The Added Value of Geographical Information Systems
in Public and Environmental Health

The GeoJournal Library

Volume 24

The Added Value of Geographical Information Systems in Public and Environmental Health

edited by

MARION J.C. DE LEPPER

National Institute for Public Health and Environmental Protection,
Bilthoven, The Netherlands

HENK J. SCHOLTEN

Department of Regional Economics, Free University & GEODAN bv,
Amsterdam, The Netherlands

and

RICHARD M. STERN

World Health Organization, European Centre for Environment and Health,
Bilthoven, The Netherlands

SPRINGER-SCIENCE+BUSINESS MEDIA, B.V.

Library of Congress Cataloging-in-Publication Data

The Added value of geographical information systems in public and
    environmental health / edited by Marion J.C. de Lepper and Henk J.
Scholten and Richard M. Stern.
        p.   cm. -- (GeoJournal library ; v. 24)
    Includes bibliographical references and index.
    ISBN 978-94-017-3771-5   ISBN 978-0-585-31560-7 (eBook)
    DOI 10.1007/978-0-585-31560-7

    1. Environmental health--Data processing.   2. Public health--Data
processing.   3. Geographic information systems--Economic aspects.
I. Lepper, Marion J. C. de.   II. Scholten, H. J. (Henk J.)
III. Stern, Richard M.   IV. Series.
RA566.A33   1994
614.4'2'0285--dc20   93-39559

ISBN 978-94-017-3771-5

The views expressed in the publication are those of the contributors and do not necessarily represent
the decisions or the stated policy of the World Health Organization or of Kluwer Academic Publishers
B.V.

Printed on acid-free paper

Foreword

Health for all by the year 2000 is the blueprint for change agreed to by the Member States of the World Health Organization. In Europe, this blueprint is built on 38 regional targets, many of which have the underlying aim of uncovering new knowledge and of using existing knowledge more effectively. The targets related to a healthy environment have the ultimate goals of safeguarding human health against environmental hazards, and of enhancing the quality of life by providing clean and safe water, air, food, and working and living conditions. Allied to these goals is the need to reduce the sense of jeopardy that many people feel about what they perceive as 'the risks of everyday life'.

These goals are an integral part of the European Charter on Environment and Health, adopted by 29 European Member States and the Commission of the European Communities in December 1989. The Charter stresses the shared responsibility of everyone to protect the environment, to be given adequate and accurate information, and to be involved in decision-making. It outlines the principles for public policy as well as what needs to be done to transform them into action. In this, strong information systems have a vital role to play by helping to monitor the effectiveness of measures taken, of trends analysed, of priorities set and of decisions made.

In keeping with the letter and spirit of the Charter, Member States need to develop multisectoral policies that effectively protect the environment from health hazards, assure community awareness and involvement, and support international efforts to curtail hazards affecting more than one country. Similarly, the machinery for implementing such policies must be developed, especially through the monitoring, assessment and control of a wide range of potential environmental hazards. To do this, specific agents and their potential effects must be studied to provide information on new risks and to better quantify old ones, thus providing the necessary input for priority setting and risk management, and for the development of integrated monitoring systems for priority substances.

Studies of the health of populations often reveal great unevenness in disease rates in time and place, which occasionally reach statistical significance. In addition to temporal peaks that may be related to epidemics, local variations are observed, frequently in the form of clusters of rare events, or of elevated or depressed rates of incidence (or mortality) in geographically restricted regions. Such results are often displayed as maps of health status, which greatly influence public opinion. This book has been developed to stimulate research and training in this area. In particular, it is directed towards developing the study of disease or adverse health outcome in individuals and populations based on the availability of geographic linkage between medical information in the health records, usually collected in routine surveys and compilations of public health statistics, and the address of the individual concerned, and information on spatial variations of potential exposures or environmental hazards. The study of such small area health statistics at large scale enables us to learn about variations in health status on a subregional, national or subnational level. It also indicates to governments the need for questioning, at the very least, the origin of such variations, especially when they imply a locally reduced standard of public health. As geographically coded information becomes available at a smaller and smaller scale, as is now the practice in many European countries, especially with the inclusion of local postal codes in addresses in hospital discharge records, local authorities, national governments, the public and the press are becoming aware of the degree of variation in health risk at small scale. When a high degree of risk is found among a population with close geographic association to what could be a potential source of environmental risk, public concern becomes acute.

In response, public authorities turn to the scientific community for advice. In particular, the scientific community is asked for an estimate of the probability that a locally high health risk is not due to chance, but is the reflection of a local increase in risk. The methods necessary for such studies in analytical environmental epidemiology are complex; they also demand high standards of scientific rigour and are in a state of rapid development.

This book is the first attempt to bring together the experience of environmental epidemiologists, statisticians and map-makers in a single volume. It is also one of the first outcomes of the WHO European programme on environment and health information systems. This programme, which has been generously supported by the German Ministry of Environment, Nature Conservation and Nuclear Safety, the Italian Ministry of Environment and the Department of Health in London will be carried out by a large number of participating institutions in Europe, assisted by several WHO collaborating centres and coordinated by the newly established European Centre for Environment and Health with offices in Copenhagen, Bilthoven, Rome and Nancy.

The wide use of this book is expected to improve the ability of the European Region to identify areas of environmental concern and to help improve public and environmental health planning policy.

Jo E. Asvall
WHO Regional Director for Europe

Acknowledgements

This volume would not have been possible without the technical and adminstrative support of the National Institute of Public Health and Environmental Protection (RIVM), Bilthoven, the Netherlands. The editors are also indebted to colleagues from the RIVM and from Geodan BV in Amsterdam for their scientific contributions. We wish to thank Han van Veldhuizen and André van der Veen for their work in preparing camera-ready copy for the manuscript, and to express our gratitude to both Alice Uppelschoten and Carmen Molina for all their efforts in facilitating the production of the book.

Marion J.C. de Lepper
Henk J. Scholten
Richard M. Stern

September 1994

Contents

List of figures

List of tables

List of maps

EDITORIAL INTRODUCTION

Richard M. Stern, Marion J.C. de Lepper and Henk J. Scholten

For a number of reasons, attention has recently been directed to the issue of the impact of environmental pollution on public health in the European Region of the World Health Organization (WHO). In western Europe, questions of planning strategies for energy production and use and of transportation management and development have been related to the potential health effects of the resulting air pollution from stationary and moving sources, prompted by the occurrence of both summer- and winter- type episodes that in severe cases, have resulted in observable health impacts, especially among sensitive populations. In central and eastern Europe, attention is now focused on the issue of environmental neglect of the past 4-5 decades, also with an emphasis on industrial air pollution as a potential source of local and transboundary health impact.

In the above context, the question has been raised as to the possibility of detecting the effects of the environment on public health based on existing information. WHO has been collecting national statistics on a large number of public health conditions since the early 1970s based on the strategy of monitoring indicators to follow progress in the Member States towards the goals of health for all by the year 2000 (HFA 2000). In the first instance one could query this HFA database concerning possible indications of environmental impact on public health. Unfortunately, one would not expect that there are a priori specific environmental health indicators, and local and national health statistics instead reflect a combination of interrelated factors, i.e., lifestyles, access to and quality of public health care, stresses originating in the indoor, outdoor, occupational and social environments, and human biology. In fact, recent analysis of tumour rates in Europe among various population strata has shown that, when corrected for the most potent of all environmental chemical carcinogens, tobacco smoke, the data do not suggest any measurable effect of the enormous increase in the production and accompanying release to the environment of tens of thousands of chemicals worldwide or locally during the past 4-5 decades.

Examination of HFA data for Europe suggests that there is a marked difference in public health status between countries in western Europe and these in central and eastern Europe. One might argue that this was due to different standards or criteria for monitoring and was a cultural artefact. Examination of the trend over the past two decades suggests a real effect, since even if there were differences in methods, both eastern and western approaches should have been able to accurately monitor relative changes even if the absolute levels were not commensurate. In particular, the HFA indicators for, e.g., life expectancy at birth and standardized mortality for all causes at all ages shows that great progress has been made consistently in the west, with little progress in the east in the period 1972-1988. One can therefore raise the question as to whether or not this east-west gradient is due to the environment or at least has an environmental component that might include air pollution.

A quantitative prediction of the total impact of environmental pollution on health is difficult to make because of the absence of detailed knowledge concerning ambient air concentrations, population and demographic details and other fundamental issues such as human dose and exposure response relationships. This is especially true for the estimate of potential effects in the recent past in central and eastern Europe, where accurate exposure (i.e. ambient concentration) data has either been unavailable or non-existent. The trends in western Europe, especially for concentrations of SO_2 and suspended particulates, have been downwards over the past two decades, a situation which is not thought to have prevailed in the east. On the other hand, there are areas in a wide belt in central and eastern Europe, extending from Germany through the five new federal states to Poland and Czechoslovakia, where environmental quality guidelines, intended to indicate the levels below which most populations will be protected from adverse health effects, have been periodically exceeded, as can be seen from, e.g., a map of 98 percentile values for ozone concentrations for Europe.

This and other such information available at present is contained in the report (The impact on human health of air pollution in Europe), recently submitted by the WHO Regional Office for Europe to the United Nations Economic and Social Council of the Economic Commission for Europe. The major conclusion of this report, based on current knowledge, is that "Millions of Europeans live in areas with air pollution severe enough to cause each year thousands of premature deaths and many more chronically ill and disabled". More specific statements concerning the magnitude or localities of the health impact of air pollution or the specific populations or subpopulations affected cannot be made.

Actual measurement of the attributable effects of environmental pollution on public health are difficult because of our present inability to distinguish between the contributions of individual risk factors to disease, since with only few exceptions, a wide variety of risk factors can lead to the same health outcome as measured by specific public health indicators of mortality or morbidity. It may well be the exceptions that provide clues to specific environmental risk factors: e.g., chronic exposure to asbestos is a unique causal factor for increased incidence of pleural mesothelioma found in parts of Turkey and northern Italy. Similarly, acute exposure to hypersensitizing agents may well also indicate a causal link between air pollution episodes and health effects (e.g., soya bean protein dust and asthma epidemics in Barcelona). These and similar observations of clusters of disease in time or space found by examination of routinely collected public health statistics may be the technique of choice for identifying priority areas for environmental study and management. Details of such analyses and examples of recent case studies can be found in Elliott et al. (1992). Furthermore, clusters of deaths, especially among the elderly and sick, have been associated with a large number of specific urban air pollution episodes (e.g., in London) and have initiated significant abatement programmes in the past that have resulted in a measurable reduction in public health effects.

The issue that must be dealt with is how to proceed towards developing priorities for environmental management based on health concerns, as we know with certainty that there are regions with a great potential for increased mortality and morbidity, yet where there is no currently available set of information that will indicate how and where to initiate management initiatives.

The appropriate questions to answer are: a) What are the priority areas in the WHO European Region for environmental pollution abatement initiatives based on criteria defined by potential health impact? and b) Is there any evidence from local public health statistics to support the hypothesis that these high impact areas demonstrate increased health risk? The two issues are to be considered separately, as one can argue that reduction of environmental pollution is necessary wherever guideline values are exceeded, while on the other hand it is necessary to determine to what extent the choice of priorities and resource use can be based on health arguments alone. From an epidemiological point of view, it is necessary to have an estimate of the expected effect of any intervention if one is interested in designing population-based studies intended to detect improvement in public health indicators resulting from such remedial efforts.

The European Charter on Environment and Health, which serves WHO as the blueprint for managing the environment as a resource for health, was endorsed by the Member States of the European Region based on their recognition of the potential for a significant impact on public health status of present and future environmental conditions. A major and recurring theme of the Charter is the urgent need for better information and better use of information in the decision-making process at all levels, from questions of individual lifestyle management to parliamentary development of the machinery for public health policy planning.

Furthermore, an awareness of the potential for conflicting interests and the requirement of efficient use of limited resources has led to the expressed need for the development, among other things, of sharp tools for establishing priorities for action based on a wide range of criteria that must cater to the interests of all parties in the public health arena.

The importance of the spatial component of the distribution of environmental risk factors or of unusual disease outcomes potentially having their origin in external factors suggests that appropriate information tools might be found in systems that deal with map-related databases and the manipulation of spatial information, an end towards which Geographical Information Systems (GIS) have been designed. Furthermore, recent developments in introducing a spatial component in environment and health data and its collection and management suggest that working with a GIS environment would add significant value to the data and the results of data analysis.

Encouragement for such an approach was provided throughout the development of the Environment and Health Information System Programme of WHO/EURO, where a GIS component had been suggested as early as 1988. It had been assumed that a well structured programme could provide evidence for the added value of GIS approaches in environment and health information and management. It was envisaged at the outset that the added value of GIS methods would immediately become evident by giving support, where appropriate, to priority setting, and for the proposing of disprovable hypotheses that relate observations of inhomogeneities in risk factors with variations in health outcomes.

An even more important encouragement was provided through the results of the introduction of a GIS programme at the National Institute of Public Health and Environmental Protection (RIVM), Bilthoven, the Netherlands. The RIVM took the decision to use a GIS programme to restructure their data infrastructure, and this long-term process showed already in the short term some striking results in the strategic products of data integration and data presentation. The RIVM has become one of the important GIS sites for environmental applications in Europe.

In order to lay the foundation for its upcoming programme, WHO/EURO organized in close cooperation with the RIVM a consultation on "The development of a health and environment geographical information system for the European Region", convened at the RIVM in Bilthoven, December 1992, attended by 32 experts from 13 countries and representatives from a number of international organizations. This volume is a direct outcome of that consultation, containing a selection of background papers presented, together with a number of solicited contributions intended to introduce important issues which were not treated in detail.

The members of the consultation, although they expressed high hopes for the demonstration of the added value of the GIS approach to environment and health information, suggested the need for great caution in proceeding with GIS programmes. Much concern had been expressed for potential difficulties in combining health and environmental data in a single database due to the possibilities of fortuitous coincidences and the political persuasiveness of maps. The editors are very much aware of the fact that spatial variations observed in health outcomes do not need to reflect variations in external causes. When and if there are spatial components reflecting variations in risk factors, these are multiple and complex. The traditional demonstration of the utility of disease mapping has been with respect to infectious diseases, and this is still useful in the developing countries and perhaps parts of Europe. But in general, as pointed out in the first contribution, the main issue is of an appropriate scale. It must be realized that "environmental data", i.e., data on the spatial distribution of putative risk factors, is not available at an appropriate scale for most of Europe (let alone the rest of the world). Collecting new data to disprove hypotheses generated by inhomogeneities in health data is a costly and resource-consuming task. One of the prime responsibilities of the spatial health researcher will therefore be to develop methods for priority setting. A suggested approach has been to start with evidence for uneven distributions of risk rather than from variations in health.

With these caveats in mind, in this volume we would like to demonstrate the value added of GIS. It draws together descriptions, assessments and opinions of GIS development and application by epidemiologists, geographers, spatial analysts, computer scientists and managers. This book is divided into seven parts containing 22 chapters prepared by different authors.

In Part I the need of information in public and environmental health is discussed. The enormous quantity of data in these two fields makes it necessary to look for indicators of public health and environmental quality. At the same time, it is important to have meta-information systems available that give information about the data: a type of Yellow Pages for data.

In Part II attention is directed to the fundamental features of a GIS, and the application of GIS is demonstrated. A description of the analysis of a possible location of a hospital gives insight into the way GIS is used. The description of the CEC CORINE project puts the data discussion into a GIS framework.

In Part III, three different sections address the critical issue of spatial analysis.

Part IV contains six examples of the application of GIS in public and environmental health. (The article by Lloyd is included not because of its conclusions but as an example of difficulties encountered when initiating investigations based on the observation of unique health outcomes rather than environmental risk.)

Part V deals with the added value of GIS by focusing on decision-making in the health sector on the one hand and the long-term potential of GIS for epidemiology on the other hand.

Both Parts VI and VII are concerned with the implementation of GIS. In addition to an elaboration on some of the conditions that have to be met in order to guarantee the effective utilization of GIS systems, Part VI also draws attention to the lessons learned from the CEC CORINE and the RIVM Critical Load experience.

In the final chapter of this volume the conclusions and recommendations of the Bilthoven Consultation are presented. The structure of the conclusions and recommendations fits into the concept of the GIS house, discussed in Chapter 4 and illustrated throughout the rest of the book.

The European Charter, which provided the framework for the European Health and Environmental GIS Programme (HEGIS), is given in Annex 1.

We expect that the development of HEGIS will enable the identification of local variations in time and place of risk factors and health status. As will be pointed out in Chapter 1, as more detailed information concerning the spatial variation of health status is collected and made available by local and national agencies, a picture of the heterogeneous nature of public health in Europe is beginning to emerge. In one notable instance within the European Community, the study of avoidable death (CEC Concerted Action Programme, edited by Holland, 1989) shows that wide local variations exist for a number of diseases at subnational levels throughout the Member States of the European Community. Similar variations can now be seen for a number of other diseases. We do not yet have evidence that these variations reflect general variations in risk factors, although there are certain clear indications that poverty, lifestyle and cultural differences contribute significantly to the patterns of heterogeneity, whereas the role of the chemical environment is not well defined.

One can anticipate that the value added of HEGIS will lie along two major paths. On the one hand, the use of GIS as a database management system will provide rapid data analysis leading to local calculations of types of risk factors, which will exhibit unique spatial correlations for environmental situations of note should they exist. We will be able to compare the variation in health outcomes across Europe with those found across a single country, state or in subregional groups of countries. For example, examination of the distribution of standardized death rates for the countries of the European Region in 1988 shows that there is a bimodal distribution, with the countries of central and eastern Europe forming a separate distribution from the rest of Europe in terms of both trends and absolute rates for cardiovascular disease, but only for trends in terms of standardized death rates excluding mortality from cardiovascular disease, accidents and suicide. One can then ask whether the individual countries show similar bimodal distributions in the public health indicators of individual administrative districts at various levels of aggregation or are homogeneous within themselves. The importance of this type of information lies in the resulting strategy for improvement and intervention: countries that look average may be uniform or they may contain regions of extremely high and low risk that contribute to an acceptable average aggregated level. The development of health and environmental planning policy will be different for homogeneous and heterogeneous countries and subregions.

References

Elliott, P., J. Cuzick, D. English, & R. Stern, (eds) (1992). *Geographical and environmental epidemiology: methods for small area studies*. Oxford, Oxford University Press.
Holland, W.W. & the E.C. Working Group on Health Services and Avoidable Deaths (1991). *European Community Atlas of Avoidable Death*. Oxford, Oxford University Press, 2nd ed.

Part I

NEED OF INFORMATION IN PUBLIC AND ENVIRONMENTAL HEALTH

1 ENVIRONMENT AND HEALTH DATA IN EUROPE AS A TOOL FOR RISK MANAGEMENT: NEEDS, USES AND STRATEGIES

Richard M. Stern

Abstract

Accurate, adequate and accessible data are a necessary tool for reducing existing environment and health risks and preventing the introduction of new and uncontrolled risks accompanying industrial development and societal growth. Without such data, environmental management in the WHO European Region will proceed based on risk perception alone and will be unduly sensitive to public pressure for priority setting. The systematic and unified collection of geographically localized environment and health data is a necessary activity which must be developed within the Region, perhaps based on subregional programmes in geographically or socially coherent areas. A concerted effort must be made to catalogue current data holdings and their method of access so that public data are made available to the entire Region for use in environmental and health management of common problems and local and regional priority-setting. Epidemiological studies using geographically linked exposure and health data are the method of choice for hypothesis testing with respect to the effects of past exposures and emissions on human health and predictions of the effect of the current state of the environment on the future state of public health. Studies of unevenness of health outcome will indicate where major research into the causes of existing inequalities in public health status and the presence of "hot spots" of disease or functional impairment is necessary. The unified monitoring of the environment at an appropriate scale is necessary to aid in the development of a regional programme to study the potential relationships between spatial variations in risk factors and in disease rates.

1.1 Introduction

The prime goal of environmental and health management is to reduce existing risks and prevent the introduction of new uncontrolled risks. Since alterations to human health are often associated with or caused by sudden or gradual changes in the environment, a prerequisite for health risk management is the identification of environmental hazards and their effects, coupled with suitable monitoring programs to provide the data necessary for priority setting and decision making. The problem of monitoring of environment and health data and their interpretation is made difficult because the association between environment and health is extremely complex, especially where health standards are already high and the more obvious causes of disease and functional impairment have been controlled.

It is extremely important to recognize that the current definition of health includes the state of wellbeing. In addition, the goals of the WHO health for all by the year 2000 (HFA2000) strategy include freedom from fear and disease or impairment. Similarly, the definition of environment encompasses a wide range of factors which describe our social

3

M. J. C. de Lepper et al. (eds.), The Added Value of Geographical Information Systems in Public and Environmental Health, 3–24.
© 1995 *Kluwer Academic Publishers.*

and physical surroundings. Although many states of human mental and physiological health (e.g., stress, birth weight, cancer) are known to be sensitive to environmental factors, the identification of the relative contribution of many of the chemical and non-chemical factors to average or individual health status is frequently unknown and may vary greatly as a function of place and time. Although many chemical risk factors that are either toxic (e.g., heavy metals, long-lived organic substances such as pesticides) or non-toxic (e.g., chlorofluorocarbons) may be observed to have a direct and measurable effect on the environment or on ecological systems, the extent of their ultimate indirect effect on human health is frequently unknown. Similarly, non-chemical risk factors such as population density, noise pollution, poverty and fear, etc., are known to play a role in determining the state of health, but their individual contributions are difficult to quantify.

Both the state of human health and the state of the environment are subject to naturally and to human-caused variations that are indistinguishable as to origin, although the direct cause of rare events can frequently be determined. Local, national or regional variations in lifestyle, the quality of and access to primary health care, and individual susceptibility to disease may each lead to variations in health status that far outweigh variations due to environmental risk factors.

Although health risk management and environmental risk management are frequently relegated to different authorities, the problems are for the most part overlapping. In order for the complex process of risk management to be set in motion, a risk must first be perceived: this is true regardless of whether the management is to deal with a personal risk or with global problems. The clue on which the perception of risk is based is usually the observation of an unexpected health outcome or distribution thereof or the identification of a recognized environmental hazard. If risk management is to be an orderly process, it requires the systematic use of proper tools to accomplish the necessary steps of hazard identification, risk quantification, evaluation and comparison with other risks, and finally decision-making with respect to priorities, use of resources and protocols.

The first challenge in Europe is the improvement of the quality, reliability, ease of access and utility of currently held data on a local, national, regional and international basis before any suggestions to develop additional monitoring programmes, even based on well defined data use, are considered in detail.

An extremely wide range of problems face the environmental and health manager. The primary need within any country is for reliable information on risks and on health status. Since many of the issues involve transboundary problems (e.g., acid rain, export of dangerous products including consumer goods and waste), or are global in nature (greenhouse effect and ozone depletion), it is reasonable to consider that the data needs of the WHO European Region and of the individual countries overlap, and a search should be made for a unified approach to environmental and health information for use in management of current risks and future risk reduction strategies. The first task in such a programme must be developing strategies for improving the use of the vast amount of environmental and health data currently held within the Region.

One necessary tool for this process is the cataloguing of sources of adequate and accurate data on health and environmental status. An extremely broad programme of environmental and health monitoring has been established in many sections of the Region: the immediate task is to identify which data already exist, and to establish mechanisms for the better use of such existing data for management purposes. From a political and scientific point of view, making the right decision for the wrong reason due to an inability to make use of existing information is as unsatisfactory as making the wrong decision or avoiding the process of decision-making entirely: the use of public resources must be justified and justifiable. As will be made clear in the following discussion, great care must be taken in the development of methods for additional monitoring of both the environment and health if the existing and new information is to be effectively used for risk management in the public sector. In some cases, it will turn out that a problem perceived by the public as being of major concern will involve trivial risk when measured absolutely, but must still be dealt with because the risk as perceived is totally unacceptable. The problem is even more complicated in the subregions where information has not been generally available to the public and the need for rapid development of environmental management programmes, with full public support, may be seen to compete with the need for rapid industrial development.

A first priority is the acquisition and evaluation of existing high-quality data on the individual classes of risk factors so that one can begin the difficult tasks of priority-setting. This is especially clear when decisions have to be made with respect to the allocation of limited resources towards improving the environment, raising the level of primary health care, educating populations with respect to health-risk behaviour and dealing with high-risk subpopulations.

It is hoped that the European Charter on Environment and Health (WHO, 1989) can be used as a guideline in developing local strategies for management of the environment as a resource for health. This must be based on public information, responsible action at all levels and a management of both our knowledge and our ignorance concerning the relationship between environmental hazards and human health risks. The Charter is seen as the blueprint for merging HFA2000 strategies with those for environmental management in the decades to come.

1.2 Information needs

Every country should be aware of the health status of its population, if for no other reason than to be able to use the information to establish priorities to allocate resources, to develop and evaluate prevention programmes, to educate the population and to be able to intervene in major shortcomings in health status. Monitoring programmes should be in place to provide local information on vital statistics (population size, age and gender distribution, cause-specific mortality) and on illness status, morbidity, time trends and geographical distribution of the incidence of specific diseases. Information on hospital admissions or discharge, health insurance, birth records, workdays lost and drug sales frequently can be maintained in central or local registries: separate registries can be created for monitoring important diseases such as cancer, cardiovascular disease, congenital malformations and contagious diseases or diseases known to be caused by exposure to widely used chemicals and specific environmental contaminants (e.g., *Salmonella* poisoning).

Similarly, each country should be aware of the state of its environment, if for no other reason than to identify baseline status in case of contingencies, and to be able to develop and measure the effect of environmental control policies. Monitoring programmes should be in place to provide local information on time trends and geographical variations in water quality, outdoor and indoor radiation, ambient air in rural and urban regions, food, soil and noise pollution. In addition, information on emission of selected air pollutant and generation and treatment of solid and hazardous waste should be available at the same grid density. Surveys should also be made of environmental resources (land, soils, water, fauna, flora, habitats), topical issues (such as forest damage, lake protection and migratory species), and data routinely collected on climatic and selected background information (temperature, precipitation, natural disasters and agriculture, especially fertilizer and pesticide use) and economic variables (sectoral employment, and energy production and use). Such a programme of European Environmental Statistics has been developed by the United Nations Economic Commission for Europe (ECE).

Because personal lifestyle, socioeconomic status, place and quality of residence, access to and quality of primary health care and individual susceptibility play a significant role in determining average and individual health status, data on these factors should also be made available. Such data can be maintained by regional or central statistical offices and has been traditionally collected by national census or household surveys in an number of countries. Although this type of information is usually utilized by social medicine, it is of extreme importance that it be made available for the purpose of environment and health risk management, since these social risk factors may be the origin of inequalities and unevenness found in the health data.

Employment will play a role as a major risk factor for individuals in a wide range of job situations. Because workplace exposures may be much higher than those for the general public, individuals engaged in certain occupations may be at high risk for certain specific diseases or functional impairment. Organizational and communication difficulties frequently prevent the integration of occupational health data in general health statistics and of occupational exposures with environmental exposures. Registers of occupations associated with known risk for exposures or health outcomes are maintained, although the data are most useful when coupled with public health records: the recording of occupational information at the same time health status is determined is of the utmost importance. Very little is currently known about how occupation affects public health, especially in one-industry or one-employer communities.

1.3 Studying risks to the environment and to health and their geographical variation

It is necessary to distinguish between effects on health and effects on the environment and between three time scales: determination of effects of previous exposures and emissions on the current status of health and the environment by comparison between exposed and unexposed individuals and areas, respectively; determination of the effect of current exposures and emissions based on epidemiological and ecological studies; and estimation of the effect of future actions by means of health risk and environmental impact assessments.

Since all attempts to manage current and future health and environmental risks are based on the results of studies of the effect of past human activities, a major issue will always be the ability of data to support hypotheses on the relationship between the state of the environment and health status: what has been called the menace of everyday life.

Strategies for the collection and use of data lie between two extremes. On the one hand, "environmentalists" argue that there exist environmental quality or exposure guidelines for most priority substances and sufficient toxicological information on most others to consider that both public health and the environment are protected by the simple means of enforcement of such guidelines together with overall strategies for reduction of environmental concentrations wherever possible. Thus, health monitoring for "environmental" reasons is not necessary and environmental health management requires only adequate environmental monitoring networks. This strategy is satisfactory for management of, for example, criteria air pollutants, if for no other reason that it is extremely difficult to demonstrate the direct health effects of air pollution at most ambient concentrations. On the other hand, the other extreme demands the proof of health impact before the introduction of regulations.

A number of examples can be cited to emphasize the need for geographically linked health data, without which the management of environmentally related health risks is left to chance. With the exception of only a few genetically related susceptibilities, the human response to chemical and other insults should be universal and the problems discovered in one country should exist wherever similar risk factors are found, although local priorities may be different. The unified collection of geographically linked health data would provide the basis for hypothesis generation and testing as the fundamental principle of environmental health risk management.

One example that demonstrates unevenness is the recent concerted action exercise on the part of ten countries of the European Community to evaluate avoidable death. This is defined in relationship to 13 diseases for which mortality could be expected to be avoidable provided that the condition is detected sufficiently early and/or adequate medical services were available. Analyses were made within several regions for each country to permit intra-country comparisons based on national standards and from country to country based on European standards: 360 individual areas were defined to permit appropriate mapping of the results. The observation of unevenness of mortality is exemplified by the data for cervical cancer (ages 15-64): standardized death rates varied from less than 1 per 100,000 to over 17 per 100,000. The differences are statistically significant, were found to occur in all diseases studied and tended to cluster in particular countries. Within a given country, there were also important local variations (e.g., within England, cervical cancer varied from 4 to 8 per 100,000) although not every country showed significant variation for all diseases.

The main object of such an effort is the demonstration of such unevenness in health risk (up to a factor of 20 for some diseases) and hence support for the argument for the need to initiate research to find the origin of the observed variations. Possible non-random sources for the variation fall into three main areas; differences in death certification or coding practice; variation in incidence or prevalence in different areas; and failure of preventive measures or curative health care. The existence of such unevenness has serious implications for other types of studies.

Considerable effort is placed in the study of occupational mortality, e.g., by the Office of Population Censuses and Surveys (OPCS) in the United Kingdom. One persistent result is that, for many occupations which at first sight would appear to provide a significant risk factor for a disease, e.g., lung cancer for the metalworking trades such as welding, social class variations in absolute lung cancer rates are greater than those due to occupation (a factor of 2-3 compared with 40–60%). Furthermore, in efforts to obtain sufficiently large cohorts to achieve statistical significance in the results, groups are recruited from many different parts of the country. If the unevenness of avoidable mortality (e.g., a factor of 4) reflects variations in the state of primary health care, then this unevenness should also be evident in strong local variations in occupational mortality, reflecting not occupational risk factors but local reference rates.

Although such general data as obtained from registers of vital statistics can be useful for developing hypotheses that might be studied, the verification of such theories relating unevenness in environmental exposures with unevenness in disease outcome relies on close linking between exposure and health data. A common descriptive name for such an approach is small-area health statistics. In the United Kingdom such a type of study is possible because of the existence of 6- and 7-digit postal codes. Identifying individual cases (e.g., hospital admissions, diagnoses, death certificates) via a postal code preserves confidentiality but permits a determination of actual (or relative) exposure levels to a wide range of environmental risk as well as establishing common social variables.

An example of such a small-area health study is the recent observation of the relationship between aluminium in drinking-water and the prevalence of Alzheimer disease (pre-senile dementia) in the population under 70 years of age. By carefully relating drinking-water quality in counties of residence with disease incidence among residents, verified by computerized tomographic scanning (the diagnostic technique of choice), it was possible to show that there was a 50-70% excess risk in the age group 40-65 associated with more than 0.11 mg Al/l (during the past 10 years) compared with users of water with less than 0.01 mg Al/l. The conclusion of a 95% certainty of a relationship between Al in drinking water and Alzheimer disease was possible only because of the ability to collect data on approximately 10,000,000 inhabitants in 88 county districts in England and Wales together with water quality data for each of the districts over the past 10 years. (The CEC standard is 0.20 mg Al/l based on esthetic grounds only: the implications are significant, although the results of this study are still controversial, since the role and mechanism of environmental aluminium in Alzheimer disease is not understood.)

In a recent study in western Germany, lead levels in shed (milk) teeth collected from over 2,000 children aged 6-7 were mapped according to place of residence during the previous two years. A direct correlation was found between "hot spots" of lead in teeth and high concentrations of lead in dust fall as mapped on a 100-m grid. It was also shown that living within 1 km of a traffic intersection or metal processing plant resulted in significantly elevated concentrations of tooth lead in the children studied.

The question remains as to the ultimate utility of routinely collected geographically linked health data in terms of relating the environment to health. The concept of small-area health studies is based on the principle that selecting geographic proximity is a device to reduce dilution effects in the study population and hence any unusual local clusters of rates or cases in time or space are more likely to be due to some common external factor, e.g., environmental exposures. The approach can be used to study the statistical significance of unusually high local rates and for priority-setting and hypothesis generation, but not to determine causality. Hypothesis-testing must always be based on ad hoc epidemiological studies. In spite of public perception, most local variations in health status studied in the past were found to be due to chance alone.

1.4   The need for high resolution and global environmental data: indicators

Studies of health effects require environmental data on the same scale as observed for unevenness in the health data: perhaps as fine as a 100-m grid for studies of urban air pollution and a 10-km grid for studies of the effects of stationary point sources such as major industrial facilities.

Monitoring for management of the state of health and the environment via a set of indicators could be developed based on existing monitoring programmes in many cases. After both chemical and non-chemical risk factors are carefully weighted, a number of key areas could be chosen to adequately define the state of health and the environment on local, national regional and subregional levels, such as:

(i)   energy;
(ii)   emission inventory;
(iii)   air quality;
(iv)   water quality;
(v)   food;
(vi)   occupation;
(vii)   soil;
(viii)   general environment;

(ix)   health services;
(x)   lifestyle;
(xi)   health policy;
(xii) environmental policy;
(xiii) environmental health.

One could consider obtaining some priority subsets of such indicators relatable to areas with populations of the order of 500 000 (or less if possible). Key data sets could be maintained on an absolute basis, including quality control programmes, so as to provide a comparison between the countries of the Region. The rest of the data could be maintained on a relative basis, providing information on variation within each country: presumably gradients could be followed on a relative basis as well. These indicators would provide a simple means of examining east-west and north-south gradients and urban-rural factors commensurate with, e.g., the Healthy Cities project taking into account chemical and non-chemical risk factors and psychological as well as physiological health status.

1.5   Strategies for identifying priority data

In order to have any hope of coping with future change, we must have a reliable knowledge base, not only to be able to predict the future based on the behaviour of the past but to be able to reduce uncertainty. In many cases, especially with respect to public perception of risk and public priority-setting, it is the nature of the uncertainty that determines the management decision. One good example is the question of radiation and its health consequences. One result of incorrect priority-setting within the nuclear industry has been that radiation risks at any level are unacceptable to the public. This forces management decisions for risk avoidance to be made that have no basis in fact and that can be shown to be irrational. At Three Mile Island, some 10 million litres of wastewater slightly contaminated with radioactivity have accumulated during clean-up. This water could be discharged into a nearby river without exceeding any federal regulations, but this has not been done because of community objection: the effect on the most exposed individuals would be equivalent to 4 minutes of extra exposure to the background radiation. The solution acceptable to all parties is the evaporation of the wastewater, at an additional cost of $5,000,000: the resulting public exposure is not any lower than from direct discharge to the river, but is apparently more acceptable. The cost of averting a single radiation-induced cancer death by the policy decision of not to discharge the wastewater to the river is estimated to be $25,000,000,000. (The cost of preventing a single death from cervical cancer by screening and education is estimated to be $25,000.) The lesson is that the ability of environment and health managers to make rational priority-setting is based on the development of a sound information base and their ability to influence public risk perception is based on adequate and appropriate information and education strategies.

Environmental risks due to chemicals can be conveniently divided into several broad classes: cancer risks, non-cancer risks, ecological effects and welfare effects. Although in many countries cancer risks are not separated from non-cancer risks, the one advantage of treating cancer risks separately is the possibility of using a uniform database and risk model, which permits (semi)quantitative ranking, both in terms of potency and in terms of the populations at risk. Ecological effects result from both habitat and environmental pollution. Welfare effects include damage to property, goods, services or activities to which monetary values are or can be assigned. What remains elusive are the intangible characteristics of environmental risks that the public often finds just as important: problems of voluntary versus involuntary risks and equity are also important but difficult to quantify and use in priority-setting.

Priority-setting for the development of management tools can be based either on current risks, as reflected by monitoring of the actual state of the environment (in the broadest definition), future risks (as determined by the spectrum of current hazards and modelling of the actual risk levels inherent in human activity for which there is no current risk measurement (probabilistic risk assessment)) and public perception of risk. Because of the extreme divergence of these sectors, it is necessary to consider setting of priorities within each as a separate exercise. In particular, it is necessary to examine a list of vehicles and media (air, water, food, separate chemicals, biological agents, etc.) to arrive at a set of priority chemical problems that require managerial decisions.

It is of interest to examine the result of a recent exercise of the US Environmental Protection Agency (USEPA) along these lines following such principles, based on a mixture of quantitative data and, where missing, expert judgement. In cases where assumptions were found to be incompatible across the four sectors, an effort was made to determine the effect of the bias so introduced. The basic effort was based on an estimate of the total effect that would accumulate from the problems as they exist now, taking account of time lags between emission and exposure and between exposure and damage. Risks to both total population and small, highly exposed sub-populations were taken into account separately. The resulting Preliminary Priority Listing of Chemical Risks in North America is outlined below:

Cancer risk (population based)

(i)   worker exposure to chemicals (250 cancer cases annually from formaldehyde, tetrachloroethylene, asbestos, methylene chloride);

(ii)   indoor radon (estimated to be over 5,000 cancer cases annually, with some individuals at very high risk);

(iii) pesticide residues on foods (herbicide, fungicides, insecticides, growth regulators) (about 6,000 cancer cases annually based on exposure to 200 separate cancer-causing substances) and indoor air pollutants other than radon (tobacco smoke, benzene, p-dichlorobenzene, chloroform, carbon tetrachloride, tetrachloroethylene, trichloroethylene) (3,500-6,500 cancers annually, mostly from tobacco smoke);

(iv)   consumer exposure to chemicals (formaldehyde, methylene chloride, p-dichlorobenzene, asbestos) (100-135 cancers per year from these four; estimate of 10,000 other chemicals in consumer products);

(v)   hazardous/toxic air pollutants (20 substances, classes of substances, waste streams) (approximately 2,000 cancers per year from 20 substances).

In the remaining list, the priorities were: drinking-water, pesticide application, radiation other than indoor radon (excluding medical applications and natural background levels), other pesticide uses, active hazardous waste sites, non-hazardous industrial waste sites, new toxic chemicals (no data), non-hazardous municipal waste sites and contaminated sludge.

Non-cancer risk (based on population times potency times ambient dose)

(i)   criteria air pollutants;
(ii)   hazardous air pollutants;
(iii)   indoor air pollutants (not radon);
(iv)   drinking-water;
(v)   accidental release of toxic substances;
(vi)   pesticide residues on food;
(vii)   consumer product exposures;
(viii) worker exposure to chemicals.

Of the remaining list, the priorities were: radon-indoor air, radiation (not radon), ultraviolet radiation/ozone depletion, indirect discharges (point source discharges) to drinking water, non-point sources, estuaries, coastal waters and oceans, municipal non-hazardous waste sites, industrial non-hazardous waste sites and other pesticide risks.

Ecological risk

(i)   stratospheric ozone depletion and CO_2 and global warming;
(ii)   physical alteration of aquatic habitats and mining, gas, oil extraction and processing wastes;
(iii)   criteria air pollutants, point source discharges, non-point source discharges and in place toxics in sediments, pesticides;
(iv)   toxic air pollutants.

Of the remaining list, most have only local impact, e.g., contaminated sludge, inactive hazardous waste sites, municipal waste sites, oil spills and groundwater contamination.

Welfare effects

(i)   criteria air pollutants from mobile and stationary sources including acid precipitation;
(ii)   non-point source discharges to surface waters including pesticides;
(iii)   indirect point source discharges to surface waters;
(iv)   to estuaries, coastal waters and oceans from all sources;
(v) CO_2 and global warming;
(vi)   stratospheric ozone depletion;
(vii)   other air pollutants, odours, noise;
(viii) direct point source discharges from industry etc., to surface waters.

Other priority items were hazardous waste sites, discharges to wetlands from all sources, other pesticide risks including agricultural chemicals etc. and biotechnology.

Public perception priorities

(i)   chemical waste disposal;
(ii)   water pollution;
(iii)   chemical plant accidents;
(iv)   air pollution.

Other lesser priorities were given to oil tanker spillage, exposure on the job, eating pesticide sprayed food, pesticides in farming and drinking-water.

It can be seen that there is little agreement across the four sectors with respect to the priorities of the major environmental issues in the United States, although criteria air pollutants, stratospheric ozone depletion and pesticide residues on foods are clearly major issues with multisectoral importance. Major health risks (e.g., radon) are not related to risks in the other sectors. Several chemicals are major concerns in several sectors (lead, chromium, formaldehyde, solvents, some pesticides). The public list of priorities focuses on less important issues, especially water pollution. The only real risks at the present time within the general area of (surface) water pollution are related to ingestion of seafood contaminated with pathogens from inadequate sewage treatment and contamination of fish from agricultural and urban runoff. Drinking-water is a significant problem in some areas, where lead contamination from pipes and fittings, well contamination from waste sites and the production of toxic substances by chemical treatment of water with a high humus content can represent a high risk for some populations. There does not exist a current risk from direct chemical contamination, although one is perceived by the public. The question of radioactive contamination of the atmosphere is not considered, as there is no current risk from radioactive release: probabilistic risk assessment of reactor failure and subsequent release from containment is not considered. Other future risks are either not considered or relegated to low priority because of lack of information. Radiation would be a significant problem if natural and medical sources were included in the risk calculation. Many complex and remote problems such as depletion of the ozone layer and global warming are not understood by the public and therefore are not given high priority.

It must be pointed out that a wide range of competing non-chemical risks have not been considered in the USEPA North American priority-setting exercise: social variables, lifestyle, consumer activity, access to and quality of primary health care services, microbial contamination of food and food safety in general, communicable diseases (as due to environmental risk factors), etc. In a more global approach, as is indicated for the European Region, such risk factors must be considered in terms of priority-setting as well, in particular, as they apply to the WHO extended definition of health, which includes the state of wellbeing.

In any consideration of priorities and approaches to environmental health risk information management in Europe, the efforts of international and multinational organizations must be taken into account. The Uited Nations (via UNEP, OPCS, International Register of Potentially Toxic Chemicals, ECE, WHO), OECD and CEC all have major programmes for the collection of data, although in many cases the purposes are for monitoring and not management. Although the information bases and databases maintained by these organizations are not necessarily adequate for the needs of a management programme, the organizational frameworks in which such information programmes have been created could be of great help to the region for any future development programme.

An initial task should be the production of a review of the state of the environment (and where appropriate, health) within the European Region, to produce the type of priority document developed by the USEPA for North America, with emphasis on two factors: the strong variation in subregional status and priorities and the range of characteristic dimensions for various thematic problems. It would certainly be appropriate to consider a survey of the special needs of, e.g., the Baltic or central and eastern subregions and to develop a pilot programme based on a limited set of relevant indicators.

Environmental thematic problems have a wide range of characteristic dimensions: global (greenhouse effect, 20,000 km), continental (acidification, 3,000 km), fluvial (water pollution, 1,000 km), subregional (the Mediterranean, the Baltic, 1,000 km), local (municipal incineration, 100 km), point sources (nuclear facilities, 10 km; toxic waste sites, 1 km; industrial noise, 100 m). Similarly, the time scale of these problems (including their generation, detection and solution) ranges from centuries to days.

The Netherlands has recently performed a national survey of environmental problems and made estimates of the resources necessary to deal with the major set of problems over the next two decades. Choosing the scale levels of global, continental, fluvial, regional, local and other, a distribution was made among a broad set of socioeconomic sectors: waste and wastewater removal, government, public utilities, households, transport, industry and agriculture. With the exception of global problems, expenditure was divided equally (within a range of 40%) over the various scale levels, although the relative fraction assigned to each activity varied by more than a factor of 30.

Similarly, the limited range of environmental themes and issues: biospheric ozone, aridification, disturbance and nuisance, indoor environment, waste and wastewater removal, eutrophication, ozone layer, dispersion and acidification, exhibited extremely wide variation in their relative contribution to each sector, but with the exception of waste and wastewater disposal, which required twice the average expenditure for countermeasures, the total costs were approximately evenly divided among all other sectoral problems (within 40%). The cost per environmental theme, however, showed almost a 50-fold variation, with dispersion and acidification requiring the most resources and the ozone layer and aridification requiring considerably less.

In addition to this economic assessment, a health effect assessment for the Netherlands was made as follows:

(i) Chemical agents (oral): nitrates (in drinking water), cadmium (in soil) and pesticides were identified as having reliable estimates of health effects, whereas those for PCDD/PCDF in mothers' milk were unreliable. Planned measures for nitrate were inadequate and would lead to a worsening of the situation in future.

(ii) Chemical agents (inhalation): summer episodes of ozone, winter episodes of SO_2, NO_2, NO_2 from boilers, CO in indoor air, passive smoking (all with moderate to good reliability in the health effect estimates); volatile organic carbon in indoor air, benzene from automobile fuels, polyaromatic hydrocarbons (PAH) from traffic and odour from industry (poor reliability of effect estimates).

(iii) Physical agents: radiation, ultraviolet light, noise (mostly in dwellings and from traffic) (with moderate to good reliability of effect estimates).

This Dutch list is very short compared with either the list of environmental themes or the detailed risk factor list prepared by the USEPA and indicates that health concerns will provide priorities for only a limited number of environmental management decisions when considerations are made on national levels of priority-setting. Local management scenarios might exhibit a significantly different spectrum of problems in various parts of the European Region.

If and when cancer becomes the health effect based on which management decisions in the Netherlands are to be made, the list of priority environmental factors becomes very restricted. Although the reliability of the estimates is poor, the following factors appear for consideration (numbers of annual cases are based on the Netherlands population) as follows: nitrate (nitrosamines), passive smoking, volatile organic carbon compounds (anywhere from zero to a maximum of 20 cases each); PAH in food (less than 1); PAH in air (5 per year after 20 years under the present load); benzene (0.5); ionizing radiation (500) and ultraviolet radiation (8,000, of which 80 deaths). Additional measures would have a considerable effect (50%) on PAH in air, benzene and passive smoking. For comparison, it was noted that there would be 7,500 deaths annually due to tobacco consumption.

1.5.1 The way forward: the European Charter

In December 1989 ministers and other senior representatives of the environment and/or health ministries from 29 countries in the European Region met in Frankfurt am Mainz to ratify the first European Charter on Environment and Health. The Charter is an extension of the European health for all policy and targets adopted by the Member States in the WHO European Region in 1984 (WHO, 1985). The Charter outlines the entitlements and responsibilities of individuals and corporate and governmental entities with respect to informed participation in the process of managing the environment as a resource for health. It outlines a series of principles for public policy, ranging from regulation at the source, the use of best available technology and the adherence to the polluter-pays policy. A series of strategic elements are proposed, including the basic principles of comprehensive strategies, the need for research, information and epidemiological surveillance, the development of international programmes of interdisciplinary issues for the region including global disturbances, urban development, drinking-water supplies, water quality, microbiological contamination of food, impacts of energy and transport options, indoor and outdoor air quality, persistent chemicals, hazardous waste, biotechnologies, contingency planning for disasters and cleaner technologies. These priority elements are to be dealt with through intersectoral planning, health promotion and international cooperation.

Two major themes of the Charter are the need for information and its dissemination and the need for a better understanding of the local effects of environment on health. It is understood that knowledge of the true state of environment is essential for rational decision-making by individuals to carry out their personal responsibility for making their local environment the best possible for themselves and their neighbours. At the other extreme, there is a need for aggregated information to enable governments to make decisions with respect to regional, national, subnational and local policy towards the same goal. The concept of thinking globally and acting locally applies to all levels of scale. Similarly, it is necessary to understand the extent and origin of local variations in health status in terms of identifying and removing risk factors to provide greater equity and to avoid dealing with non-issues based on perceptions of risk where none exist.

Based on the above examples, it can be concluded that, in the overall European context, national priorities for health-related environmental management in any one country can be expected to be but a small subset of the overall catalogue of problems facing the Region as a whole. It would be irresponsible to ignore available evidence that the situation in central and eastern Europe will warrant extensive and perhaps Herculean measures that will severely tax the technical and economic resources of the Region in the decades to come. Such evidence can already be found in several health indicators, when comparison is made between east and west, as shown in Figures 1.1-1.2 and Maps 1.1-1.2.

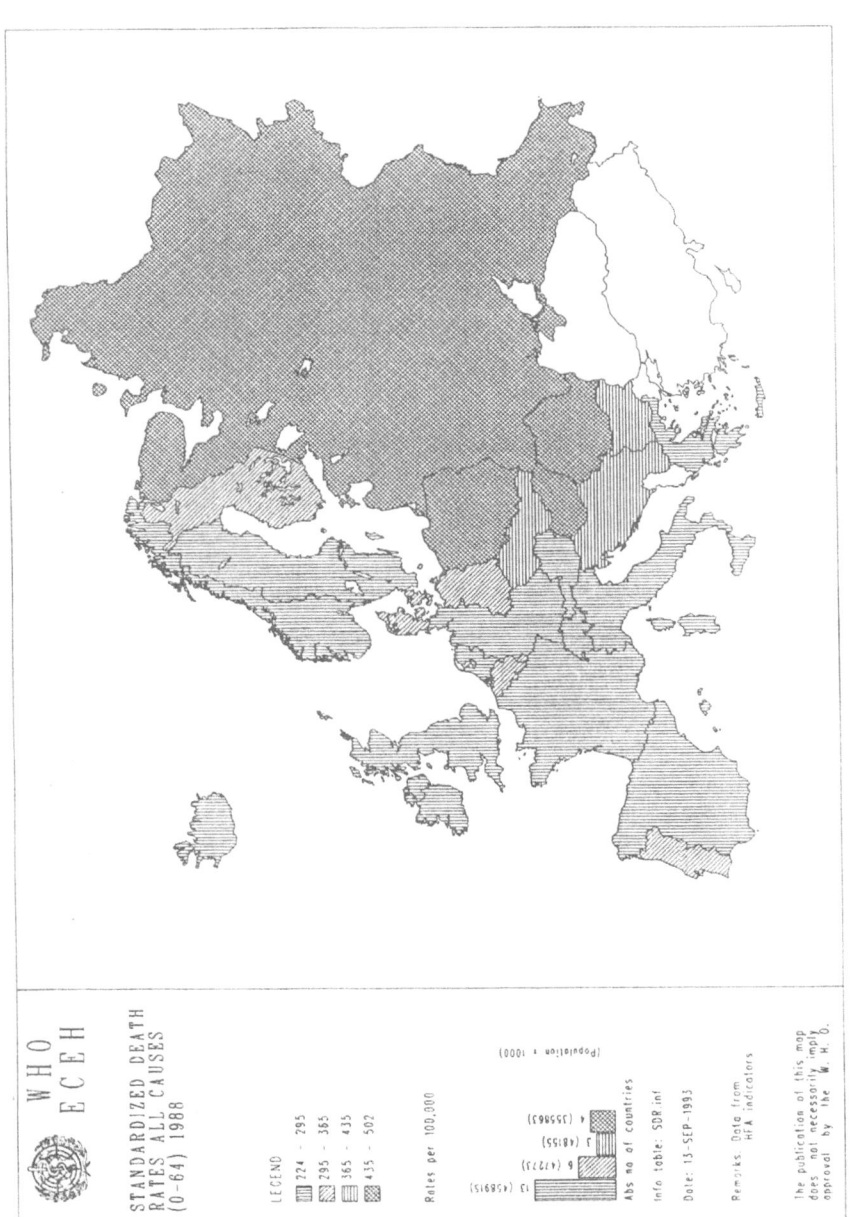

Map 1.1: Map of the distribution of age-standardized death rates per 100,000 population, all causes, ages 0-64 years, for the countries of the European Region, in 1988. Division into four uniform rate categories, indicating the number of countries and total population in each category

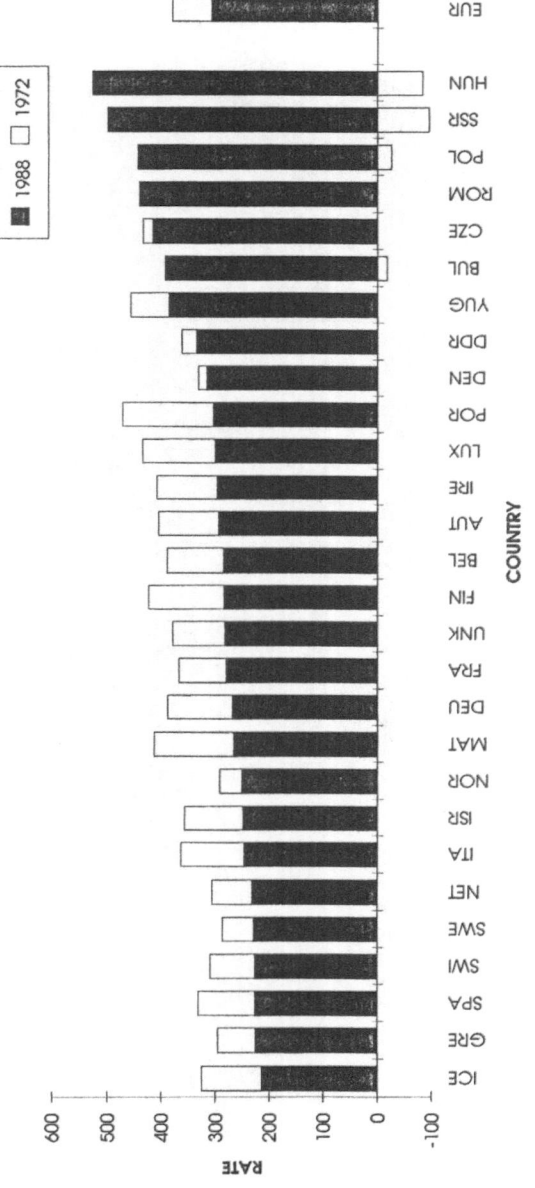

Figure 1.1: Ranked bar chart, age standardized death rate per 100 000 population, all causes, ages 0-64, for the countries of the European Region, in 1988. The differences (1972-1988) are shown as well (increases are negative) (data for USSR for 1980-1988)

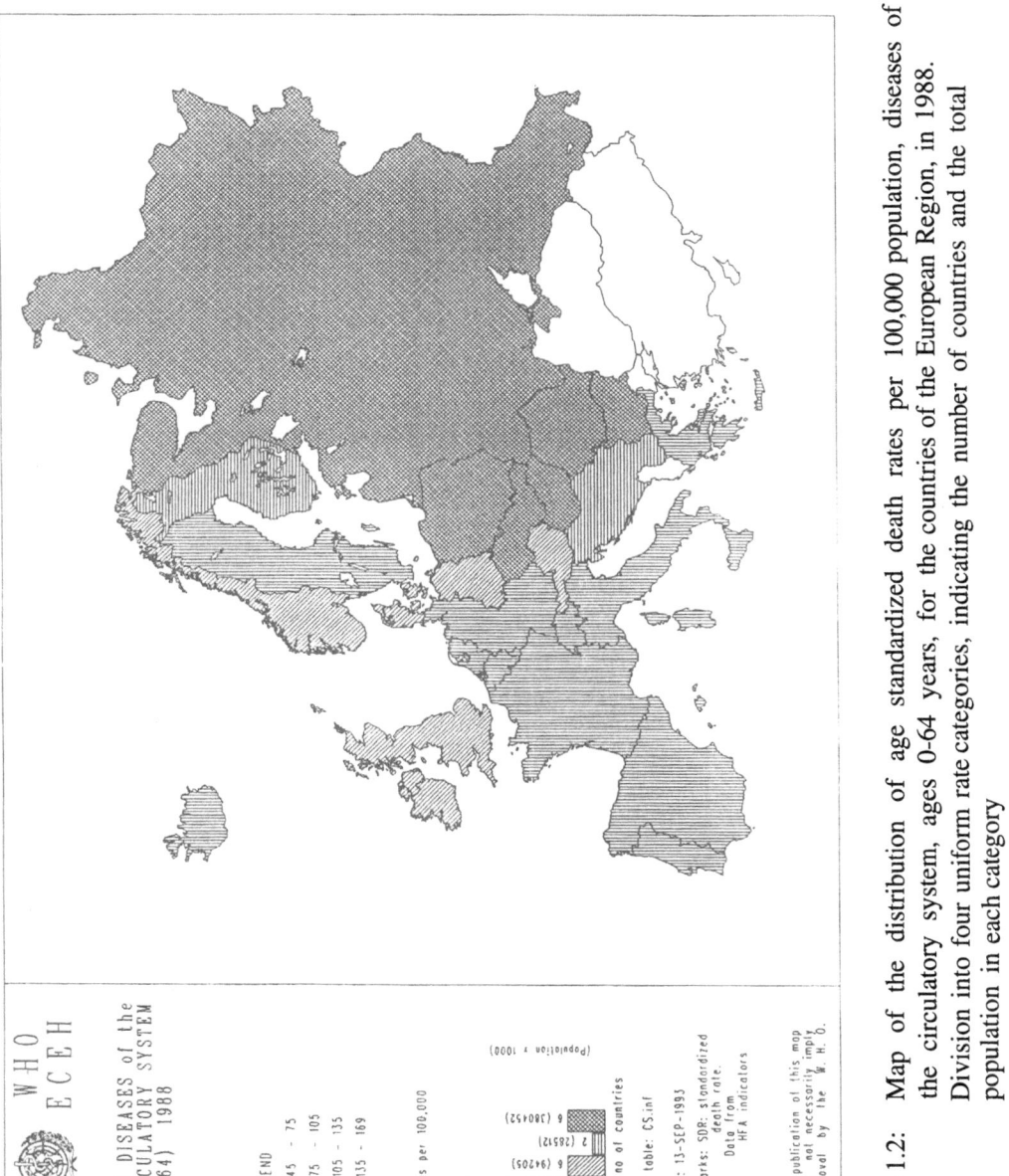

Map 1.2: Map of the distribution of age standardized death rates per 100,000 population, diseases of the circulatory system, ages 0-64 years, for the countries of the European Region, in 1988. Division into four uniform rate categories, indicating the number of countries and the total population in each category

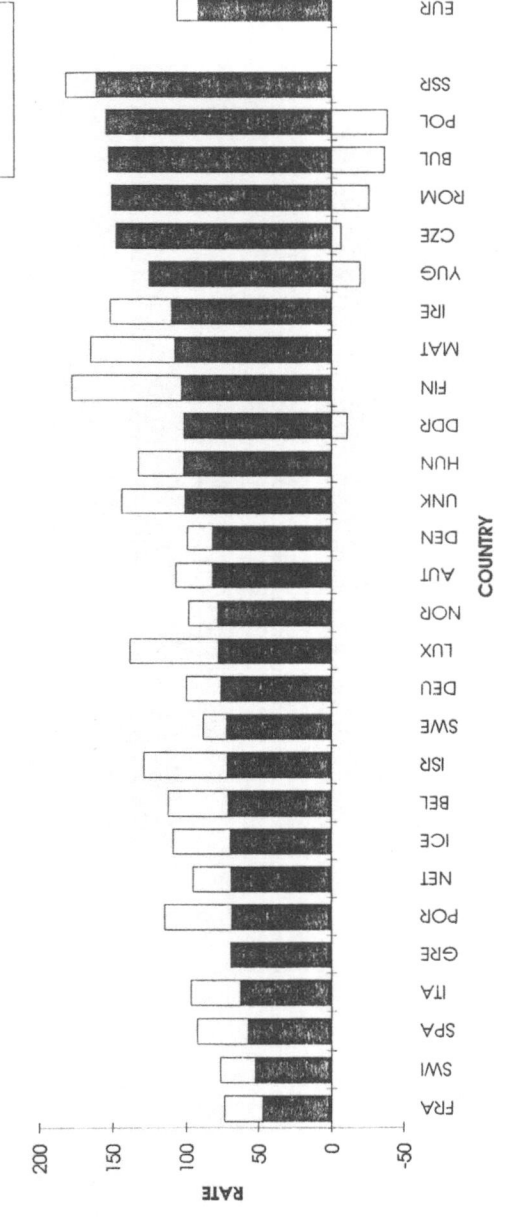

Figure 1.2:   Ranked bar chart, age standardized death rate diseases of the circulatory system, ages 0-64 years, for the countries of the European Region, in 1988. The differences (1972-1988) are shown as well (increases are negative) (data for USSR for 1980–1988)

In an epidemiological model relating to mortality or other health data to risk factors, significant contributions can be expected from lifestyle, the quality of and degree of access to primary health care, environmental factors (occupation, outdoor, indoor and other special situations) and human biology. Variations in any and/or all of these factors can be expected to lead to variations in health status over regions (of any size) for which aggregated data is available. It is expected that, throughout Europe, mortality data (e.g., as in figures 1.1-1.2 and Maps 1.1-1.2) is of high quality and hence the subregional differences expressed are real.

1.6   Conclusions

Detailed examination of the issues involved in consideration of the relationships between environmental conditions and human health reveal a number of activities that involve the collection, storage and manipulation of geographically coded or referenced data. The introduction of a spatial component occurs at a wide range of scales, from global perspective through continental, fluvial, subregional, national, subnational and local, ranging in scale from tens of thousand of kilometres to tens of kilometres. From the input and storage point of view, a relational database management system is appropriate for dealing with all the various levels and layers of information that will be of interest to a programme manager at any level. From an output and information-package point of view, maps are a convenient way of displaying spatial and temporal variations. But maps and a database management system are but two components of a true GIS: the missing function is the analytical tool box. The real utility of a GIS for environment and health has yet to be tested: it can only be tested through the additional step of performing localized and spatial analysis on several variables, to arrive at new and otherwise inaccessible information. It is hoped that the results of the first pilot programmes of a HEGIS programme (e.g., population-based assessment of the health impact of air pollution, pesticide use, etc.) will demonstrate the added value of GIS in this context. At the very least these studies will emphasize the difficulties in obtaining and integrating environment and health data at comparable scales, and in identifying competent and responsible government bodies with jurisdictional responsibilities or authority over appropriate geographical areas.

References

WHO (1985). *Targets for health for all*. Copenhagen, WHO Regional Office for Europe. (European Health for All Series, No. 1).

WHO (1989). *Environment and health: the European Charter and commentary*.
    Copenhagen, WHO Regional Office for Europe. (WHO Regional Publications,
    European Series, No. 35).
WHO (1991). *The implementation of the European Charter on Environment and Health*:
    report on a Working Group, Düsseldorf, 28-30 August 1990. Copenhagen, WHO
    Regional Office for Europe.

Richard M. Stern
World Health Organization
European Centre for Environment and Health
Antonie van Leeuwenhoeklaan 9
NL-3721 MA Bilthoven
The Netherlands

2   INDICATORS OF PUBLIC HEALTH AND ENVIRONMENTAL QUALITY

Erik Lebret

Abstract
This chapter gives a short outline of the relationship between exposure to environmental contaminants and the risk to public health involved. First, the concepts of health and environment will be introduced to provide a common frame of reference. Then, the relationship between environment and environmental contaminants with health will be described in broad terms. General indicators dealing with the different dimensions of public health and of environmental quality are presented. Some general considerations and criteria are suggested and the suitability of potential indicators for use in geographical information systems is discussed. It is concluded that, at present, many indicators lack one or more of the desired properties of indicators for environmental health. Therefore a set of indirect indicators is proposed.

2.1   Introduction

People from many different disciplines are active in the field often described as environmental health. For each of these disciplines, such words as environment and health may have different meanings. Moreover, indicators to describe the status of the environment and of health may be of use for some applications in some disciplines, but not for others. The same indicator may even have a different meaning in another discipline (e.g., environment). In a typical interdisciplinary activity such as geographical information system (GIS) applications in the field of public and environmental health, common concepts and a common frame of reference are a prerequisite for communication between specialists. This chapter therefore starts with the introduction of some concepts and definitions that are relevant to the environmental health arena. Some of these concepts are also incorporated in the public health chapter of the National Environmental Outlook 2; 1990-2010 (Kramers et al., 1992).

2.2   Concepts of health and environment

The concept of health is defined in the Constitution of the World Health Organization as: "Health is a state of complete physical, mental and social wellbeing, and not merely the absence of disease or infirmity" (WHO, 1946). Health status is judged by physicians and by individuals themselves in the light of given physical and mental capabilities. In this context, the health effects of environmental contaminants include not only the occurrence of death or clinical disease, but also effects on (vital) functions that do not directly lead to disease. Another important consequence is that the content of the term

M. J. C. de Lepper et al. (eds.), The Added Value of Geographical Information Systems in Public and Environmental Health, 25–39.
© 1995 *Kluwer Academic Publishers.*

health is not constant but may vary depending on social and cultural values; it can change over time from one culture to the other.

A number of endogenous and exogenous factors determine health status. These are summarized in Figure 2.1. As can be seen, environment may have a different meaning to different people. For some it may mean environmental pollution, for others it may incorporate lifestyle, dietary habits and the like. The figure also shows that environmental pollution, or environment in the narrow sense, constitutes only one group of determinants of health status.

Endogenous factors include the genetic make-up of a person as well as attributes acquired during life (e.g., immunity to a infectious disease, allergies). These factors determine how an individual will respond to the exogenous factors. A clear example is skin pigmentation: a dark pigmented skin can cope better with ultraviolet radiation and is less prone to develop skin cancer than is a lighter pigmented skin. Lifestyle (smoking, alcohol and drug consumption, physical activity, etc.) is an important determinant of health. Tobacco consumption, for instance, is a known risk factor for lung cancer, cardiovascular disease, chronic obstructive pulmonary disease and low birth weight of babies. Dietary factors have positive as well as negative health effects. For many nutrients, a sufficient but not excessive intake is required for good health. The work environment includes such factors as occupational exposure to pollutants, risk of accidents, physical activity, but also such sociopsychological factors as job satisfaction, stress from unemployment risk and the like.

The role of the social environment appears in many epidemiological studies as an important determinant of health, although it is often unclear which (group of) traits of the social environment are responsible. Social networks, income, education, access to medical care, lifestyle, diet and occupational factors are all involved.

Also the health care system is an important determinant of health. Vaccination programmes, health education, access to medical facilities, health care insurance and many other characteristics of the health care system all affect the health status of a population.

Exposure to environmental pollutants is only one of the exogenous factors; the remainder of this section will focus on these factors and their relation to health.

The relation between emission of a pollutant into the environment and the manifestation of health effects is illustrated schematically for contaminants in the compartment air in Figure 2.2. Pollutants are emitted into microenvironments, i.e., locations with a more or less homogeneous pollutant concentration. The resulting pollutant concentration in a microenvironment is determined by dilution, exchange with other microenvironments and physical and chemical processes (decay, deposition).

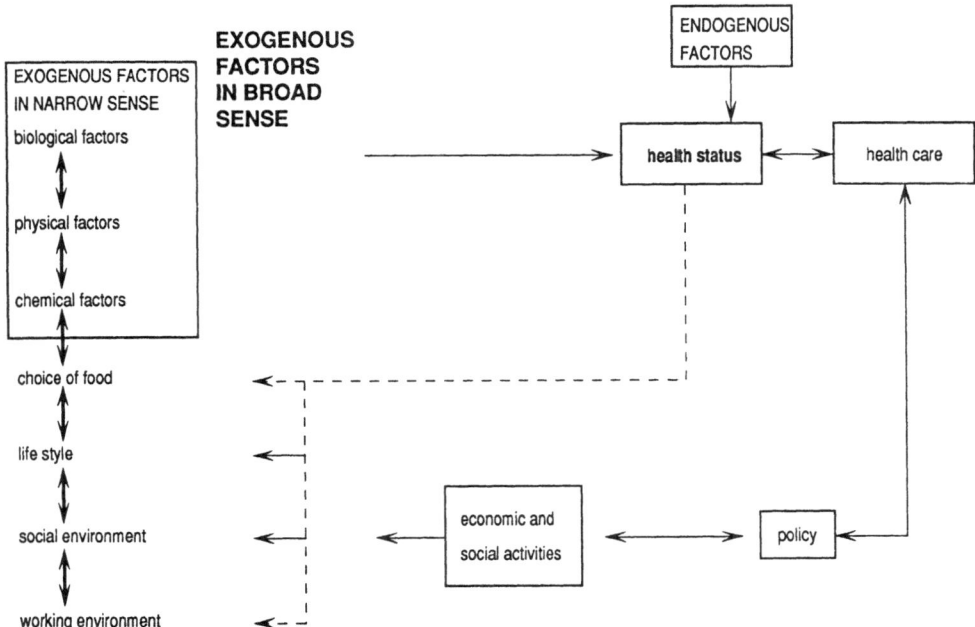

Figure 2.1: Determinants of health status according to Lalonde

Figure 2.2: Chain of events from the emission of a pollutant to the manifestation of health effects

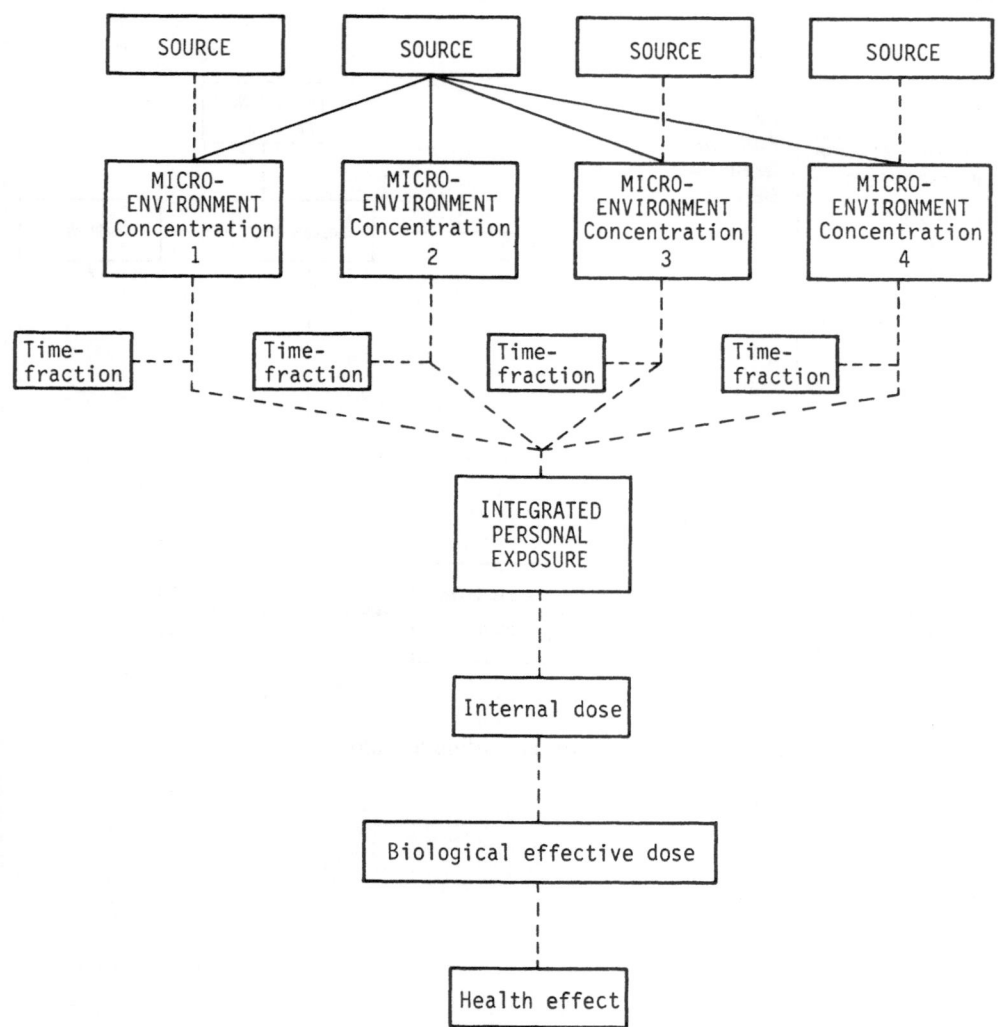

Depending on the air pollutant of concern, microenvironments may include the home, inner-city streets, cars, the work environment and the urban background. Pollutants may disperse from one microenvironment to another. For an air pollutant such as nitrogen dioxide (NO_2) or carbon monoxide (CO), microenvironment 1 might be the home or a room in a home with unvented gas appliances (gas stove, water or space heaters) as sources inside the microenvironment and inner-city streets as microenvironment 2 (with traffic, electricity plants, etc. as sources) that affects the quality of microenvironment 1. Also inside cars (microenvironment 3) there might be a source (faulty exhaust). Otherwise, the pollutant concentration will depend on the exchange with microenvironment 2.

Integrated personal exposure is an important concept in this scheme. Here, exposure is the event when a person comes into contact with a pollutant. Thus, if a person is not present in a microenvironment, no exposure can take place there, regardless of the pollutant concentration. This may appear trivial, but has important implications, since the population spends various amounts of time in the different microenvironments. In modern industrialized society, approximately 70% of time is spend at home and another 20% or more is spend indoors (work, school, homes of relatives, social gathering places). On average, only a few percent of the time is spent outdoors. The time spend in a microenvironment and the concentration in that microenvironment determine the contribution to the individual's personal exposure. Ambient air pollution levels may have a moderate impact on the integrated exposures (due to the limited time fraction spent outdoors), when indoor levels differ from outdoor levels.

Breathing rate and the physical properties of the lung will determine the dose of air pollutants, and pharmacokinetic and toxicodynamic processes influence the effective dose at the target organ. Intrinsic susceptibility, finally, will determine whether a person will develop clinical symptoms.

To complicate matters, it should be noted that several environmental contaminants have exposure through several other compartments and pathways, e.g., through soil into food, drinking-water and soil ingestion (infants). These may all contribute to the integrated personal exposure and to the dose. For reasons of simplicity, these pathways were not incorporated in the scheme in Figure 2.2.

2.2.1 Indicators of health status

Given the broad definition of health, health has many aspects. These cannot all be covered by a single indicator, and hence, a range of indicators exists that deal with different dimensions of health. In a positive approach, the health status of a population can be regarded in terms of life expectancy, or the quality of life. In a negative approach (more common in environmental health), health status is described in terms of mortality, morbidity, functional disorder or annoyance. Some of these indicators are briefly described here.

Table 2.1: Characteristics of some selected approaches to health monitoring (from WHO/CEC 1989)

	Sample Population	Data Providers	Potential For Quantification Of Environmental Impact
National Registers	Entire population of country	(Para)medical personnel Hospital records Laboratory records	Large, provided linked with exposure data (e.g. from occupational registries)
Local Registries	Entire population of smaller territorial/ administrative entities. Sometimes problem regions	(Para)medical personnel Hospital records Laboratory records	Large, provided linked with exposure data (e.g. from occupational registries) Conclusions can be generalized to whole country if sample population is representative
Sentinel Networks	Population covered by data providers	Selected practitioners, laboratories, hospitals	Only if relevant exposure data are concurrently collected or are available from other sources Otherwise specific epidemiological studies are needed
Periodic Health Surveys	Ideally: randomly drawn from population	Specially trained survey team	Only if relevant exposure data are concurrently collected or are available from other sources and dependent on sample size

Life expectancy is usually age- and sex-specific. Present life expectancy at birth (LEB) can only be estimated. The LEB is aggregated over all determinants of health. In this century, LEB has increased considerably in many European countries, among other things, due to techniques to prevent and treat infectious diseases. In Europe, life expectancy for women is six to nine years longer than it is for men.

The quality of life indicates to what extent the prolonged life is spent in good health, e.g., expectancy of life without limitations due to morbidity or infirmity. For example, in the Netherlands, expectancy of life without limitations at birth was 59.5 years for men and

58.2 years for women in the period 1981-1985. Estimated additional life with limitations was 13.3 years (men) and 21.3 years (women) (van Ginneken et al. 1989).

Mortality is commonly expressed as total or cause-specific mortality per 100,000 inhabitants. For comparison of populations with differences in age structure, mortality data must be age-standardized. A population with a high percentage of elderly people will have a higher cancer mortality than a younger population. The registration of such vital statistics is fairly well established in the European Region.

Morbidity data on the incidence (number of new cases of a disease over a specified period of time) or prevalence (total number of people in a population with the disease) are less readily available than mortality data. Indirect measures of morbidity are the utilization of medical services, including hospitalization, consultation of general practitioners, use of medication, sick leave and the like.

Table 2.2: Examples of appropriate health indicators to be monitored (from WHO/CEC, 1989)

Age Group	Cause-Specific Mortality	Cause-Specific Morbidity
0 - 11 months	Infectious diseases (e.g., diarrhoeal and respiratory diseases) Congenital malformations sudden death	Infectious diseases (respiratory diseases, diarrhoeal diseases) Congenital malformations Functional deficiences (e.g., growth retardation) Rickets
1 - 14 years	Accidental deaths from traffic accidents and poisonings Infectious diseases (e.g. malaria)	Traffic injuries and poisonings Rickets Allergies Infectious diseases
15 - 64 years	Cardiovascular diseases (inc. cerebrovascular disease) Certain forms of cancer (e.g. lung cancer, mesothelioma, leukaemia) Infectious respiratory diseases Traffic accidents and accidental poisonings Liver cirrhosis Ill-defined causes of death	Cardiovascular diseases (inc. cerebrovascular disease) Cancer Liver cirrhosis Traffic injuries and accidental poisonings Drug-induced adverse effects chronic respiratory diseases (e.g., chronic bronchitis, emphysema) Communicable diseases related to hygienic conditions Allergies, asthma Certain nutritional diseases
65+	Certain cancers Cardiovascular diseases Respiratory diseases Liver cirrhosis	Certain cancers Cardiovascular diseases Respiratory diseases Liver cirrhosis

Table 2.3: Examples of environmental risk factors to be monitored (from WHO/CEC 1989)

-    Ambient pollutants including emissions from anthropogenic sources

-    Risk factors in the workplace

-    Emissions from energy production, industry, pesticide use, etc.

-    Alcohol consumption

-    Smoking status

-    Patterns of drug use

-    Hygienic quality and availability of drinking-water

-    Continuous or accidental release of hazardous industrial chemicals (solvents, chlorinated hydrocarbons) into soil, water or air

-    Environmental radiation from natural sources (e.g., radon) or due to the accidental release of radioactive materials into the environment

-    Traffic density and driving patterns (cars, motorcycles, bicycles, etc.)

The suggested examples for monitoring of environmental risk factors are not listed in any order of priority.

Function disorder of certain organs can be observed even before disease becomes manifest. Examples are loss of pulmonary function or kidney function. Birth weight and the growth or decline with age of body length and organ functions can also be considered as such. Usually such data are only available from special surveys and do not provide country-wide coverage.

Annoyance reaction to environmental pollution in the population should be considered a health effect according to WHO's definition. In some cases, annoyance is influenced by the perceived health risk and not so much by the real risk of the exposure to the contaminant involved. However, in many cases annoyance is a direct reaction to noise, odour or exposure to irritants.

Tables 2.1, 2.2 and 2.3 show examples of monitoring strategies and examples of health indicators and environmental risk factors to be monitored, as suggested by a WHO/CEC consultation in 1989 (WHO/ECE, 1989). The availability of health indicators depends on whether they are routinely monitored or whether results from surveys in representative population samples are available (Table 2.1). Health care infrastructure is important in that respect: access to data in countries with a national medical care system and central administration is more feasible than in countries with independent (private and/or local) health agencies. Sufficient coverage of the population is, in most countries, obtained only for mortality and, to a lesser extent, morbidity data. Quality and comparability of indicators across countries is of course a prerequisite for any meaningful comparison of data using GIS or other techniques.

The relation between different type of indicators can be illustrated by a simplified example dealing with the biological responses in a population at a given (high) exposure to air pollution (Figure 2.3). The illustration shows that, given a certain exposure, the majority of the population will exhibit some physiological response, the health significance of which is difficult to evaluate. A smaller proportion of the population will react with pathophysiological changes and yet a smaller part will have clinical manifestation of disease. Mortality will occur in the most sensitive individuals and people with an already otherwise compromised health. Similar reasoning with respect to biological responses can be applied to other types of pollutants.

2.2.2 Indicators of environmental quality

Environmental quality is treated here within the context of direct effect on human health. Moreover, it is restricted to exogenous factors in the narrow sense, i.e. chemical, physical and biological agents. Indirect long-term risks from global warming and the effects on ecosystems are excluded. Environmental quality is commonly expressed in terms of one of the following indicators.

Source strength (emission rate), presence of sources or type of emission (continuous or intermittent, point or line source, high or low stack) indicates environmental quality. These usually need to be combined with models to determine transport and decay in water, air or soil to provide a quantitative estimate of concentrations and thus environmental quality.

Concentrations of contaminants can be measured in monitoring networks for water, soil and air. This normally takes place at fixed monitoring sites that are considered to be representative for a region or a population. The selection of the site is critical for the feasibility to use the data for desired purposes. A site in the centre of a city is well situated to describe air quality in that city but not to estimate the deposition of cadmium on arable land in order to assess contamination of food.

Figure 2.3: Schematic representation of the biological responses in a population on exposure to a pollutant (adapted from American Thoracic Society, 1985)

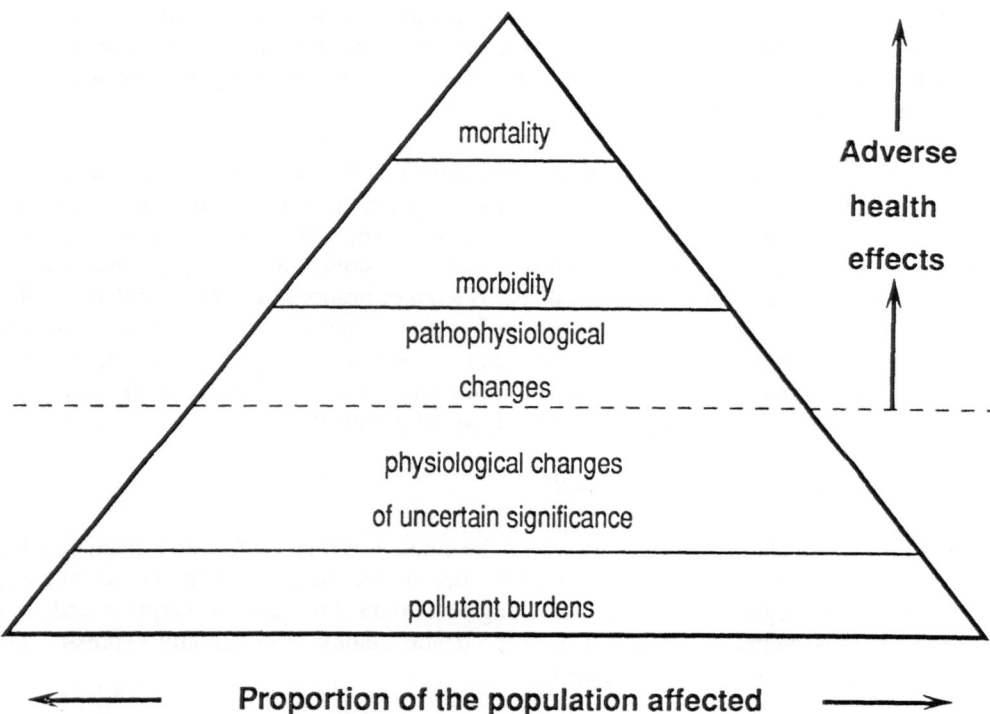

Exposure measures can be obtained in a direct (only feasible for air pollutants) or indirect fashion. Contaminant levels are determined in foodstuffs and beverages and in the air of microenvironments. These values are then weighted by the quantity of food or beverage consumed, or the amount of time spent in different microenvironments. Usually, only a few selected food products or microenvironments can be measured. This method integrates exposure through several pathways and routes.

Biomarkers, i.e. pollutant levels or metabolites thereof, can be determined in body materials such as blood, urine, saliva, or obduction/biopsy material, e.g., kidney or fat tissue. This provides an indicator of the body burden of pollutants. The method not only integrates exposure through pathways and routes but also reflects the differences of uptake efficiency of pollutants from the environment into the body.

Biomarkers can also reflect early effect indicators. The difference between an exposure biomarker and effect biomarker can become very small.

It follows from Figure 2.2 that environmental quality should preferably be measured in at least the dominant microenvironments and pathways, since these most closely reflect the integral personal exposure and thus potential health effects. In practice, however, only emission rates or contaminant levels are available and for just one compartment or for one microenvironment for a significant proportion of a country. More detailed measures of environmental quality are only available from (small-scale) surveys.

2.3   Combining health and environmental quality indicators in a GIS

GIS can be a helpful tool to display and analyse geographic differences in health status. These differences occur due to geographical differences in the endogenous and exogenous determinants of health. Such differences exist between countries and within countries. Similarly, environmental quality may differ from one area to the other due to differences in emission, meteorology and topography and geological and hydrological situation. Since the purpose of using a GIS for the representation of data on health status and environmental quality is to make geographical comparisons, there is an strong need for high-quality data with a high degree of specificity. This includes: high coverage of the population involved and uniform and accurate classification of the causes of mortality, of morbidity and of functional disorders. Also, general data on the population, such as age structure, is important for calculation of valid rates for areas or subpopulations. Similar reasoning applies to data on environmental quality. A special concern in data quality is the natural variation in disease incidence. Particularly in rare diseases, the yearly rates in small populations, or areas, are heavily influenced by chance variation. Thus, the time-frame over which the data are collected should be taken into account when comparing disease rates between areas.

2.4   General considerations and criteria for health and environment data in a GIS

When data of sufficient quality are available on health status and environmental quality, there is the potential to link the data in a GIS to study similarities in geographical patterns. While doing so, it should be realized that the health determinants as represented in Figure 2.1 are usually closely interrelated. The health care system, health policy, socioeconomic activities and exogenous (and sometimes endogenous) factors in the broad sense may exhibit comparable geographical patterns, both between and within countries. This can easily lead to what is called the ecological fallacy: a health risk is attributed to one exogenous factor, while in reality it is caused by one (or more) other determinants.

A further impediment is caused by the lack of specificity of both health and environmental quality indicators. Most diseases have multifactorial causes with many determinants for which no data is available. Most commonly available environmental quality indicators, i.e., emissions and concentration data, lack specificity for the phenomenon of interest because they poorly reflect the exposure that may cause the health effect. Moreover, present environmental quality indicators may not be associated with present health status in the case of a latency period between exposure and the manifestation of disease (e.g., cancer).

As Tables 2.1, 2.2 and 2.3 show, the list of potential indicators is long. This raises the statistical problem of multiple comparisons. The chance that a region has an increased cancer incidence for any one of some 80 types of cancer is over 50% when a P-value of ≤ 0.01 is used ($1-(0.99)^{80}$) (Neutra, 1990). This phenomenon warrants restrictions on the number of indicators to be used for a core set with desired properties.

Presently, only data on cause-specific mortality seem to have adequate coverage and standardization. To a lesser extent, this is also the case for some morbidity data. On the exposure side, sufficient coverage is usually restricted to emission rates and ambient air pollution (which is a poor predictor of population exposure). For data on food contamination, there is the additional problem that food is transported over great distances and is often not consumed in the region where it is grown (and contaminated). With the present (lack of) completeness and data quality and specificity, it appears that the possibilities of linking health and environment data in a GIS is limited to the upper part of the triangle in Figure 2.3 (mortality and morbidity). These are, of course, the most serious effects, but the ones occurring less frequently are also due to environmental contaminants. Moreover, these effects have other well known risk factors that should be taken into account, but for which data is usually not available in sufficient detail.

In the selection of (a core set of) indicators to be used and in the interpretation of findings, the following criteria deserve careful consideration:

(i)   completeness of data;
(ii)   accuracy of data;
(iii)   precision of data;
(iv)   signal-to-noise ratio;
(v)   time-frame between cause (exposure) and manifestation of health effect;
(vi)   specificity of environmental quality indicator for relevant (e.g., past) exposure;
(vii)   specificity of health indicator;
(viii) adequacy of the control for confounding factors;
(ix)   effects of multiple comparisons.

2.5   Indirect indicators for environmental health

Instead of directly linking the data on health and environment in a GIS, with its numerous limitations, indirect indicators for environmental health are a possibility to consider. This can be done by comparing measured or modelled pollutant concentrations with uniform health-related quality guidelines as a benchmark. These can have the form of acceptable daily intake for foodstuffs or WHO's *Air quality guidelines for Europe* (WHO, 1987) can be used. In this case, the indicator to be used may have the form of a percentage of time that a guideline is exceeded, or the percentage of food or harvest with contaminant levels above a certain level or the percentage of homes or streets with air pollution levels above the guideline. These indicators may then be displayed and analysed at an appropriate aggregation level by using GIS.

This approach (although not in a GIS framework) was successfully applied in the Netherlands "Concern for Tomorrow" report to assess the health risk of indoor air pollutants (van der Wiel et al., 1990). This way, it could be shown that indoor air quality exceeded health-based quality guidelines in a majority of the houses, based on the distribution of indoor sources in the housing stock combined with results of specific surveys. On an international scale, this approach has been adopted currently in the data-gathering phase for the WHO report *Concern for Europe's tomorrow*.

2.6   Conclusions

Health is a multidimensional phenomenon with a wide range of determinants including exposure to environmental pollutants. This exposure take places through different microenvironments and pathways and is, in general, poorly reflected by concentration data collected at fixed monitoring sites. Few indicators rank high on the set of criteria presented here. Most lack complete coverage, quality or specificity for the relation of interest, i.e., the role of exposure to environmental pollution in the etiology of disease.

It is suggested that a core set of indicators should include indirect indicators of environmental health. This approach has distinct advantages. Firstly, it takes into account the functional disorders and other health impacts that are not routinely monitored, since the quality guidelines aim at preventing such effects. Secondly, this approach is not affected by confounding effects of other health determinants that may obscure the true relation between exposure to environmental contaminants and health. Thirdly, this approach allows one to assess the risk to future health effects from the present exposure to environmental contaminants.

References

ATS (American Thoracic Society (1985). Guidelines as to what constitutes an adverse respiratory health effect, with special reference to epidemiologic studies of air pollution. *Am Rev Respir Dis*, **131**: 666-668.

Ginneken, J.K.S. van, A.F.I. Bannenberg & A.G. Dissevelt (1989). *Gezondheidsverlies ten gevolge van een aantal belangrijke ziektecategorieën in 1981-1985. Methodologische aspecten en resultaten.* Leiden, NIPG-TNO/CBS.

Kramers, P.G.N., E. Lebret & P.J. van de Mheen (1992). Effects on public health; how ill does the environment make us? *In: National environmental outlook 2; 1990-2010.* Bilthoven, National Institute of Public Health and Environmental Protection

Neutra, R.R. (1990). Counterpoint from a cluster buster. *Am J Epidemiol*, **132**: 6-13.

WHO (1946). Constitution.

WHO (1987). *Air quality guidelines for Europe*. Copenhagen, WHO Regional Office for Europe (WHO Regional Publications, European Series, No. 23).

WHO/CEC (1989). *Health monitoring in the prevention diseases caused by environmental factors*. Geneva, WHO (PEP/89.23).

van de Wiel, H.J., E. Lebret, W.K. van der Lingen, H.C. Eerens, L.H. Vaas & M.J. Leupen (1990). Assessing future trends in indoor air quality. *Toxicol Ind Health*, **6**: 103-115.

Erik Lebret
National Institute of Public Health and Environmental Protection (RIVM)
Department of Epidemiology
Environmental Epidemiology Branch
P.O. Box 1
NL-3720 BA Bilthoven
The Netherlands

3 META-INFORMATION SYSTEMS FOR ENVIRONMENT AND HEALTH

Jan A. Bakkes[a]

Abstract
Drawing upon national experiences regarding environmental meta-information and a WHO consultation on a European Metadatabase on Environmental and Health Information Sources (Munich, 8-10 May 1989), the role and elements of a meta-information system in relation to an envisaged European health and environment geographical information system are described. Three essential elements are identified: (i) absolute restriction to meta-information, that is: referral service only; (ii) a human interface; computers are not essential; (iii) a good thesaurus. In addition, the potential for completing national meta-information systems through the forming of an international network is briefly discussed.

3.1 What is a meta-information system?

A meta-information system is a catalogue of information. Its minimum aim is to put the users on the trail of what they need. The information sources to which the catalogue refers can, in theory, be of any sort: ongoing research, people with specified expertise, written documentation, computer models, data sets and so forth. If the meta-information system refers to data sources, it may be called a metadatabase, but this specific phrase is sometimes also used to loosely indicate all systems that hold information about information rather than the information (the data, documentation etc.) itself. The rationale behind putting effort into meta-information systems is that the bottleneck in many studies is not the existence of information but its availability.

The catalogue may operate in several ways: as a database kept by the user or made available elsewhere (e.g., by a host service). It may be disseminated in the form of hard copies such as books or optical disks or it may be an information service in which questions are dealt with by an information officer who is, in turn, supported by a variety of systems.

More essential than the medium used for storage and dissemination is the ability of a good meta-information system to service the clients who cannot exactly specify what they are looking for. Imagine yourself looking for supportive data for your research proposal or for experience that you can use in making a sample design for the underwater soils of your city. In our experience many of the queries fall into this category and

---

[a] The views expressed in this chapter are those of the author and do not necessarily reflect the policies of the National Institute of Public Health and Environmental Protection or the views of other staff members of the Institute.

M. J. C. de Lepper et al. (eds.), The Added Value of Geographical Information Systems in Public and Environmental Health, 41–50.
© 1995 *Kluwer Academic Publishers.*

therefore a human interface is called for - maybe to be complemented in the future with techniques of artificial intelligence. Dealing with vaguely formulated questions also calls for sound and multiple indexing and robust retrieval methods.

Conventional examples of meta-information systems are:

(i)   the well-known bibliographic databases as operated by commercial hosts, such as Chemical Abstracts, hosted among others by the European Space Agency;

(ii)   the catalogues of data sources of international organizations or projects, such as the ACCIS guide to the United Nations Information Sources on the Environment (ACCIS, 1988) the survey by Gesellschaft für Strahlen- und Umweltforschung for the UNEP HEM Office (Fritz 1990) and the EUROSTAT inventory of national data publications published by Statistical Office for the European Communities in Luxembourg;

(iii)   national environmental source catalogues or referral services, such as UMPLIS and ECOTEC (Da Vinci Consulting, 1989), CIMI (Bakkes, 1990) and the US EPA (1989).

Although national environmental compendia (e.g., CBS of the Netherlands, 1990, GEPAT, 1987; Statistisk Sentralbyrå, 1992; Glówny Urzad Statystyczny, 1991) are meant by their compilers to be sources of actual data rather than national catalogues of data sources, it has been established that such compendia are also regularly consulted in the latter role.

The typical user of general meta-information systems can best be characterized as the professional who for a moment steps beyond his or her own specialization (knowing information sources within one's domain seems to be no problem; at least that is not what referral services are typically approached for). The professional may, for instance, be a model-builder working on the exposure to ultraviolet B radiation or summer smog and now looking for precise data on the proportion of time members of risk groups spend indoors and outdoors. Increasingly, commercial information brokers are the ones who do the asking, on behalf of the actual users. In our experience, consultancy firms have always been very regular customers, appreciating the opportunity to have a *tour d'horizon* before starting a new project.

3.2 The added value of meta-information systems

Within the context of a geographical information system (GIS) an obvious but helpful distinction can be made between two types of meta-information:

(i)   information about what is inside the GIS;
(ii)   information on what is available outside the GIS;

This distinction has been made in order to point out that a thoroughly kept catalogue of the data that is inside an (geographical) information system does not constitute the added value of a metadatabase. In itself, it will be a more than average job to keep track of data and sources in the actual data system, which is especially true on a European level, in which case it will probably consist of various national and international components stacked on top of each other.

The added value of an European meta-database for health and environment lies in disclosing sources of information beyond what is contained in the related GIS system. Because we have only vague understanding of the interactions between environmental quality and human health, the subject area of an envisaged European health and environment GIS cannot and must not be defined very strictly. On the other hand, it seems improbable that we can afford to store and keep up to date in such a system all datasets of possible interest - even if we could specify our data requirements. And as our understanding develops, other expertise and data on other variables will appear to be of interest. Therefore, it will be very useful to have a tool that permits to efficiently locate datasets, expertise, reports or current research of possible interest - not only in the starting phase of the development of a GIS but also as a permanent means of looking beyond the contents of the GIS holding the actual data.

3.3 The CIMI formula

The recommendations in this contribution have been arrived at by imagining how the national Reference Centre for Environmental Information in the Netherlands (CIMI) would have to be complemented in order to fulfil the role of a national component of an European metadatabase on environment and health and what this would imply at the international level. Background information on CIMI is given below.

3.3.1 Basic idea

CIMI is an acronym for Centrale Ingang Milieu-Informatie of the Netherlands. It has been developed during the past ten years and has now reached a stable and fully operational form. The basic idea of the CIMI has become clearer in time: large sets of information and knowledge should be left in the hands of the scientific environment that created it, thus achieving optimal support and up-to-dateness. At the same time, knowledge about and access to environmental information as well as expert knowledge should be increased and facilitated on a structural basis. Although the above may seem self-evident, there was an initial (i.e., late 1970s) inclination towards the creation of a central, large, computerized dataset. Only very gradually, when it came to considering practical consequences, was this idea abandoned - a line of development towards a meta-information system that seems in fact rather common. First of all, it was feared that the centralistic option would lead to updating problems. Furthermore, institutional willingness to cooperate was not assured, and also the sheer volume of the data was expected to create difficulties. By 1986, the CIMI had arrived at a very conscious restriction to meta-information, as opposed to the actual data or expert answers. Of the three considerations against the more centralistic options, at least the first two (regarding updating and cooperation) are still relevant in my opinion. As to the third aspect (data volume), it will be interesting in the next few years to examine the balance between the tendency to make large datasets directly available for GIS on the one hand and the increasing use of meta-information systems on the other.

3.3.2 Scope

The CIMI provides references to sources of the three following types of information on the environment.

(i)   Expertise, i.e.: knowledge and experience of individual experts or of institutions. Sources at the national level are well covered, including environmental expertise hidden within organizations that do not have a well-known mission in the environmental field, such as the Institute of Soil Fertility, consumer organizations or local government. About 600 institutions are included.

(ii)   Datasets with a predominantly spatial character, e.g., results of environmental monitoring networks, flora and fauna inventories, groundwater mapping, etc. Nationwide systematic sets are all coveredand at least all flora and fauna inventories at a more detailed level. About 120 surveys and datasets are described at present.

(iii)   Literature, both articles and monographs, from or about the Netherlands. Coverage is as complete as it can reasonably be, including "grey" literature. The present growth rate is 5000 references per year; half of this relates to monographs.

The subject area of the CIMI reflects a rather wide concept of environment. It ranges from land-use statistics to the economics of the environment and from enforcement plans to the bird census. In a geographical sense, the scope of the CIMI is limited to the Netherlands: information from Dutch sources or regarding the environment in the Netherlands.

3.3.3 System architecture

The meta-information corresponding to each of the above types of information requires its own specific data format and is therefore stored in a different CIMI database. The retrieval information to be assigned to it, such as keywords and classification codes, is harmonized via a common thesaurus, i.e., a structured set of index terms. The distribution over different databases permits addition or removal, as the system develops, of system modules corresponding to specific subject areas. Thanks to the common terms used in indexing, the different modules can be searched with the same search profile. Modular growth of the reference system has been kept in mind also in the design of the thesaurus: it contains relatively simple terms, from which more complex terms can be constructed during computer searches. This is called post-coordination. The thesaurus is composed of a nucleus of limited size (about 1500 general terms) to which annexes are added as specific perspectives are added to the system. The addition of index terms for foreign geography would be an example of such a new perspective. This set-up makes the system flexible and not very vulnerable with regard to ageing: new complex terms can in most cases be formed from existing simple termsand if it is decided to expand the system to a new subject area, this can be done without great problems. Both have been done: a large system module on current research has been abandoned and the addition of a new module on simple computer models is being prepared.

The present thesaurus (Enderman et al., 1990) is a product of cooperation between the CIMI and other institutes. This has provided a good starting-point for the present efforts aimed at coordinated databases (current research and literature) and at shared production efforts (bibliographical databases).

3.3.4 Users and services

The CIMI is intended for professional or semi-professional users of environmental information. The telephone service of the information officers is the cornerstone of the services provided. The information officers handle the intake of information requests and, if necessary, reformulate questions during a short intake interview. They provide referral information (names, addresses, literature references) with the aid of the CIMI computer system or on the basis of personal knowledge. They redirect the customer if CIMI is not the appropriate source to handle the question. They may also offer additional searches in other databanks (databank brokerage).

Some easily accessible products are added to the type of services provided by the information officers. The literature database is available on-line on a publicly accessible host. It also forms the basis of an abstract bulletin. Periodically a summary catalogue is compiled on the basis of the expertise databank and issued in print (CIMI, 1991).

3.3.5 Resources

The CIMI - at the time of writing - is part of the National Institute of Public Health and Environmental Protection. It is essential for a reference centre to be embedded in an organization that plays an active and central role in the subject field. Otherwise it would soon become a sterile affair. The profiles of the sources of information to which CIMI refers are drafted in cooperation with these sources. Only information holders who are prepared to answer questions referred to them by the CIMI will be incorporated in the system. The information in the reference system is verified and updated annually with the aid of the information holders. The cost of the prototype was about USS 2.5 million. The present system costs about US$ 900,000 annually. The bibliographic databank is by far the largest and the most costly. The technical set-up is kept rather conventional (PCs, two minicomputers, BRS retrieval software).

3.4   Recommendations

It is assumed that a meta-information system on a European scale will be built mainly from national building blocks.

3.4.1 Delimitation of a national meta-system

At the national level the CIMI can stand as a model - in the first place not with regard to the scale of operations but with regard to its essential formula. In comparison with the CIMI, some elements have to be added in order to reach a minimum coverage in environment-related health information.

The essential formula consists of the following:

(i)   Absolute restriction to meta-information. The information volume must be kept manageable. Actual data are in better hands with the sources with regard to updating and user support. A clear and restricted role is more easily accepted by information holders.

(ii) Well oriented information officers are the cornerstone of the system, although databases of considerable size may be involved. Many questions are loosely formulated and need interpretation with knowledge of its context. Moreover, it is at present beyond the available resources to computerize the many and recent small facts that are valuable in putting clients on the trail of the information they are probably after, and if we could do so, timeliness would be a problem.

(iii) A good thesaurus, although a backstage instrument, is very important. If the meta-information system can be built from different modules (e.g., modules on different topics or country modules), this greatly improves the flexibility of the system. If the information in the different modules has been indexed on the basis of a common thesaurus, coordinated retrieval should always be possible. This is also true if not all modules are kept by the same institute. A common thesaurus is absolutely necessary if different institutions want to cooperate by filling the same databank.

Other aspects suggested by the experience of the CIMI are the following.

In addition to the service via the information officers, dissemination via other media may have specific advantages. Paper publications are important because of ease of access; on-line access appeared to add much to the credibility of the reference centre among professional information traders, including envisaged partners for collaboration. One aspect of this is that collaborating partners typically want to see the result of their effort immediately.

It is useful to locate the meta-information system physically at a large and central environmental institute. As the information officers are the cornerstone of the system, their orientation to what is going on must be as good as possible. A central place such as RIVM also adds to the credibility of the reference centre, balancing to some extent that it "only" offers information about information.

The budget must be stable and not too modest. Although a meta-information system is relatively cheap when compared with storage and dissemination of actual information, the costs of information input must not be underestimated. Whether a meta-information system is actually used by those for whom it is intended depends very much on its imageand this is largely influenced by it being up-to-date or not. Updating must therefore be regular and frequent and not be influenced by hiccups in the flow of resources.

3.4.2 Scope of a national meta-system

It is assumed that information needs in the field of environment and health in terms of broad categories of sources of information, are not very different from the field of environment in general. Under this assumption, the marketing survey prepared by the CIMI (Research and Marketing, 1986), held among professional users of environmental information, indicates the following priorities for meta-information on environment and health:

(i)   national literature;
(ii)   expertise of individual experts or of institutes;
(iii)   geographically referenced datasets of national and sub-national coverage;
(iv)   current research, at project level.

Topics (i) and (iv) require rather large and therefore costly databanks. It is therefore doubtful that such national databanks will be set up at short notice in the framework of an international project in countries where they do not yet exist. To some extent, existing commercial databases offer an alternative by their good coverage of the international scientific literature. Topics (ii) and (iii) remain as the possible substance of national meta-information systems throughout the region.

3.4.3 The international level of the meta-system

Apart from coordination of the development of the meta-information system as a whole, two specific tasks remain to be dealt with by the central point of the system.

(i)   Development and maintenance of an international thesaurus for environment and health. It could be developed from the CIMI thesaurus and the Multilanguage Descriptor System (Commission of the European Communities, 1983), one of the precursors of the CIMI thesaurus.

(ii)   References to information kept by international bodies. A revitalization of the concept of lead databases for United Nations subsidiary bodies, as developed by ECE Geneva, would be useful.

The national source catalogue and the query handling by the information officers of the national meta-information system imply a certain form of quality statement about a source. Also, coordinated referral is expected from national systems (always refer to source x for data type y). With this in mind, handling of queries about sources in a given country should be handled by the national meta-information system, not by the central point. The central point of the meta-information system should therefore not have a super-catalogue listing thousands of sources within countries.

3.5   Conclusion

The foundation of a concise meta-information system seems logical as one of the preparatory steps towards a health and environment GIS for the European Region (Bakkes, 1990; see chapter 22). There will be a need to maintain the meta-information system permanently. The method of CIMI can be used for the meta-system. Important features are: a restricted role, a special thesaurus and main service via information officers by telephone and fax. Computer requirements are modest. Of the possible information types to be covered, datasets and expertise seem most important in this context. It is assumed that the system will largely consist of national meta-information systems, working together via a common thesaurus. Handling of queries about sources within countries should be left exclusively to national systems.

References

ACCIS (Advisory Committee for the Coordination of Information Systems) (1988). *ACCIS Guide to United Nations information sources on the environment.* New York, United Nations.

Bakkes, J.A. (1990). Outline of a potential environment and health metadatabase for the Netherlands. Prepared for the WHO-EURO consultation on a Metadatabase for Environment and Health, München, 1989. *Inf Services Use*, **10**: 223-228.

CBS of the Netherlands (1990). *Environmental statistics of the Netherlands, 1990,* Condensed English version of the *Algemene Milieustatistiek.* The Hague, CBS.

CIMI (Centrale Ingang Milieu Informatie, 1991). *Gids Nederlandse milieu-expertise 1991-1992,* (Environmental expertise in the Netherlands). The Hague, CIMI.

Commission of the European Communities (1983). *Multilanguage descriptor system for the European inventories on the environment.* Pilot edition, September 1983. Published by Peter Peregrinus.

Da Vinci Consulting (1989). *An information system for access to environmental data in the European Community. Feasibility and planning report, January 1989.* Contains information on ECOTEC, UMPLIS and CIMI.

Enderman, J.C. et al. (1990). *Milieuthesaurus. Systeem van gecontroleerde termen voor het ontsluiten van milieu-informatie. CIMI, TNO-CID, VROM/DVEB-Bidoc.* [Environmental thesaurus. System of controlled terms for disclosing environmental information]. Also on diskette. Bilthoven, 1990. VROM Milieuthesaurus.

Fritz, J.S. (1990). *A survey of environmental monitoring and information programmes of international organizations.* Draft, May-August 1990. Munich, UNEP HEM Office, c/o GSF Munich.

GEPAT, ed. (1987). *Compêndio de estatísticas do ambiente.* Lisbon, Ministry of Environment and Natural Resources: Compendium of Environment Statistics.

Glówny Urzad Statystyczny (1991). *Raport o stanie, zagrozeniu i ochronie srodowiska 1991* [Report on conditions, hazards and protection of the environment], Warsaw.

Research & Marketing, Heerlen (1986). *Onderzoek inzake de behoefte aan en het gebruik van milieu-informatie t.b.v. het projectburo CIMI te Leidschendam.* [Survey regarding the need for and use of environmental information, for the CIMI project].

Statistisk Sentralbyrå (CBS of Norway) (1992). *Natural resources and the environment 1991.* Oslo-Kongsvinger, Statistik Sentrabyrå.

US EPA (1989). *Information resources directory.* Washington, DC, US EPA.

WHO (1991) *Development of a health and environment geographical information system for the European Region.* Report on a WHO Consultation, Bilthoven, 10-14 December 1990. Copenhagen, WHO Regional Office for Europe.

Jan A. Bakkes
National Institute of Public Health and Environmental Protection (RIVM)
RIVM Information Manager for Environment
P.O. Box 1
NL-3720 BA Bilthoven
The Netherlands

PART II

THE COMPONENTS OF GEOGRAPHICAL INFORMATION SYSTEMS

4   AN INTRODUCTION TO GEOGRAPHICAL INFORMATION SYSTEMS

Henk J. Scholten and Marion J.C. de Lepper

Abstract
This chapter introduces geographical information systems (GIS) by describing some fundamental concepts important to them in more detail. It aims to help the newcomer to GIS and to provide some understanding of what GIS are and what they can be used for. The concept of GIS will be described in a number of ways: formal definition, examination of distinguishing features in regard to other information systems, the type of questions it can help answer, its ability to carry out certain specific operations and, listing the components a GIS comprises. In addition, the subject matters of this chapter are both the latest developments in the fields of hardware and software in a GIS environment and future directions and trends in these fields. As the concept of GIS not only includes hardware and software, this chapter will, in addition, briefly consider types of data and data storage and the different types of user groups and their differentiated GIS needs.

4.1   Defining geographical information systems

About 30 years have passed since a number of geographers conceived a system for storing and organizing spatial information in a computer and since R.F. Tomlinson proposed the basic concept of a geographical information system (GIS). Supported by the remarkable development of computer hardware and software since then, research and development in GIS has made great progress all over the world. Paralleling advancements in the technology has been the growth in applications.

The use of GIS grew dramatically in the 1980s and has become commonplace in business, universities and governments, where they are applied for a variety of purposes. Consequently, many different definitions of GIS have been developed. The amount of literature - and discussion - on what constitutes GIS is vast and growing. The GIS literature contains many terms that are used as synonyms for GIS, which include spatial information system; geo-data system; geographical data system; and land-use information system (Clarke, 1986; Parker, 1988). This plethora of meanings arises in part because GIS is a very young science and has important relationships with other physical and social science disciplines that involve the processing of spatial data. These include remote sensing, photogrammetry, cartography, surveying, geodesy, environmental science, regional science, planning and, of course, geography. All these disciplines are attempting the same sort of operation - the processing of spatial information.

M. J. C. de Lepper et al. (eds.), The Added Value of Geographical Information Systems in Public and Environmental Health, 53–70.
© 1995 *Kluwer Academic Publishers.*

A number of definitions of GIS appear in the literature (Burrough 1986; Marble, 1983; Calkins & Tomlinson, 1984; Berry, 1986). Despite efforts to distinguish different types or forms of spatial information systems, it is commonplace to use the term GIS to refer to all types of spatial information systems. We will illustrate the important differences between spatial information systems that fall under the GIS umbrella.

Many widely used computer programs, such as spreadsheets and statistical or drafting packages, can process simple geographic or spatial data. Why, then, are they usually not thought of as a GIS? The generally accepted answer is that a GIS is only a GIS if it permits spatial operations on the data by means of linking datasets together using location as the common key. Certain complex spatial operations are possible with a speed and set of functions with a GIS that would be very difficult, time-consuming or impractical otherwise.

GIS differ from computer graphics because the latter are largely concerned with the display and manipulation of visible data. What's more, computer graphics do not pay much attention to the non-graphic attributes that the visible entities might or might not have and that might be useful data for analysis.

In short, a GIS does not hold maps or pictures, it holds databases. The geographical database structure (topological structure) is central to a GIS and is the main difference between a GIS and a simple drafting or computer-mapping system, which can only produce good graphic output. This brings us to a general definition of GIS: "An organized collection of computer hardware, software, geographic data and personnel to efficiently capture, store, update, manipulate, analyze and display all forms of geographically referenced information" (Dangermond 1990).

GIS is not just hardware and software, or even these two plus data. As Dangermond (1990) puts it: "Every geographical information system is composed of five elements: computer hardware, computer software, data, personnel to run the system and a set of institutional arrangements to support the other components. Each of these components must be coped with successfully." Hardware is used to store, process and display the digital map data; software performs GIS operations. Expertise i.e. people who provide the intelligence to use the system, is the most important of all these components of a GIS.

4.2   Types of spatial information systems

Although GIS do have a lot in common with other spatial information systems, major differences also exist between the three main groups of spatial information systems that are distinguished below.

4.2.1 Computer-aided design (CAD) systems

CAD systems are graphics systems used by industrial designers, architects and landscape architects to support and display their work. CAD has replaced the drawing board. Although early CAD systems were purely automated drawing systems, later packages have provided enhanced facilities for qualitative and quantitative design analysis as well as database facilities in which information can be stored and in which a large number of symbols can be used. CAD systems allow the possibility of automated drawing, the manipulation of drawings (changes of scale, location, zooming in, rotating and editing) and the presentation of this information in a professional format. Graphics software development has been under considerable influence from the world of CAD and GIS presentation software has tried to incorporate CAD features. The developments in GIS have tended to lag behind those taking place in the CAD environment. It is important to recognize that the basic concepts in the two worlds are different. GIS is totally involved with the concept of the database, whereas CAD is more concerned with the design process and the accompanying use of symbols. Both the hardware and software in the CAD environment are focused on presentation, whereas this principle applies less within the framework of GIS.

Automated mapping (AM) was developed initially as another application of computer graphics technology alongside CAD. However, AM was extended to permit the storage and retrieval of associated locational and attribute data linked to the graphics, thus creating Automated mapping/facilities management (AM/FM) for use as a specialist application in the management of utilities.

4.2.2 Land-use information systems (LIS)

The objective of this type of information system is to function as an administrative system for the management of geographical data on land use, and in this sense, LIS is similar to AM/FM in several ways . Many different demands are often made of LIS that have a direct influence on the way in which data should be stored. In the example of real estate information, concern over the legal obligations related to the precision of data may be of paramount importance. However, with regard to information on pipelines or cables, for instance, there may be less worry about legal matters but much more concern that the exact location of a pipe or cable can be established quickly. In this type of information system, the central focus is the development of a very detailed database. Moreover, there are tools available in this type of system that allow data with a very high degree of accuracy (double precision) to be stored, managed, integrated, updated and displayed. Relatively few spatial or geographical analyses are carried out within this type of system.

4.2.3 Geographical information systems (GIS)

While LIS provide powerful tools for local planning authorities and public infrastructure agencies operating at a very detailed, micro scale, GIS tend to support analysis, planning and evaluation at a more macro scale. They are used in a variety of contexts to assist the research required to formulate and evaluate central and local government policy with respect to different aspects of physical and environmental planning on the one hand and economic or strategic planning on the other. Such information systems adopt both a database approach and a set of tools to assist data collection, management, updating and presentation. The distinguishing feature of GIS in comparison with LIS and CAD is the availability in GIS of the tools of spatial analysis.

These three categories of spatial information system are certainly not mutually exclusive as far as data are concerned. Many public utilities and local authorities have, for example, begun to use CAD systems to illustrate their information because CAD systems fulfill the demands for precision. However, the problem that confronts this group of users is the addition of information and its relationship to associated attribute information. A local authority or municipality may input its inventory of street lamps into a CAD system that possesses a number of qualities that are indispensable in the design process, for example. However, it is clear that this information may be related to other land-use information, and thus the integrated use of the information requires a database approach, for which a LIS or GIS is required.

The conclusion is straightforward: when the goal is to make maps, one can suffice with a mapping package, which is rather cheap and easy to handle. When one is interested in the spatial analysis of the data, one needs a GIS environment, which can integrate all the data and can provide some necessary analysis tools.

4.3   The objectives of geographical information systems

GIS represent a technology designed to achieve particular objectives. In recent years, a variety of GIS products to assist the management and manipulation of spatial and non-spatial data have arrived on the market, and users worldwide have begun to gain familiarity with these new systems. Experience suggests that there can be no doubt but that the application of GIS is making significant contributions in facilitating the availability, integration and presentation of information.

As a technology, a GIS is not necessarily limited to the confines of one independent system. It may well have several components, each with a particular objective. A GIS must therefore accomplish the following four main tasks:

(i)   Firstly, there is the storage, management and integration of large amounts of spatially referenced data. A spatially referenced database contains two types of information, locational and attribute data. Locational or spatial data are two- or three-dimensional coordinates of points (nodes), lines (segments) or areas (polygons). Non-locational or descriptive data, on the other hand, refer to the features or attributes of points, lines or areas. Data is obtained from a wide variety of sources, and one of the most important features of GIS is the facility to integrate data, e.g., converting data values to a common spatial framework.

(ii) The second main task of GIS is to enable spatial retrieval. These spatial query possibilities are very flexible and very powerful. It is possible to retrieve data for defined areas (polygons) such as a municipality or postal code area, but there are also a number of possibilities that would appear to be extremely relevant for public and environmental health purposes (Openshaw & Charlton, 1990), such as: radial retrieval on a single point, radial retrieval for a search region defined as a buffer around a site's perimeter, retrieval of data lying in a corridor focused on a linear map feature (e.g., an overhead wire), retrieval of data for a user-defined area and retrieval of data for search areas defined by manipulating existing digital map databases (e.g., relationship of cancer rates with geology patterns).

(iii) The third main task of GIS is to provide methods that enable analyses that relate specifically to the geographical component of the data. The analysis techniques may be simple or more sophisticated. At the simplest level, for example, data about different spatial entities such as soil type (per square kilometre) and land use (by local administrative area) can be combined by overlay analysis. At an intermediate level, GIS may allow statistical calculations of the relationship between datasets to be computed, or distances between entities may be used to determine the route that must be followed to move as quickly as possible from one location to another. The implementation of GIS creates the opportunity to have an enormous amount of (point)data available which requires new, explorative methods (Openshaw 1990). The most sophisticated analysis occurs when modelling is introduced. In this context, there are a variety of analytical opportunities. It is possible, for example, to use atmospheric modelling techniques to discover which areas might be affected by pollution resulting from an explosion at a particular hazardous installation (e.g., Chernobyl), given certain wind and weather conditions. Alternatively, modelling methods can be used to determine the impact of locating a large public facility (e.g., a hospital) at different sites in a city region.

(iv)   The fourth main task of GIS involves displaying data on map forms of high quality. Maps no longer have to be drawn by hand; they are an implicit product of all the work that is carried out within a GIS. However, for many different purposes, other forms of display (e.g., graphs and tables) may also be required, often for use in combination with maps. The health for all system is a nice example of an alternative method of displaying spatial information in an interactive and user-friendly way.

4.4   Technical aspects of geographical information systems

4.4.1 Types of data storage

A spatially referenced database contains two types of information, locational and attribute data. Locational or spatial data are two- or three-dimensional coordinates of points (nodes), lines (segments) or areas (polygons). Non-locational or descriptive data, on the other hand, refer to the features or attributes of points, lines or areas. Data are obtained from a wide variety of sources, and one of the most important features of GIS is the facility to integrate data, e.g., converting data values to a common spatial framework.

It is possible to distinguish further between three forms in which locational data can be incorporated within a GIS: raster, vector and quadtree storage.

Raster or grid storage

This form of storage for locational data involves a regular grid of cells being laid over an area. Attribute data are collected for each grid cell which may measure, for example, 500 m by 500 m. This means that the whole area is covered by a group of cells, each of which has an attribute value. Within a grid-oriented system of this type, it is often the case that only limited use is made of the attribute data. Satellite photographs, in which a considerably smaller grid size is used, provide raster information. In a satellite photograph, a single value is attached to each cell. In this way, factual data can be collected in a very efficient way. Several authors indicate how remote-sensing data is frequently contained in raster-based GIS.

Vector storage

The storage of locational data in vectors gives a very precise representation of reality. In this way, points, lines and areas are incorporated as Cartesian coordinates in the computer. While lines can only be represented as a series of cells in a raster structure, in the vector storage method, the exact middle point of a line can be identified. Whereas an area is represented by a group of cells with an angular boundary in the raster structure, the precise boundary of an area can be included when a vector structure is used. This precise storage of x- and y-coordinates usually generates larger datasets. Within a vector system, the relationship between locational and attribute data is of great importance. Each element is related to a record in the database with the same, unique identification number.

Quadtree storage

The quadtree storage method falls between the grid and vector storage methods. In the quadtree storage method, the data are stored in grid cells of variable size. Within larger homogeneous areas, a large cell size is used, whereas towards the edges of the area, the cell size diminishes to form a precise picture. An area is therefore covered by considerably fewer quadtree cells of varying size than is the case in the regular grid storage method. The speed with which analyses can be carried out with a quadtree structure is high, while the original precision of the data is retained rather well.

4.4.2 Hardware in a GIS environment

The initial GIS products date from the 1970s, a decade in which the typical hardware configuration comprised a central computer surrounded by memory and storage disks and a number of peripheral devices. Time-share systems enabled a large number of terminals with lines attached to the mainframe to be used at the same time. At the beginning of the 1980s, this centralized approach was extended by connecting minicomputers to the central mainframe in order to carry out certain processes. This was the period in which GIS came of age. Very large databases began to be assembled and the need for processor capacity increased enormously. In the middle of the decade, the personal computer (PC) arrived, although at that moment its impact for GIS meant in many cases little more than an extra terminal. Nevertheless, it goes without saying that the PC has become central to the popularization of GIS.

In the second half of the 1980s, the PC played an important role in simple GIS tasks such as the automatic production of maps. It is exactly this function that has brought GIS to the attention of many people, while the basic concept of a common central database tends to be forgotten. Attempts to transfer fully fledged GIS from the mainframe onto the PC have been commercially successful, even though in many cases their performance leaves something to be desired. Of much greater importance in the mid-1980s was the further development of minicomputers and work stations connected to a network. At this time, the larger organizations making use of GIS realized that the central mainframe option was not the solution to a number of GIS tasks. It was recognized that each separate task required its own processor capacity or its own working environment. Throughout the 1980s, it became clear that hardware vendors were meeting these demands perfectly well by means of further optimization of minicomputers using servers and the increased processor capacity of work stations and that hardware costs were decreasing significantly. We may expect in the 1990s a further exploding growth of processor power. The differences between PCs and work stations is disappearing. An even more important development is the growth in the storage capabilities. Gigabytes are already available at the servers in the network, but they become also available for a PC user. This happens by the use of compact discs. Large geographical data files of more than 500 megabytes can now be bought for just 30 dollars! Both developments will be responsible for a further decentralized use of GIS without losing the benefits of the centralized data gathering and data management.

4.4.3 Software in a GIS environment

Identification of the main tasks of a GIS environment allows us to specify a corresponding set of software demands.

Data input and verification

Data input covers all aspects of transforming data captured in the form of existing maps, field observations and sensors into a compatible digital form.

Data storage, database management and query

Data storage and database management concerns the way in which the data about the position, linkage (topology) and attributes of geographical elements (points, lines and areas representing objects on the earth's surface) are structured and organized, both with respect to the way they must be handled in the computer and how they are perceived by the users of the system.

One of the most significant advances in GIS software development came at the end of the 1970s, with the introduction of the concept of the relational database. In the relational model, different datasets are linked together by the use of common key fields. For example, attribute data available for two different sets of spatial units (areas) will require a third set of information to show how the two spatial databases fit together. This type of structure can also be used to construct spatial databases in which lines are linked together to represent polygons. The creation, maintenance and accessing of a database requires a data base management system (DBMS). In order to process very large quantities of information, a relational DBMS is a standard tool of most GIS software packages. It facilitates data manipulation and analysis.

Data analysis

The software required to perform certain analytical tasks varies according to the nature of the problem, the quantity of information available and the objectives of the organizations involved. A variety of basic analytical tools are now available within most of the GIS software packages. The overlay procedure, for example, has been widely used for combining different datasets in order to identify areas or sites with required characteristics. Buffering, address-matching and network analysis are additional tools adopted in planning applications of GIS. However, the development of analytical functionality within GIS has tended to neglect the important benefits that modelling procedures can contribute through data transformation, integration and updating; simulation; optimization; impact assessment; and forecasting (Birkin et al., 1987; Openshaw, 1990; van Beurden & Scholten, 1992). The specialized and complex nature of modelling algorithms and the restrictions imposed by processor capacity have both been influential in keeping the modelling component separate. One of the challenges confronting the next generation of GIS is to improve the integration of modelling and GIS so as to provide researchers, decision-makers and planners with enhanced model-based decision-support systems.

Access and presentation software

One of the most important functions of GIS development is frequently the provision of access to information by a variety of different users. In the context of planning, individuals in different departments of the same organization (public works, social services, transport, parks, etc.) may require access to the same database. Similarly, users in different national, regional and local organizations may wish to access the database simultaneously.

The technological advances in computer hardware over the past 20 years have had a direct impact on the presentation of information in GIS. There have been striking advances in automated mapping, i.e., the development of programs enabling maps to be produced automatically. The experience gained over this period has meant that very professional products are now the norm and relatively cheap, micro-based mapping software (ATLAS*Graphics, GIMMS, MAPINFO, for example) provides high quality output.

4.5   Type of user and kind of need

When an organization takes the decision to implement a GIS, one of the first steps that ought to be taken is an analysis of the needs of users and the decision-making process that the GIS is intended to influence and be a part of. This step is most critical to GIS success because this will ask the potential users of the system about what information they need in order to perform their work. Such an analysis will result in an identification of different user types and their differentiated needs.

A number of categories of users can be identified on the basis of the objectives of the organization to that the users belong. A number of types of organizations can be classified that differ from one another according to the type of activity that each performs.

The type of GIS that is adopted and applied therefore varies between each category of organization and between organizations of different size and function within the same category. However, it is possible to identify particular groups of individuals across the spectrum of organizations whose occupational characteristics with respect to GIS are distinctive.

Different types of organizations

Four main types of organizations may be distinguished. Firstly, there is the research institute, where research may be carried out in order to find solutions to problems or answers to questions posed by external paymasters. Data collection and manipulation take place and descriptive, explanatory and predictive analyses are undertaken. Secondly, there are administrative institutions such as public utilities or property registration agencies. Here, the objective is to manage information in such a way that the process of acquiring and manipulating data is made as simple as possible. The management of a waste-disposal system of sewerage pipes for a local administrative area is one example where accurate and quick answers are required on the basis of the information stored in the GIS to questions such as: "Where are the oldest parts of the system?"; "What is the total length of pipe involved?"; "At what depth are the pipes buried?"; and "How many houses are connected to these pipes?".

The third type of organization is the government agency, whose objective is to formulate policy recommendations. For this purpose, concept design and evaluation takes place. Commercial enterprises are the fourth type of organization. Their aim is to maximize their profits by selling goods and services. Information is collected and manipulated within an integrated modelling or GIS environment, for example to establish optimum locations for new retail outlets.

Different types of user groups

Within every organization - whether it can be categorized in, for instance, the first or the third type of the above specified types of organization - different groups of people perform different tasks. There is, in other words, a differentiation of functions. Different functions and different needs for and use of information are interrelated. It is clear that, because of these differences, the user demand on GIS varies. The following groups of users can be distinguished: information specialists, researchers, research coordinators, policy-preparers, decision-makers and third parties.

In each of the above specified four categories of organizations, information specialists are required to acquire and manage data, computer hardware and software. The information specialist usually works with the raw data and requires a large GIS (e.g., ARC/INFO, ARGIS, SYSTEM9) in which the database is fundamental and which is flexible and able to be connected to other systems. Researchers, on the other hand, tend to be confined to their own institutes or to commercial companies. They work either with raw data or data that has been partially processed or transformed. They demand user-friendliness from a GIS, analytical features (as with SPANS, for example) and appropriate interfaces that allow the transfer of information to other packages for modelling and other purposes. Research coordinators are concerned with the interrelationship between the different products of the organization and therefore work with manipulated data and require a simple, user-friendly GIS. Policy formulation is usually the responsibility of a government agency, and the main requirement of a GIS in this context is that it should be easy to use. The same applies to the decision-makers in administrative, government and commercial organizations, whose job it is to translate information into policy statements and to third parties, who simply utilize information provided by government agencies or research institutes.

The successful implementation of a GIS therefore depends on a careful preliminary assessment of the type of GIS required to meet the demands to be satisfied by the various users of the system. Ideally, there needs to be a detailed functional specification that outlines the needs that a GIS has to meet. While GIS packages are now available for the above-identified range of GIS users, there remain considerable difficulties, particularly in large organizations, in establishing the configuration of hardware and software and in enabling the easy exchange of information between the machines and packages involved. It is important to ensure that there is a GIS management framework that takes into account the separation of functions between user departments and the central data store.

Differentiated GIS needs

In general, a GIS enables four main functions: preparation, analysis, display and management of geographical data. Preparation includes such functions as data collection, digitizing point data and editing. The analysis function examines data in order to create new data, the purpose of that is producing information. All operations which produce graphic output fall under the display function. Management is the handling of permanent alphanumeric and geographical data. From the foregoing it clearly follows that not all users wish or need to make similar use of these main functions. The distinction between different types of user of spatial information have important consequences in practical terms in the application of GIS software. Table 4.1 shows the different user groups and their varying demands.

The tools for processing geographical data differ widely in level of sophistication. Meijer (1989) refers to this as a continuum from low-end to high-end GIS. At the one end are simple microcomputer mapping packages solely for presentation of data; at the other end is the mainframe implementation of complex tools for all kinds of geographical analysis. The emergence of new hardware and software systems has greatly contributed to the possibilities for different users to choose the right GIS tools. It is not the purpose here to review a number of GIS software packages, nor to identify their advantages and disadvantages, but merely to describe in more general terms the main characteristics of low-end and high-end GIS tools.

High-end GIS refers to a full integration of all tools in one system which is, among other things, capable to handle both raster (grid) and vector information and to perform such types of analysis as overlay techniques, buffering etc. Low-end GIS consist of tools that have only part of this functionality. Notwithstanding the fact that a high-end GIS has a much larger functionality than a low-end GIS, not all users are professionally trained GIS users and will probably be better off with the right GIS tools at the lower end of the GIS spectrum.

From the foregoing sections on different users and their differentiated GIS demands follows that, in most cases, three characteristics are most important: user-friendliness, interfaces and output quality.

Table 4.1: User groups and their demands

Type of user	Information demand	User demand	Type of GIS	Development
Information specialist	Raw data	Data Management Analysis	Large Flexible	Links to other packages
Researcher	Raw data and pretreated data (=information)	Analysis Well accessible Specific	Compact Manageable GIS software	Macro languages Problem-oriented
Management/ decision maker	Strategic information Policy models	Easy to use Evaluation/ Key information	"Small and beautiful"	User-friendly interfaces
Target group/ third parties	Information to users	Well accessible	"Small and beautiful"	User-friendly interfaces

Source: Scholten & Padding (1990).

User-friendliness is most important for the decision-maker who has hardly any knowledge of computers. Even the reading of a manual should not be necessary for completely self-explaining programs. The decision-maker should be able to use the GIS tools almost without any specialist training. A transparent menu-driven program is most suitable. User-friendliness is far less important for the information specialist who is familiar with working with commands and will probably only work with menus if this would speed up the work.

Interface with other software - sometimes high-end GIS - is, however, of utmost importance for the information specialist. Both the locational and attribute information must be easily transferable to and from other systems for a proper functioning of low-end GIS. Those engaged in research, policy preparation and policy-making are less interested in the graphics interface; in a true GIS working environment, the information specialist takes care of the conversion. These user groups will not create the locational information they need themselves. The digitizing will often be obtained from third parties. The locational information they work with will in general not be derived from the data, but a priori given, such as administrative boundaries. The database interface is critical to these users as they will have all kinds of attribute information in packages such as dBaseIII+, Lotus123 or a statistical package such as SPSS/PC+. Finally, transferring data to the graphic GIS tools should be very easy.

The third aspect of low-end GIS, the quality of the output, refers to the product of the GIS process, which is used in reports and must therefore be of high quality. Complex maps are made by the information specialists for which high-speed media such as electrostatic plotters are required. For all the other users, the relatively low-priced pen-plotters and laser printers will do.

4.6 Conclusions

In the definition of GIS, the accent clearly lies on data gathering and management, data integration and analysis and data presentation. These components are the determining factors for the structure of the GIS. Symbolically represented, the structure can be seen as a house, the GIS house. The most salient feature of a GIS is the presentation, the final product. Final products made with the aid of a GIS - an ecological map or a catchment area of a hospital - determine this presentation side. They can be regarded as the final results of the data processing. Seen in terms of the terminology of a GIS house, the presentation can be regarded as the roof of the house. It must ultimately support the decision-making.

In contrast to the presentation side, we have the geographical database as the basis of a GIS, the foundation of the GIS house. It is the geographical part of the data collection that is important for the operating processes within the various organizations. Topographical data for the GIS come under the heading of these collective location data, as do ports or coastlines, local authority boundaries or certain functional subdivisions that are important for an organization. A number of features of these geographical databases are important in discussing the GIS house concept. First of all these data are frequently not specific to a certain organization but only describe the area of study of an organization. Secondly, these data are highly defined and standardized. Thirdly, these data have been measured with a certain degree of accuracy and are kept updated. Lastly, these data, as explicitly stated at the beginning, are expensive when it comes to gathering and management. These different points make it obvious that agreements need to be reached on cooperation, on the one hand within an organization and on the other between organizations with the same geographical area of study.

Between the output of the GIS and the geographical core dataset are the alphanumeric databases. Alphanumeric data and geographical data are integrated and processed into information in the form of tables, graphs and maps, for instance, by all kinds of processes. The databases are the building bricks of the house. The data processes such as the modelling of water flow or the modelling of noise, however, are the mortar you need if you want to end up with the information requested.

The GIS house is a concept for information supply (Figure 4.1). Around the GIS house there are a number of other factors at play. After all, for implementing the processes hardware and software are needed. Agreements have to be reached and standards devised for the hardware and software but also for the data. An organization is needed that makes available personnel, space, facilities and financing. And last, but by no means least, there are the people themselves who can work with an information system but who are also responsible in managing it.

A sophisticated GIS has the ability to perform all the above mentioned tasks and can answer the following type of questions:

What is at...? The first of these questions seeks to find out what exists at a particular location. A location can be described in many ways using, for example, place name, postal code or geographical references such as latitude and longitude.

Where is it...? This question is the converse of the first and requires spatial analysis to be answered. Instead of identifying what exists at a given location, we may want to find a location where certain conditions are met (e.g., an unforested section of land at least 2000 m^2 in size, within 200 m of a road or railway station and with soil suitable for supporting buildings).

What has changed since...? The third question might involve both of the first two and seeks to find the differences within an area over time (trends).

What spatial pattern exists...? This is a more sophisticated question that may be posed in order to determine whether cancer is major cause of death among residents near a nuclear power station or to determine how many anomalies there are that do not fit the pattern and where they are located.

What if...? The last of these questions is related to modelling and is posed to determine what happens, for instance, if a new road is added to a network or if a toxic substance seeps into the local groundwater supply.

Answering these questions covers 90% of our information demand. However, it makes no sense to expect that GIS will solve 90% of our information demand.

Figure 4.1. The geographical information system house

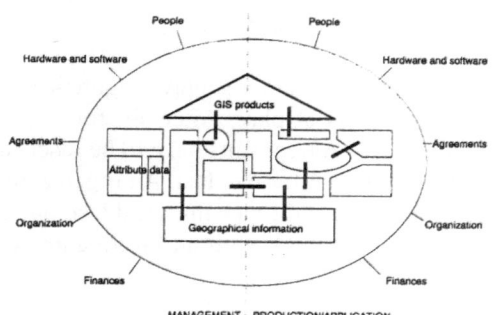

References

Berry, J.K. (1986). Learning computer assisted map analysis. *J Forestry*, 39-43.

Birkin, M., G.P. Clarke, M. Clarke & A.G. Wilson (1987). *Geographical information systems and model-based locational analysis: ships in the night or the beginnings of a relationship?* Leeds, School of Geography, University of Leeds.

Burrough, P.A. (1986). *Principles of geographical information systems for land resources management.* Monographs on Soil and Resources Survey 12. Oxford, Clarendon Press.

Calkins, H.W. & R.F. Tomlinson (1984). *Basic readings in geographic information systems.* Williamsville, SPAD Systems.

Clarke, K.C. (1986). Recent trends in geographic information system research. *In: Geoprocessing*, **3**, pp. 1-15.

Dangermond, J. (1990). How to cope with geographical information systems in your organisation. In: H.J. Scholten & J.C.H. Stilwell, ed. *Geographical information systems for urban and regional planning.* Dordrecht, Kluwer.Academic Publishers.

Fedra, K. & R. Reitsma (1990). Decision support and geographical information systems. In: H.J. Scholten & J.C.H. Stillwell, ed. *Geographical information systems for urban and regional planning.* Dordrecht, Kluwer Academic Publishers.

Marble, D.F. & D.J. Peuquet (1983). *Geographic information systems and remote sensing*. Manual of Remote Sensing, Society for Photogrammetry and Remote Sensing, USA.

Meijer, E., M. Meijer & M. van der Vlugt (1990). The integration of different GIS-platforms in practice. *In*: Harts, J.J., H.F.L. Ottens & H.J. Scholten ed. *EGIS'90 proceedings*. Utrecht, EGIS Foundation.

Nijkamp, P. (1990). Geographical information systems in perspective. *In*: H.J. Scholten & J.C.H. Stillwell, ed. *Geographical information systems for urban and regional planning*. Dordrecht, Kluwer Academic Publishers.

Openshaw, S. (1990). Spatial analysis and geographical information systems: a review of progress and possibilities. *In*: H.J. Scholten & J.C.H. Stillwell, ed. *Geographical information systems for urban and regional planning*. Dordrecht, Kluwer Academic Publishers.

Openshaw, S. & M. Charlton (1990). *Applying GIS to small area health statistics system*. Paper presented at the SASHU Technical Workshop in collaboration with WHO, 22 June 1990, London.

Parker, D.H. (1988). The unique qualities of a geographic information system: a commentary. *In: PE&RS*, 54 pp. 1547-1549.

Scholten, H.J. & M. van der Vlugt (1990). GIS in Europe, a state of the art, *In*: Worrall, L., ed. *Information systems for urban and regional policy analysis*. London, Bellhaven press.

Scholten, H.J. & P. Padding (1990). Working with GIS in a policy environment. *In*: *Environment and planning B: planning and design*, **17**: 405-416.

van Beurden, A.U.C.J. & H.J. Scholten (1990). The environmental geographical information systems of the Netherlands and its organizational implications. *In*: Harts, J.J., H.F.L. Ottens & H.J. Scholten ed., *EGIS'90 Proceedings*, EGIS Foundation, Utrecht.

van Beurden, A.U.C.J. & H.J. Scholten (1992). Analytical capabilities and possibilities of geographical information systems. *In: New technologies and techniques for statistics. Eurostat proceedings*. Luxembourg, Office for Official Publications of the European Communities, pp. 159-171.

Henk J. Scholten
Free University Amsterdam
Department of Regional Economics
De Boelelaan 1105
NL-1081 HV Amsterdam
The Netherlands

Marion J.C. de Lepper
National Institute of Public Health and Environmental Protection
P.O. Box 1
NL-3720 BA Bilthoven
The Netherlands

5 THE INTEGRATION OF INFORMATION IN GEOGRAPHICAL INFORMATION SYSTEMS

Arthur U.C.J. van Beurden and Marion J.C. de Lepper

Abstract
The integration of different types of information in a consistent manner is one of the most important issues for all disciplines attempting the same sort of operation: processing spatial information. The possibilities for data integration have grown dramatically with the emergence of geographical information systems (GIS). The key question addressed in this chapter is whether managers, planners and scientists in a wide range of applications actually need GIS and their sophisticated capabilities for successful information integration in order to achieve sensitive and intelligent decision-making.

5.1 Introduction

Decision-makers and analysts of the 1990s, in fields ranging from regional planning, environmental management, public health, land planning, natural resource management, environmental risk assessment and planning, tax mapping, ecological research, demographic research, epidemiological research to international marketing, are faced increasingly with complex issues. These professionals realize that there is often a geographical or spatial component to their questions: where is the optimal location for a new hospital; which bird, fish and animal populations are affected by an oil spill; is the incidence of disease related to proximity to toxic waste sites? Answering this type of questions requires both geographical and other information.

Access to relevant information is a critical component when trying to answer questions related to problems such as those mentioned above. A very significant problem here is that the required information may come from a variety of sources, is often only available in many different forms, may be of various types, extensive in quantity, variable in quality, referring to areal units of different size, may contain differences in temporal accuracy and so forth. It goes without saying that - in order to understand and penetrate the complexity of the problems faced nowadays - integration of this variety of information is needed as well as a means to successfully perform this integration.

This chapter will review the general principles of information integration and will look at the advantages that geographical information systems (GIS) appear to offer in the integration of information. This will be illustrated by means of a case study wherein a state agency was asked to determine the best location for a new hospital. In this chapter,

M. J. C. de Lepper et al. (eds.), The Added Value of Geographical Information Systems in Public and Environmental Health, 71–86.

methods by which the integration is performed and the objectives of the integration, or - to put it differently - based on the questions it can help to answer. These three types of data integration are: visual integration (performed by systems designed for graphics and mapping), (dis)aggregation of information (a conventional approach adopted in order to be able to answer particular queries, for which GIS can yield benefits) and geographical integration (for which GIS is essential). This chapter seeks to demonstrate that one of the key benefits of GIS lies in creating an environment within which data collected from different sources for the same set of areas or individuals can be related in such a way that new, previously unknown information can be generated. This function of adding value through data integration is one of the major strengths of GIS and can be achieved in various ways through overlay techniques with locational information or through combining attribute information using statistical or modelling techniques.

The key contention of this chapter is that more and more decision-makers, planners and scientists in a wide range of applications will realize that GIS is not only useful but in some instances even essential to successful information integration and - because of this - to sensitive and intelligent decision-making. After all, GIS is a tool that not only helps to organize and to integrate data about today's problems, it also provides the capability of performing spatial analysis with the data and of displaying results. In doing so, GIS helps both to answer spatial questions and to understand their spatial relationships; such an understanding is the basis for more sensitive and intelligent decision-making (Burrough, 1986; Scholten & Stillwell, 1990).

5.2   Spatial and descriptive information in GIS

Some of the concepts about GIS are important with regard to the integration of information. One might basically define a GIS as a system only containing a set of programs for capturing, storing, checking, integrating, manipulating, analysing and displaying information. In such a description, only the GIS "engine" (computer) and application software is meant. However, a definition of GIS should say more. GIS also implies specific characteristics of the information used in the system; it is spatial information. Data spatially referenced to the Earth involve a spatially referenced computer database and its own spatial analysis tools. Performing geographical analysis is what sets GIS apart from digital mapping systems.

A GIS is not simply a computer system for making maps, although it can create maps at different scales, in different projections and with different colours. A GIS does not hold maps or pictures; it holds a database. The database concept is central to a GIS and is one of the main differences between a GIS and a simple drawing or computer mapping system, which can only produce good graphic output. Most contemporary GIS incorporate a database management system or provide links with standard database management systems. Going beyond just making pictures requires knowledge of three pieces of information about every feature stored in the computer: what it is, where it is and how it relates to other features. A digital map database therefore always consists of two types of information: spatial and descriptive. Spatial data contains information about location, shape and relationships between geographical features, usually stored in coordinates and topology. Descriptive or attribute information is tabular or textual data describing the geographical characteristics of map features. The composite of spatial and descriptive data is generally referred to as geographical data. The data stored in a GIS, in other words, describe objects from the real world in terms of their position with respect to a known coordinate system, their attributes that are unrelated to position (such as colour, cost, pH, incidence of disease) and their spatial interrelations with each other (topological relations), which describe, for instance, how they are linked together or how one can travel between them.

All spatial data in geographical data can be reduced to four basic topological concepts: points, lines, areas and their coherence (or position or situation). A map is a set of points, lines and areas that are defined both by their location in space with reference to a coordinate system and by their non-spatial attributes. The information conveyed by a map is represented graphically as a set of map components. Locational information is represented by points for such features as wells and telephone poles; lines for such features as roads, streams and pipelines; and areas for such features as lakes, county boundaries and census tracts. A point feature is represented by a discrete location defining a map object whose boundary or shape is too small to be shown as a line or area feature. Or it could represent a point that has no area, such as the elevation of a mountain peak. A line feature is a set of ordered coordinates that, when connected, represent the linear shape of a map object too narrow to be displayed as an area. Or it could be a feature that has no width, such as a contour line. An area feature is a closed figure whose boundary encloses a homogeneous area, such as a state, a land-use area or a body of water. Attribute information describes the characteristics of features, typically stored in tabular format and linked to the feature by an user-defined identifier. For example, the attributes of a well, represented by a point feature, might include depth, pump type and owner but also area and link to the water layers below.

Topology is the method for explicitly defining spatial relationships. It is the way to add the position or situation information to mere locations. In most cases mathematical procedures are used to derive topological information. For maps, topology defines connections between features, identifies adjacent polygons and can define one feature such as an area as a set of other features (i.e., lines).

Most maps represent the real world, which has been projected from the Earth's globe onto a flat surface. The real-world coordinates are also projected and transformed to one of many new coordinate systems, which represent real locations on the Earth's surface. A projection is nothing more than a geometric transformation, characterized by a set of mathematical equations and parameters. The locations on a map and their true locations on the earth are therefore always known. Like flat maps (known from almost prehistoric times), GIS use various planar (often called Cartesian) coordinate systems to map the Earth's surface. This planar coordinate system may also contain the longitude and latitude.

Commonly used map projections are coordinate systems called Universal Transverse Mercator, Lambert Conic Conformal or Albers Conic Equal-Area. Each map projection used is based on a particular set of mathematical equations and parameters, which also relate to the way real objects are represented on the flat map. Some map projections preserve the integrity of shape; others preserve accuracy of area, distance or direction and many distort all of these characteristics. A coordinate is an x,y location in a Cartesian coordinate system or x,y,z in a three-dimensional coordinate system. Features on the earth's surface are mapped onto flat, two-dimensional maps as points, lines and areas. An x,y (Cartesian) coordinate system is used to locate map objects, which by means of the known projection can be referenced to ground locations. Each point is recorded as a single x,y location. Lines are recorded as a series of ordered x,y coordinates and areas as a series of x,y coordinates defining line segments that enclose an area. With x,y coordinates, points, lines and areas are represented as a list of coordinates and not as a picture or graph. Derived from these coordinates is topological information.

These two types of information — spatial and descriptive — are stored as files on a computer. The power of a GIS lies in linking these two types of data and in maintaining the spatial relations between the map features, whatever spatial operation is performed on them (Figure 5.1).

5.3   Information integration

For the purpose of this chapter, three different types of integration of information are distinguished, based on both the methods by which the integration is performed and the objectives of the integration, or based on the questions it can help answer. These three types of data integration are: visual integration (performed by systems designed for graphics and mapping), (dis)aggregation of information (a conventional approach adopted in order to be able to answer particular queries, for which GIS can yield benefits) and geographical integration (for which GIS is essential). This distinction does not pretend to be exhaustive but merely serves as a context within which the advantages and disadvantages of information integration using different methods can be discussed; this will be illustrated by means of a case study (Tydac Technologies, 1989).

Figure 5.1: Linked spatial data and descriptive data

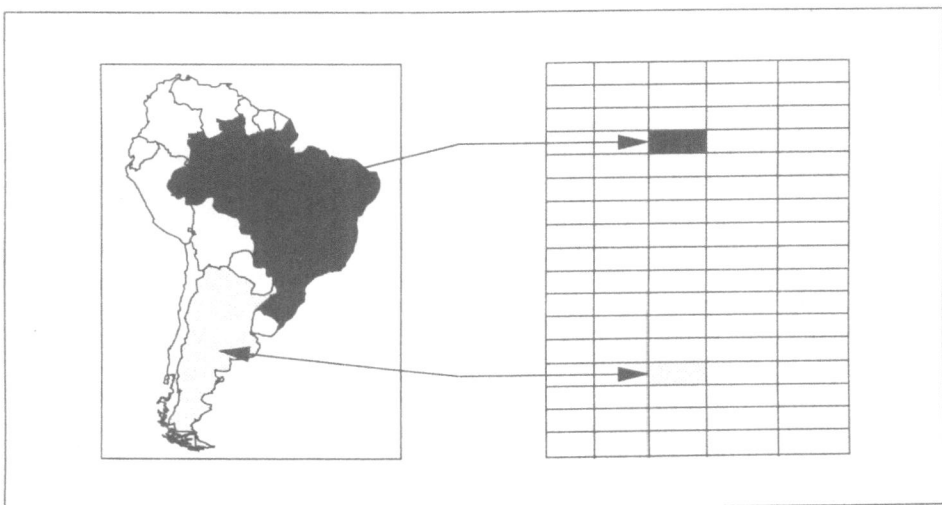

In this case study, a state agency was asked to determine the best location to site a new 60- to 100-bed hospital in the state of Maryland, USA, using a number of criteria, to provide input to planners. Based upon interaction with appropriate experts, it was determined that there are a number of critical factors involved in determining where the new hospital should be located. These factors basically fall into three categories: physical, demographic and competitive. The physical factors were those which were important to locating the hospital from a purely physical point of view. Since it makes little sense to locate a new hospital that is not accessible to a major population centre, one such factor was pure location (distance) with respect to cities or other incorporated places. Another factor was the infrastructure with respect to utilities such as water and sewers; after all, hospitals are major installations that require reliable water and sewage service. A third factor in this category was the slope of the terrain. From a construction point of view, it is important that the slope of the site selected be within reasonable limits for the construction of a major facility. The demographic information in the study is population. Obviously, population is going to be the single most important factor in determining the demand for hospital services. It is also known that older people specifically require statistically significantly more hospital services than the rest of the population does. The third category of information was competitive in nature and is also critical to the site selection process. This information was measured in terms of current hospital capacity.

The database available for the Maryland study contains all kind of data: point observations of facilities, numerous basic data for the State of Maryland on a 90-acre grid cell basis, 56 different land-use zones within the state measured on 10-acre grid cells, point datasets containing the location, name and population of the cities and towns, road network information such as state and interstate highways at 1:2,000,000 scale and so on.

Visual integration

Visual integration is basically what systems designed for graphics and mapping perform. This method of integration of information is, in principle, nothing more than simply drawing maps on top of each other or overlaying transparent copies of maps. The result is a map containing, for instance, both point and area information. An important characteristic of this kind of integration is that the composite map can not be processed, interrogated or analysed: there is no intelligence in the map. The goal of visual integration might be to get an impression, but no more than that, of where certain objects are located in relation to others. The databases used to create the map can not be accessed through the map, and it is consequently not possible to answer particular queries. Sometimes the link with descriptive data may exist in order to allow simple queries of one of the used map layers.

In the Maryland example, this could apply to the slopes. Slope was considered important as a construction issue, and areas where slope made the terrain unsuitable for the construction of a large hospital were excluded from examination. Within an agreed acceptable range, areas with a more gentle slope were preferred, since the construction costs could be expected to be reduced.

Map 5.1:   Slope map (in %) of Maryland

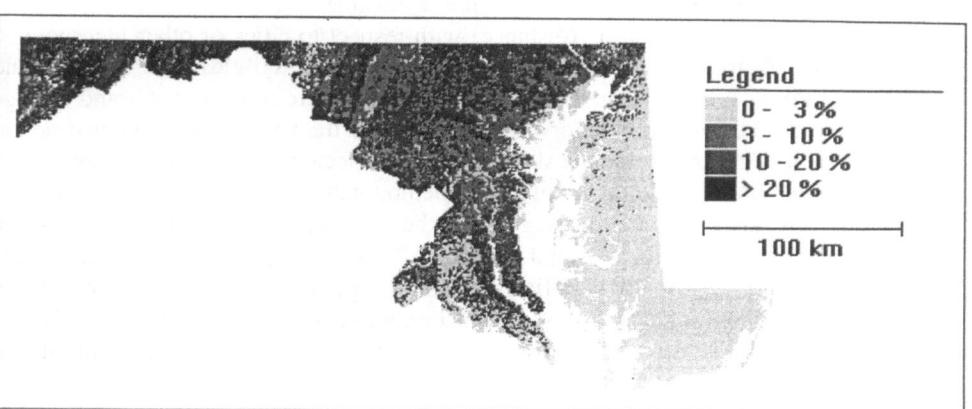

Each ZIP area (zip code zone) can be allocated a certain amount of money to build hospitals. Where slopes are steep, building costs are high. Map 5.1 shows the slopes and the known boundaries of ZIP areas are drawn in Map 5.2.

Map 5.2:   ZIP boundaries of Maryland

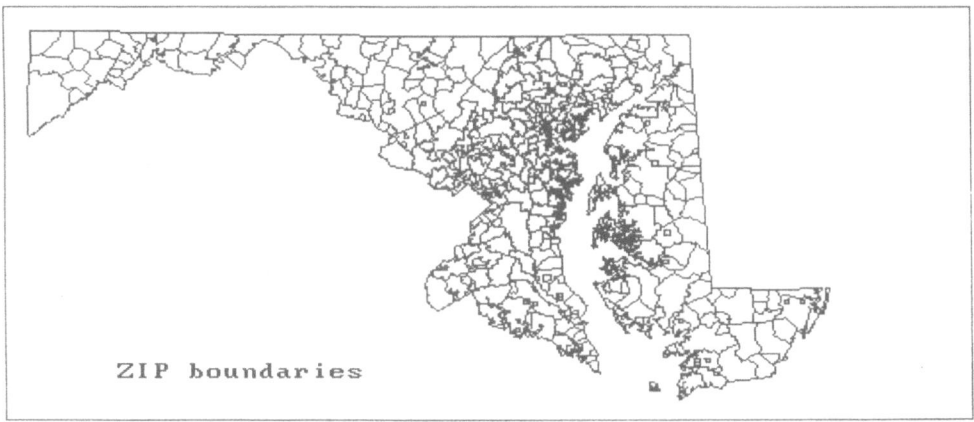

Drawing the two maps on top of each other, one can see which ZIP areas have steep slopes and which ZIP areas have less steep slopes (Map 5.3). It is, however, not possible to perform a query that will, for instance, demonstrate ZIP areas where the mean slope is less than 10%.

Map 5.3:   Visual overlay of the previous maps (Maps 5.1 and 5.2)

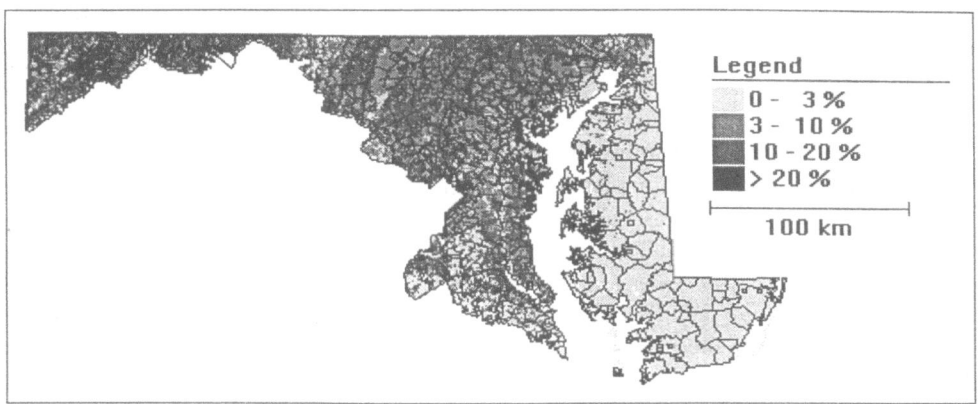

Map 5.4:   Main land-use classes in Maryland

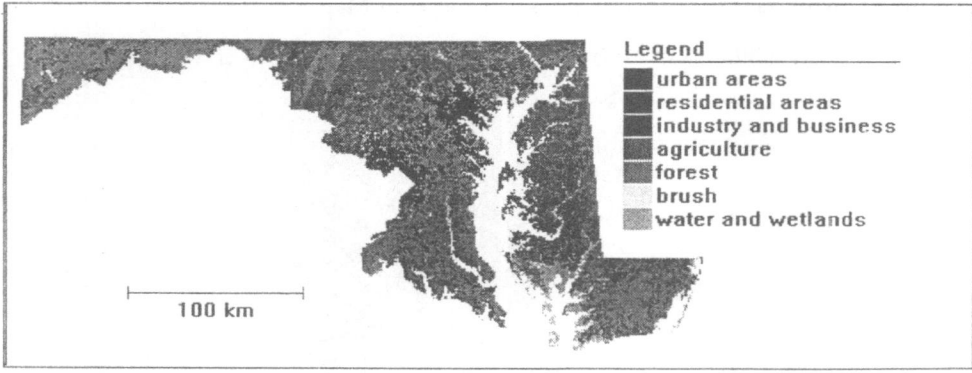

Aggregation or disaggregation

Aggregation or disaggregation of information refers to the second type, i.e. the integration of information by means of attribute-information of features.

Aggregation of information is used in order to obtain a more general idea of where how much of something is located, e.g., the summation of county census data transformed to national data. Disaggregation refers to the reverse. This approach was conventionally adopted for answering spatial queries about standardized areas. This method of integration comes down to the linkage of datasets using geography or space as the common key between the datasets. Information is linked only if it relates to the same geographical area. A major advantage compared with visual integration is that the composite map can be interrogated, processed or analyzed. A major disadvantage is that this can only be done for standardized areas. Identifying, for instance, what exists at a certain location or answering such questions as "where is it.. ?" (e.g., finding a location where certain conditions are met) is possible, but the exact location of features within the standardized area is lost.

In our case study of hospital siting, the general land-use of the candidate sites was judged important. In accordance with planning guidelines, sites were preferred in which the existing land use was high-density residential or institutional. All other land-use conditions were judged equally suitable. From this it follows that information about land-use in each ZIP area is needed. The available land-use data is recorded by satellite and represented in Map 5.4. The ZIP area boundaries are displayed in Map 5.2. Map 5.5 shows the aggregated land-use data, in which case the most abundant land-use class is assigned to a ZIP area. The total area of such a land-use class may be 95% of the total ZIP area, but it may also be 51%. Whether other land-uses occur within the ZIP area cannot be retrieved.

Map 5.5:   Dominant land-use classes per ZIP area in Maryland

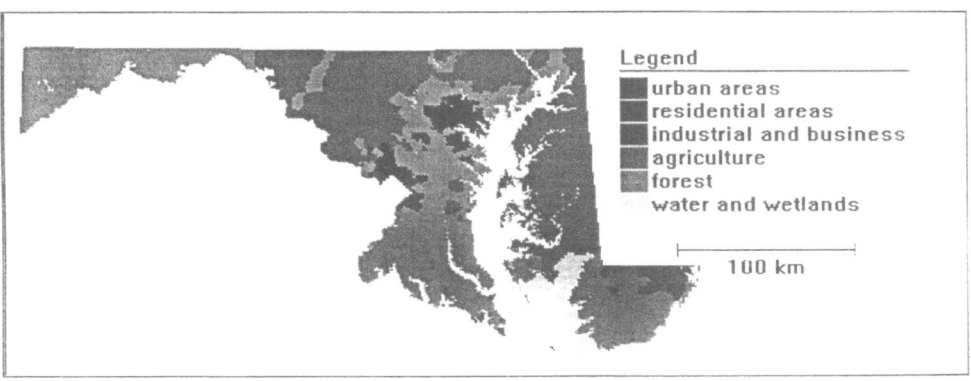

Geographical integration

Compared with the two other distinguished types of integration, geographical integration is, in principle, the only form of true information integration. It is (almost) impossible to perform this kind of data integration without a GIS. In this respect, the power of GIS lies in the link between the graphic (spatial) data and the tabular (descriptive) data as well as in the topologically structured geographical data. The connection is a one-to-many link between the geographical object (feature) and related records in the attribute table. As in other databases, the identifier (key of the feature) occurs in both the feature file and the attribute table (relational database concept). The geographical or topological link is a mathematical transformation function that can be used to convert data from any input map to match with other maps.

Once the link is established, maps may be queried and results may be displayed (attribute information), or a new map can be created based on the attributes stored in the attribute table. The GIS analysis tools may then be used in a way that is consistent with the data and between the maps. Again, here it is demonstrated that the key difference between producing a computer-drawn map and GIS is that GIS contains a database: the points, lines and areas that form part of the map objects in the GIS have attributes attached to them. The database contains information that can be used to answer the user's problem, usually represented in maps, although tabular reports are possible. Most GIS packages have added software to deal with non-spatial (i.e., attribute) data.

The main advantage of the geographical type of information integration is that a wide range of spatial analyses becomes possible. Performing geographical analysis is what sets GIS apart from digital mapping systems. GIS provides a range of analysis capabilities that will be able to operate on the topology or spatial aspects of the geographical data or on non-spatial and spatial attributes combined. Creating and storing topological relationships has a number of advantages. Data is stored more efficiently when topology is used, which speeds up the processing of data and enables the processing of larger datasets. Topological relationships further enable certain analyses, such as modelling any kind of flow through the connecting lines of a network, combining adjacent polygons with similar characteristics and overlaying geographical features. Geographical analysis allows us to study real-world processes by developing and applying models. Such models illuminate underlying trends in the geographical data and thus make new information available. A GIS enhances this process by providing tools that can be combined in meaningful sequences to develop new models. These models may reveal new or previously unidentified relationships within and between datasets, thus increasing our understanding of the real world.

A GIS typically answers questions that require spatial operations on the data. Many widely used computer programs such as spreadsheets (e.g., Lotus 123), statistical packages (e.g., SAS, Minitab, SPSS) or drafting packages (e.g., AutoCAD) can handle simple geographical or spatial data. There are, however, spatial queries that can only be answered using spatial data. The main difference between GIS and systems for computer-aided cartography is the provision of capabilities for transforming and linking the original spatial data in order to be able to answer particular spatial queries.

There are five generic characteristics of spatial data with which a sophisticated GIS should be able to deal. They are usually linked to the kind of queries or questions of users. The data characteristics deal with location, conditions, trends, patterns and modelling.

(i)   The first type of question seeks to find out what exists at a particular location: "What is at ...?".

(ii)   The second question ("Where is it?") is the same question the other way round and requires spatial analysis to be answered. Instead of identifying what exists at a given location, the purpose is to find a location where certain conditions are met.

(iii)   Questions dealing with trends in time ("What has changed since...?") might involve both of the first two and seek to find the differences within an area over time; the attribute data are therefore most important. There are, however, also geographical trends, mostly depicted as surfaces: knowledge about groundwater tables at points and accompanying models enable interpolation about other spots, previously unknown. In such cases the spatial part of the data is bound by specific spatial statistical rules, referred to as geostatistics.

(iv) Questions referring to patterns ("What spatial patterns exist?" or "Can I find a homogeneous zone?") are more sophisticated. These questions might be asked to determine, for example, whether cancer is a major cause of death among residents near a nuclear power station. Just as important, we might want to know how many anomalies exist that don't fit the pattern and where they are located.

(v) Finally, modelling type of questions ("What if...?") are posed to determine what happens for example if a new road is added to a network or if a toxic substance seeps into the local groundwater supply. Answering this type of question requires both geographical and other information.

Map 5.6:   Some hospital data for Maryland

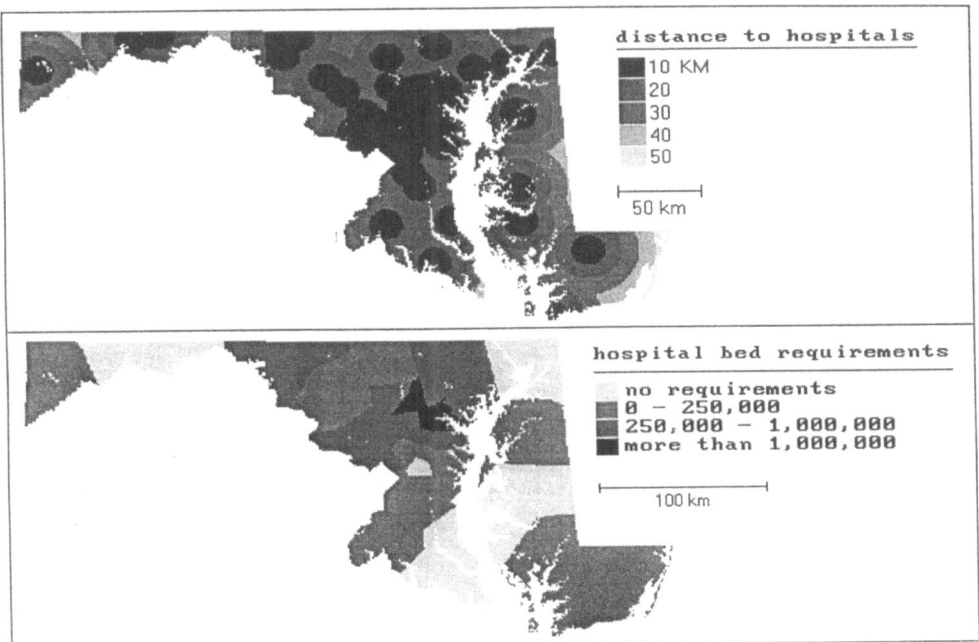

This type of information integration - whereby different datasets are combined using geography or space as the common link between the datasets, without loss of the exact locations of features - can also be illustrated by means of the Maryland study. The two maps of Map 5.6 show some data dealing with hospitals: the distance from urban centres to hospitals and the calculated bed requirements, allotted to a certain area.

Map 5.7:   Some population data for Maryland

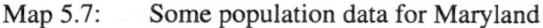

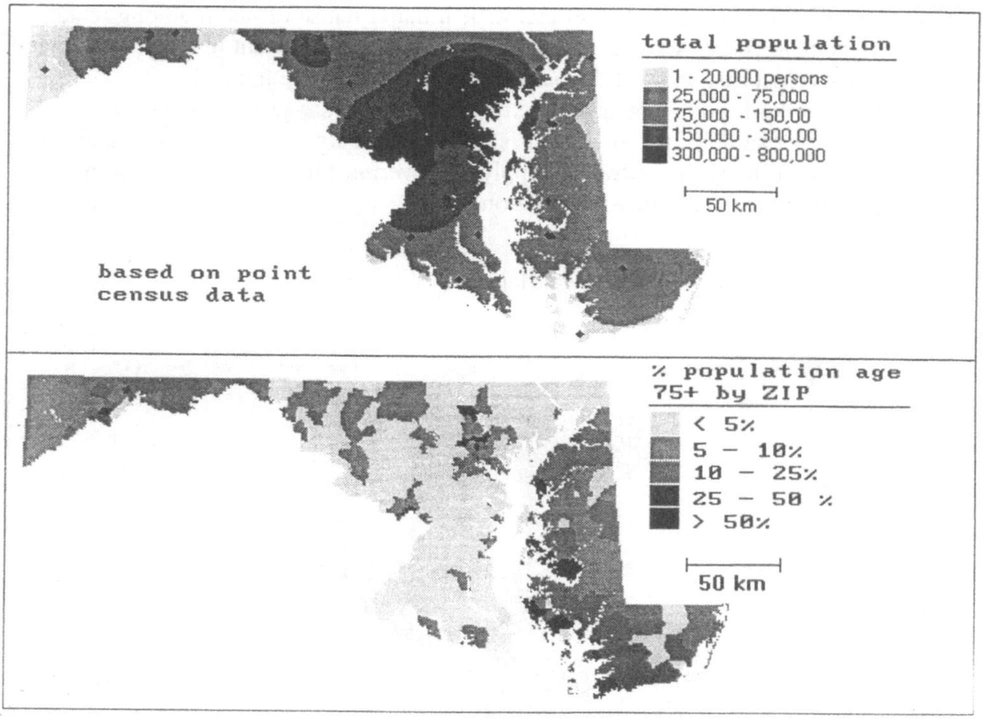

The population data (Map 5.7) contain many attributes. Most data are census data, belonging to a point. These point census data can be transformed as if people will spread equally in the area. Derived from these same data are the percentages of people older than 75 years for each ZIP area.

Other data (Map 5.8) that was needed, deal with the conditions posed in the final weighing. Steepness of slopes has already been discussed; other conditions contain information about the quality of the sewer system.

All available data is finally used in an indexed overlay (Map 5.9). Indexing overlays are most appropriate when the objective is to produce a map of relative, as opposed to absolute, measure of some geographical quantity that is composed of several determining factors. This type of analysis has the advantage of allowing the analysts to encode their knowledge of a problem in relative rather than in absolute terms. In general, an indexing overlay procedure calls for the user to specify in relative terms the importance of each individual theme of information. A weighting factor to each map and a ranking to each class in the map has to be assigned by the user. The weighting factors are quantitative factors that indicate the importance to the result of the information contained in a map relative to the other maps. The ranking factors are applied to the classes within a map. They denote, in relative terms, the desirability or lack of desirability of a certain condition. In the Maryland study, the results contain indices giving a relative indication about the advantages of one site over another. The weights can be changed to show the impact of a factor or the way the analyst uses the indices. After this indexed overlay, the optimum location for a new hospital site, according to this particular procedure, is known and the contributory conditions can be retrieved; not because the conditions are in the map, but because the system contains a geographical database.

The results of the Maryland study showed that the eastern shore of Maryland was the most suitable location for a new hospital facility. From the input information it can be obtained that this area is characterized by lack of service by and proximity to an incorporated area, relatively large population and a high percentage of the population in the older age groups.

Map 5.8:   Some additional data for Maryland

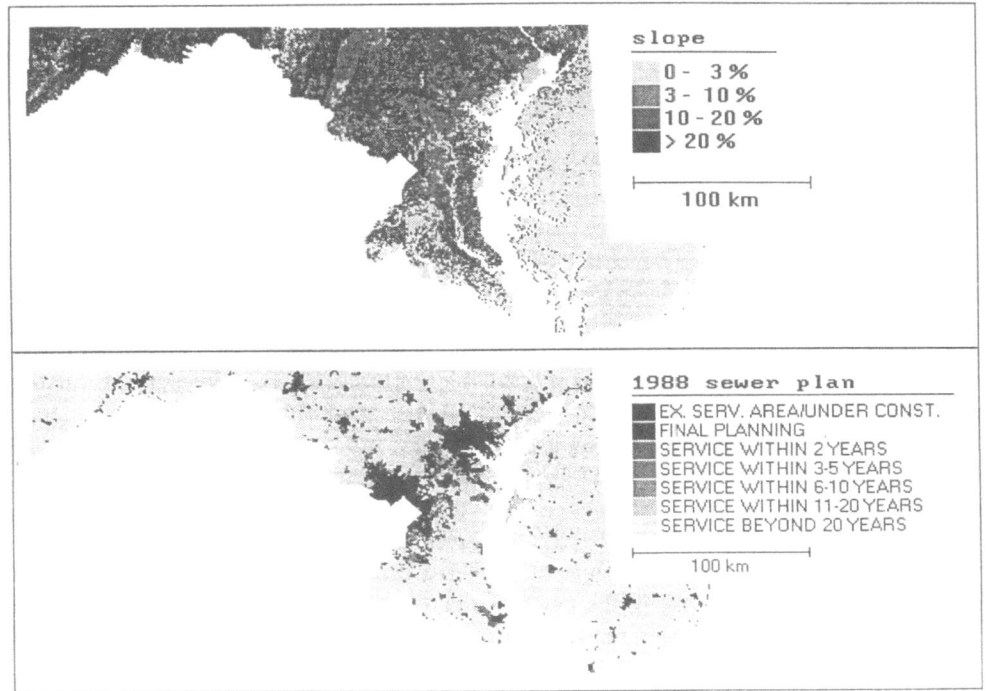

Map 5.9:   Final resultfor optimal hospital sites in Marlyland

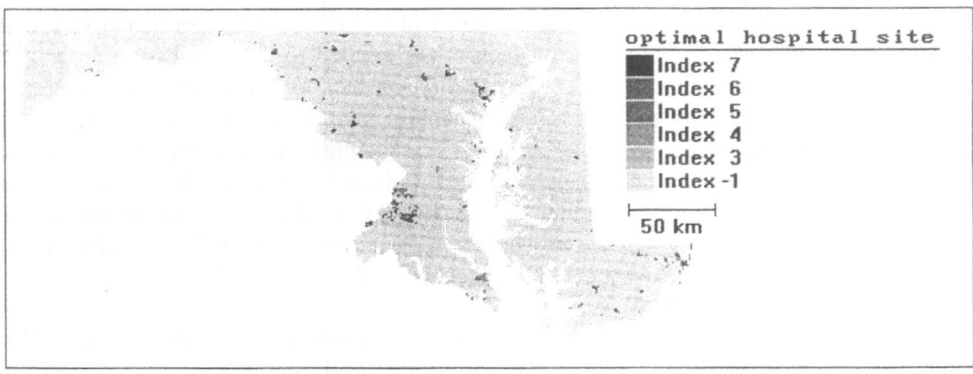

In conclusion, this case study illustrated that the power of GIS to combine data from many different sources using many different scales and data models is one of its major strengths. This data integration opens the way for powerful and varied ways of looking at and analysing data. Certain complex spatial operations are possible with a GIS that would be very difficult, time-consuming or impractical otherwise. It further demonstrated that, because the data can be accessed, transformed and manipulated interactively in a GIS, they can serve as a simulation for anticipating the possible results of planning decisions. By using GIS in a similar way that a trainee pilot uses a flight simulator, it is, in principle, possible for planners and decision-makers to explore a range of possible scenarios and to obtain an idea of the consequences of a course of action before the mistakes have been irrevocably made in the real world itself.

5.4   Conclusions

This chapter has looked at the importance of data integration for spatial processing disciplines and has reviewed the general principles of three different types of information integration more specifically with the goal to establish their effectiveness and efficiency. By means of a case study wherein a state agency was asked to determine the best location to site a new hospital, it has been demonstrated that GIS - compared with other methods - clearly offer advantages in the integration of information. Creating an environment within which data collected from different sources for the same set of areas or individuals can be related in such a way that new previously unknown information can be generated may well be one of the key benefits to be obtained from GIS. This function of adding value through data integration can be achieved in various ways through overlay techniques with locational information or through combining attribute information using statistical or modelling techniques. The Maryland case study further demonstrated that, because information can be accessed, transformed and manipulated interactively in a GIS, it can serve as a simulation for anticipating the possible results of planning decisions. By using GIS in a similar way that a trainee pilot uses a flight simulator, it is, in principle, possible for planners and decision-makers to explore a range of possible scenarios and to obtain an idea of the consequences of a course of action before the mistakes have been irrevocably made in the real world itself. Taking this into account, we feel that there is no escaping the fact that GIS is not only useful but even essential for the advancement of the health and wellbeing of the world's population and environment.

Acknowledgements

We wish to thank our colleagues Paul Padding and Henk Scholten from the National Institute for Public Health and Environmental Protection in Bilthoven for their stimulating and constructive comments on previous drafts. To TYDAC Technologies - in particular Mr Michael Arno - we are indebted for their assistance in integrating the datasets of the State of Maryland for the location of hospital services.

References

Burrough, P.A. (1986). *Principles of geographical information systems for land resources assessment.* Oxford, Clarendon Press.

Environmental Systems Research Institute Inc. (1990). *Understanding GIS*; the ArcInfo method. Redlands, CA. ESRI.

Evers, H.W. & H.J. Scholten (1991). How to integrate raster and vector, formats and standards, hardware and people. *Proceedings Second European Conference on Geographical Information Systems.* Brussels, Belgium, 2-5 April, 1991. Utrecht, Vol. 1, 316-323.

Leighton, D.H. & M.D. Kutsal (1991). GIS as an integration platform. *Proceedings Second European Conference on Geographical Information Systems.* Brussels, Belgium, 2-5 April, 1991. Utrecht, Vol. 1, 621-627.

Openshaw, S. & M. Charlton (1990). *Applying GIS to a small area health statistics system.* Presented at the SASHU Technical Workshop in collaboration with WHO Regional Office for Europe, 22 June 1990, London.

Scholten, H.J. & M.J.C. de Lepper (1990). The benefits of the application of geographical information systems in public and environment health. *In: World Health Stat Q,* **44** (3): 160-171.

Scholten, H.J. & J.C.H. Stillwell (1990). Geographical information systems: the emerging requirements. *In*: H.J. Scholten & J.C.H. Stillwell, ed. *Geographical information systems for urban and regional planning*. Dordrecht, Kluwer.

Tydac Technologies (1989). *SPANS case study — hospital siting in Maryland*. Internal case study report. Ottawa, Tydac Technologies.

van Beurden, A.U.C.J. & H.J. Scholten (1990). The environmental geographical information system of the Netherlands and its organizational implications. *In*: *Proceedings First European Conference on Geographical Information Systems*. Amsterdam, 10-13 April. Utrecht, Vol. 1, 59-67.

Arthur U.C.J. van Beurden and Marion J.C. de Lepper
National Institute of Public Health and Environmental Protection
Department of Informatics Service Centre / Geographical Information Systems
P.O. Box 1
NL-3720 BA Bilthoven
The Netherlands

6 ORGANIZATIONAL ASPECTS OF GEOGRAPHICAL INFORMATION SYSTEMS

David W. Heath

Abstract
This chapter addresses some of the important organizational issues which a geographical information system (GIS) in support of public and environmental health, especially at a Europe-wide level, will have to address. It looks at various issues of acquisition, sources, integration, project development and management, use and dissemination of data, as experienced by Eurostat, the Statistical Office of the European Communities. Eurostat receives data mainly from the official statistical services of the 12 Community Member States and makes this data widely available, e.g., to policy services of the Commission of the European Communities. It is currently engaged in developing a GIS for its own use and for the services of the Commission: the GISCO project. The remarks made herein, however, mainly reflect experience in developing and managing harmonized data from the different Member States and in using very large databases for their processing, storage and dissemination. In particular, a search for fully appropriate, completely accurate data, to be made available in a form that is optimized to well-defined known problems, would be an illusory target for large multi-purpose international data collection.

6.1 Data organization

6.1.1 Datasets and data flows

A first important distinction is that between collecting static datasets and setting up continuing data flows. Collecting specific datasets is particularly justified:

(i)   as a first step in setting up an information base in order to get some answers even if imperfect;

(ii)   for data with long-term validity where the state or relative level of different groups is the main interest (as in (i), there are likely to be difficulties in use because of data inconsistencies);

(iii)   for basic physical features that remain unchanged over the time-span of interest;

(iv)   for datasets that, for accidental reasons, are not being renewed.

For public health policy purposes, it is necessary to follow evolution over time. To do this, data flows with regular reporting have to be set up. Increasing harmonization between different suppliers of matching information can be achieved over time (or at any rate should be sought after). Harmonization of data is better achieved at source through agreed definitions and methods, rather than by later calibration. Changes can be followed either by recounting and comparing with previous count (e.g., 10 yearly population

M. J. C. de Lepper et al. (eds.), The Added Value of Geographical Information Systems in Public and Environmental Health, 87–97.
© 1995 *Kluwer Academic Publishers.*

and changes of residence). For some data, it is easier to assess changes than absolute levels (often rates of inflation can be compared between countries more easily than can absolute price levels). For most topographical data, rates of change are slow, and many have neither predictive value nor direct policy interest.

6.1.2 Data sources

Socioeconomic data are widely available in the databanks of the official statistical services. These data are cheap and well harmonized. Their weakness for geographical information systems (GIS) purposes is that fine local detail is not generally available; the data may not fit the problems under discussion and are very difficult to change in the short run. Such data are from three main sources.

(i)   Administrative records, e.g., registers or such administrative operations as unemployment benefit or customs control. These data are cheap, comprehensive, reliable and renewable, but what they measure is determined by the purposes of the administrative action that gives rise to them. They need to be converted to a version nearer to the interest of users of statistics. Even then there may be a definitional gap.

(ii)   A second important source of statistical data is censuses. These are cumbersome operations; they are costly with considerable risks of observational errors, and there are limitations on the questions that can be put. However, they can give finely localized information and more and more national statistical services are planning small-area statistical output from censuses. They provide also infrequent general benchmarks and sample frames for later selection of specialized sample surveys and for obtaining sample results.

(iii) Sample surveys are cheap and quick to carry out and process. They can have a detailed questionnaire and be addressed to a specifically relevant population. Specially trained interviewers or special procedures can cut down observation error, and the errors due to random sampling fluctuation can be calculated. One such sampling type is the sampling and processing of administrative records, e.g., health and treatment of selected patients.

As well as relevant socioeconomic data, a (health) GIS will have to contain basic geographical data. This block of data will provide the background situation for analysis. It provides a unifying framework for socioeconomic data from diverse sources. It allows presentation of results in map form (Maps 6.1 and 6.2). The source of such data is mainly national cartographic institutes. There are substantial problems in obtaining consistent Europe-wide data with a common projection system, updating rhythm and accuracy. Eurostat is starting to examine the possibilities of collaboration with national mapping agencies and their international organizations, e.g., CERCO (Comité Européen des Responsables de la Cartographie Officielle).

The data needed include permanent physical features, administrative boundaries (where more frequent updating is necessary) and basic geographical information of the type of climate, soil quality and geology. There are issues of appropriate scale to be decided.

Eurostat is already responsible for a standard (and widely used) nomenclature of administrative units, the NUTS (Nomenclature des unités territoriales statistiques) system. This nomenclature is at various levels, e.g., level 1 being the Member States and level 3 being (for example) the departments in France. The provision of wide- ranging data in a GIS for wide public use raises issues of coordination, consistency of datasets, guarantees of data quality, permanence of service, conditions of access and of pricing policy, data confidentiality, copyright, etc. For official statistics these matters, if not resolved, are at least being handled in identified ways, nationally and internationally. For the geographical component of GIS, the situation is much less clear. In some countries (e.g., Brazil and Mexico) the same office covers statistics and geography. Within the services of the Commission of the European Communities, Eurostat is investigating its capacity to act as a focal point for this new geographic information need also. A dispersed ad hoc approach is likely to be inefficient and unsatisfactory from numerous points of view.

6.1.3 Making the most of data

The socioeconomic data in a general-purpose GIS are not tailor-made to the users' purposes. The organizational consequence of this is firstly that the user needs supplementary information in order to adapt the basic material. This may be small scale, specialized complementary data from microcensuses, pilot surveys, clinical trials, etc. or assumptions based on scientific background knowledge or on analysis of the data. Users need possibilities for storing their own data and for analysing the common stock of data in conjunction with private data. A second organizational consequence is that the user needs to have ready access, ideally on-line, to adequate descriptions of the data provided. This meta-data needs to help the users identify what relevant data are available and inform the users of the characteristics of the data. The more general is the purpose of a GIS and the wider is its range of users, the greater is the need for meta-data

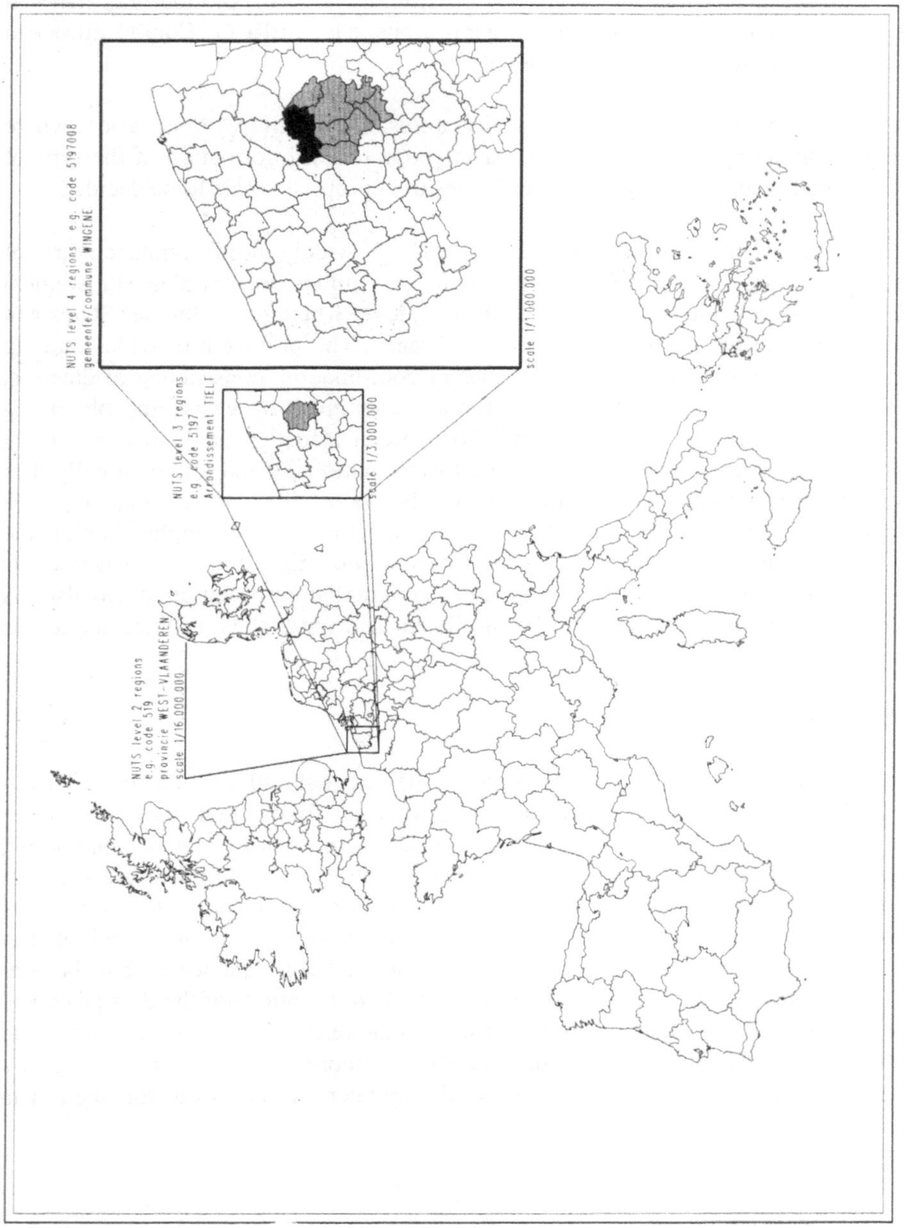

NUTS level 4 regions e.g. code 5197008
gemeente/commune WINGENE

scale 1/1 000 000

NUTS level 3 regions
e.g. code 5197
Arrondissement TIELT

scale 1/3 000 000

NUTS level 2 regions
e.g. code 519
provincie WEST-VLAANDEREN
scale 1/16 000 000

Map 6.1:   Illustration of the content of nomenclature of territorial units for statistics of the European Community

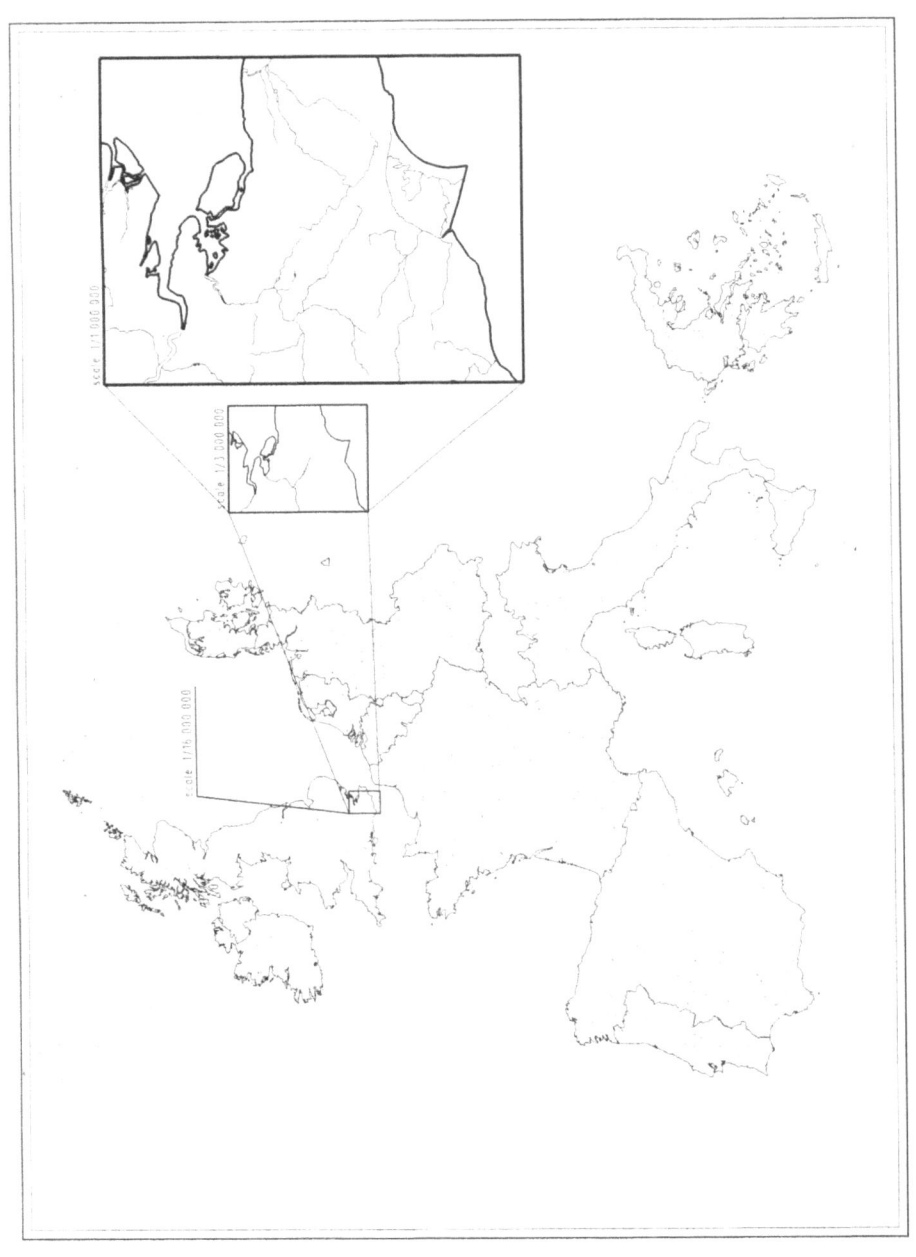

Map 6.2: Illustration of the River Network and its hierarchical structure

(Savolainen, 1990). Although suggestions have been made and work is under way to tackle this problem (intelligent front-ends to databases using expert systems, hypertext for methodological notes, etc.), this is still an unresolved problem. Even when satisfactory computing tools are in general use, there will remain the problem of ensuring discipline so that meta-data and data are entered in parallel into databases including GIS.

If management problems can be overcome, the provision of a very wide range of related data in the GIS multiplies its usefulness.

(i)   Data can be collected for a specific purpose (but thereafter be available for other purposes) or consciously intended as a generally useful multi-purpose indicator of some phenomenon. Data can have important indirect uses, e.g., the nutrition status of children from health clinic data to monitor food policy in countries with chronic food shortages.

(ii)   Although location plays this role to some extent in GIS, if a unifying framework can be found (such as in the national accounts, social or food accounting matrices), data can be checked for consistency, gaps filled by calculation or new data needs identified. Integrated surveys, e.g., the Communities' survey on the structure of agricultural holdings (where very many and widely ranging data are collected on each unit surveyed) can also serve as a unifying structure for other simpler datasets. This requires a positive management of contents and also a central responsibility (person, coordinating committee or other administrative device).

(iii)   The GIS can act as a tool in data collection by improving sample selection and by acting as a knowledge base, e.g., for an expert system to interpret remote- sensed data or for image enhancement.

The larger the GIS, the greater is the chance that it will containing data relevant to the user's problem (but the greater is the difficulty in finding these data and in using them successfully). Where the extension involves extra countries, there is also increased likelihood of missing data. The usefulness of the GIS can be substantially enhanced if estimates can be provided to make datasets complete. This can be left to the initiatives of particular data owners. It can be undertaken or stimulated centrally based on an analysis of use or requests. For this use analysis, tools must be available to identify priorities. Completing geographical cover can involve a wide range of approaches ranging from sophisticated interpolation methods (e.g., kriging) to simple disaggregation according to a key. The responsibility for the choice of method needs to be identified and its implications made public.

6.1.4 Responsibility for data and data costs

The data in a GIS account for the majority of costs and practical difficulties. One way to build a GIS is to approach it as a container for use by different data providers, each taking responsibility for his or her own data (accuracy, updates, documentation, etc.), while making them generally available to others at zero cost. In return, each has free access to other's data. This approach, variously called Spanish banquet, chest of drawers, owner-client or library (in which the librarian orders and catalogues books (= datasets) and makes them available together with a reference service) has some obvious advantages compared with one central organization having to pay for everything. The data owners generally have a natural interest in their own data being of good quality, up-to-date, etc. and are generally the best qualified to ensure this. There is still a need, however, for a central management to identify and arrange for the purchase of widely needed data not freely offered by any data owner.

This decentralized approach also requires particular attention to be paid to data integration. Aspects that need to be considered are the appropriate harmonized nomenclatures, units and methods. These may have to be adapted from general standards to take account of particularities of spatial data (e.g., local units). Remote- sensed data provide a calibration tool across national frontiers, but nomenclatures (e.g., land-use) may need adapting to what remote-sensing can distinguish. There are also difficulties related to time: often geographical breakdowns are less frequent than coarser figures and are not synchronized across countries. Standardizing adjustments can be made to levels of fine geographical data if there are important (known) variations at national or large regional level. To reach agreement on these requires considerable internal discipline with coordination structures across disciplines and procedures for assigning responsibilities for standardizing adjustments through the individual data supplier or GIS manager. Such discipline can be demanded of late joiners but is difficult to impose at the outset. The necessary adaptations can sometimes be made to data after they have been collected, by some correction or calibration procedure (e.g., where remote-sensed data on key items can be matched with data collected by traditional national systems based on different methods). In principle, it is better to work towards achieving changes in data collection procedures in order to get harmonized results.

6.2   GIS development and management

The development method for a GIS does not differ essentially from those of any major computing project and thus will involve problem statement, feasibility study, project development method, deliverables, milestones, useful early releases, planned upgrades treated as new projects, user involvement, documentation, user manuals, clear responsibilities (project owner, system supplier, project manager), pilot products and so on.

This development will generally take place within an existing computer environment and have to be compatible with general rules. As an example can be cited some features of the framework within which the GISCO project is being developed: open systems, multi-vendor equipment, maximum use of purchased software, decentralization but with substantial central mainframe applications remaining, portability, public data in reference databases, increasing confidentiality precautions for statistical data, centrally standardized choices of hardware and software, rapid technical developments (e.g., electronic mail and desktop publishing) and decentralized and centralized graphics facilities.

Although there is much talk of distributed databases, it is still difficult to point to working examples. In the case of GIS, particular features push towards a centralized, (probably) mainframe approach: the huge volume of data, the importance and large volume of standard geographical data, the (probable) need for a consistent approach to cartographic presentation and the existence of expensive, specialized (e.g., cartographic) facilities. These conditions do not require that updating be done centrally. Where possible, it is useful to decentralize responsibilities for data management, e.g., to statisticians responsible for particular subject matter areas. This requires facilities for a working computer environment to allow data to be vetted before entry into the public GIS. This environment needs to have easy access to data in the GIS for use in the checking process. It could be on a local micro computer with rapid downloading facilities or it could be a facility within the GIS for data to be stored and used but only by the owner of the data until they are passed into the public domain of the database by a formal procedure.

Users will often wish to use their own non-public data in conjunction with data in the public GIS. This can be achieved within one central GIS by levels of access rights. However, the planning for the GISCO project is different. It consists of a cloning of a centrally developed system (hardware, software, basic geographical data and interfaces to Eurostat's major reference databases) with provision for periodic updates. The clones will be provided to interested policy services (regional, environment, agriculture, transport, etc). Each of these services would then add on its own special features. It would grant access rights to outsiders interested in the add-on features on the basis of bilateral agreements.

There are other organizational aspects of GIS development. There is the need to fit into the time development of related computing activities. The more economies in development that are being achieved by taking advantage of existing facilities, the greater is the need to consider the evolutionary path of the computer part of the GIS and its upward and downward compatibility. One must also consider how far the GIS replaces, supersedes or structures other activities. Eurostat is substituting several dispersed mapping developments by an integrated development programme and centralized service within the GISCO project.

6.3   Direct user access

Important consequences for organization and for the nature of the GIS follow from the type of user served. At one extreme is a small number of regular skilled users; at the other are many occasional unskilled users. The actual situation will be in-between and call for a mixed response. For a few skilled users the need for support documentation, catalogues, meta-data, continuing training, help desks, help telephone lines, help commands and menu-type user-friendly interfaces is reduced. In order to maximize benefits from the GIS investment, one may wish, however, to provide a service allowing a wider circle of users. If the circle is wide but identifiable then training becomes important (as well as all the other facilities listed) and investment in workshops, in developing distance learning material and in staffing regular in-house or external courses becomes worthwhile. Wide use of a GIS can only take place by helping customers to help themselves. Although some skilled staff are needed to develop and carry out the user help, even more would be needed if all user requests pass through a central point. In addition, a free service is easily abused while charging brings its own problems.

Different users can have different interests (Scholten & Padding, 1990). The same GIS can serve different users better if there can be different user views (external schemata). Encouraging feedback from users is often neglected but is important; as with any product, the users provide the real test of quality. Some formalization of contacts with users is a worthwhile investment and helps in planning developments, identifying weak points and controlling data quality. Even with a good product, periodic publicity campaigns or newsletters may be needed to draw it to the attention of new potential users.

Requests for information can arise from the public via telephone, personal contacts, fax, letter, etc. Answers sometimes require only sending an existing standard product, but they may also require special work. They may be for copies of substantial parts of the data. The more wide ranging is the GIS and the more its importance and use, the more such requests will arrive. A considered policy to be agreed and applied by all concerned is necessary. What will be accepted for free answer? What sort of work will be done specially and at what charge? Administrative procedures need to be set up for calculating charges, for billing and for collecting use statistics. If the organization managing the GIS decides to provide such a public information service a specialized unit is usually necessary that has general knowledge of the GIS and is able to answer more general queries, avoiding using unnecessarily the time of subject-matter experts. Developing a range of standard products may facilitate meeting certain requests for information, but if these are also public products (e.g., publications), they may stimulate more information requests.

6.4   Dissemination

As well as providing direct access, a GIS can also produce a variety of public products. These can capitalize on the GIS investment and increase the availability of knowledge. The organization managing the GIS may have an existing publications policy, to which the GIS has to conform. There are issues of copyright and pricing policy which may apply particularly to products involving the output of mapping agencies. Where the GIS data have multiple owners, delicate negotiations may be required before composite products can be put on the market.

The actual products depend on the data and on the users. They can include broadsheets in response to current policy problems, substantial publications giving in-depth treatments of specific topics, popularized presentations designed to raise awareness of the general lay public and regular situation reports, e.g., yearbooks. Large parts of the GIS could be made available on CD-ROM provided that the copyright and charging issues can be settled. This is option is particularly attractive to educational and research users. It is difficult to reach directly the dispersed mass of potentially interested clients at the European level. Two-stage approaches through the specialized press, by co-publication or through agency agreements with national bodies can help. Marketing arrangements can be made with firms specializing in selling value-enhanced information customized to the needs of clients. A central responsibility for coordinating and organizing public products is necessary.

References

Savolainen, K.M. (1990). *European meta database on environment and health information sources*. Paper presented at WHO Consultation on the Development of a Health and Environment GIS for the European Region, Bilthoven, 10-12 December 1990.
Scholten, H.J. & P. Padding (1990). Working with GIS in a policy environment. *Environ planning B Planning Design*, **17**: 405-416.

David W. Heath
Eurostat
Statistical Office of the European Communities
Bâtiment Jean Monnet
L-2920 Luxembourg
Luxembourg

7   DATA ASPECTS OF GEOGRAPHICAL INFORMATION SYSTEMS

J. Maes and M.H. Cornaert

Abstract
Geographic information systems (GIS) are major tools for data management and information provision in the European Community CORINE Programme and for the European Environment Agency Task Force. Data are a key element in the implementation and development of GIS, acting both as a challenge and major opportunity. The key issue about data relates to their availability, quality, usefulness, accessibility and the cost for updating and conversion into digital form. On the other hand, GIS acts as an integrator and translator of data, giving an enormous opportunity for adding value to data. The capacity of GIS to bring together data from diverse sources results in a synergic stimulus where the whole is much more than some of the individual parts.

7.1   The CORINE Programme

To be in a position to conserve and improve the environment, one must know what state it is in and how it is developing. In other words, one must have access to information on: the distribution of the fauna and flora and their habitats; the sources, extent and location of pollution; the state of natural resources; and the natural risks threatening the environment and human activities. This is true at all levels of responsibility: local, sub-national, national and the European Community. A difficulty peculiar to the Community level is the considerable variation of environmental data characteristics (availability, definitions, measurement methods, etc.) between countries and often even between regions. As a consequence, in the great majority of cases, the data cannot be compared and directly assembled to present the state and development of the European environment for the benefit of Community policy.

In response to this need for information, it was decided to undertake the CORINE Programme to gather, coordinate and ensure the consistency of information on the state of the environment and natural resources. Initially planned for a four-year period in 1985, the Programme was extended by two years, until 1990.

The Programme had two aims:

(i)   to verify the usefulness of a permanent information system on the state of the environment for Community environmental policy, to check the technical feasibility of creating such a system and to identify the conditions required for its installation and functioning;

M. J. C. de Lepper et al. (eds.), The Added Value of Geographical Information Systems in Public and Environmental Health, 99–114.
© 1995 *Kluwer Academic Publishers.*

(ii)   to supply information useful for Community environmental policy on topics of priority concern (for example, biotopes, acid deposition and the Mediterranean environment).

Concerning the first aim, the CORINE Programme results show that a permanent information system on the state of the Community environment is necessary and technically feasible. Furthermore, the Programme has permitted the precise definition of the conditions necessary for the realization and operation of such a system.

The second of the Programme's aim has also been successfully attained. Data on the priority topics were collected, supplemented by a series of basic data and organized in an operational geographical information system (GIS).

7.2   The European Environment Agency and its information and observation network

In 1990, a regulation was adopted that establishes the European Environment Agency and the European Environment Information and Observation Network. However, the entry into force of this regulation has depended the choice of the Agency seat; Denmark was chosen in October 1993. In the meantime, a task force has been created within the Commission of the European Communities Directorate-General for Environment, Nuclear Safety and Civil Protection (DG XI), which is in charge of the technical aspects of the preparation of the setting up of the EEA, including maintenance and use of the existing CORINE Information System. In 1991, it was decided to extend three main CORINE inventories (Biotopes, CORINAIR and Land Cover) to the central and eastern European countries. A number of EFTA countries have also expressed their interest to join these inventories. Furthermore, in conjunction with the United Nations Economic Commission for Europe (ECE), a major exercise of data collection has been set up to publish a pan-European report on the state of the environment by the end of 1993.

The European Environment Agency and its associated network will be established with the objective of providing both the Community and its Member States with objective, reliable and comparable information at European level to enable them to take the necessary measures to protect their environment, as well as to assess the results of these measures and to ensure that the public is properly informed about the state of the environment. It is intended that the Agency will also furnish the scientific and technical support necessary for these purposes.

The Agency is the logical successor, in particular, of the Commission's CORINE Programme. The techniques and the databases established under the CORINE programme will provide a significant contribution to the new Agency.

7.3   The CORINE Information System

Data gathering under the CORINE Programme resulted in the creation of numerous data files covering geographical reference data (topography, hydrography, etc.), basic data (climate, soil types, etc.) or thematic environmental data (biotopes, emissions into the air, etc.). These files from various sources were put together into a consistent GIS. For this it was necessary to be sure of the quality of the data, to structure them in such a way as to allow data management, data search and analysis and the production of graphs and maps. An overview of the contents of the CORINE Information System is given in Table 7.1.

The main functions of the CORINE system may be summarized as follows:

(i)   Reception of digital data. Most of the data were supplied in digital form, so that the efforts of the CORINE Secretariat could be concentrated on converting the data to fit the GIS standards, rather than digitizing them (which had been already done by the project leaders or the experts in the technical working parties).

(ii)   Integration of data and setting up the databases. This phase comprised, in particular, adjustment of the national data files to administration borders and adjustments to the data from different sources (for example, only one reference coastline should be used and all the other sets of data should be made to correspond to it no matter where they come from). This phase of the work also consisted of verifying the geographical references of the thematic data and in establishing cross-references between the files.

(iii)   System and data management. This is not exclusive to the GIS but is a necessary obligation, together with updating of the data, file safeguard procedures, development of the system and its utilities, standardization of procedures for routine functions, etc.

        (iv)   Research and spatial analysis functions. Herein lies the originality and interest of the GIS, giving the possibility of combining data from different sources

Overview of the contents of the CORINE Information System

Theme	Nature of the information	Volume of information Description	Mbytes	Resolution/scale
Biotopes	Location and description of biotopes of major importance for nature conservation in the Community	5600 biotopes described, according to approx. 20 characteristics. Boundaries of 440 biotopes computerised (Portugal, Belgium)	20.0 2.0	Location of the centre of the site 1/100 000
Designated Areas	Location and description of areas classified under various types of protection	13 000 areas described according to approx. 11 characteristics (file being completed) Computerized record of the limits of the areas designated in compliance with article 4 of the EEC/409/79 directive on conservation of wild birds	6.5	Location of the centre of the site 1/100 000
Emissions into the air	Tons of pollutants (SO2, NOx, VOC) emitted in 1985 per category of source : power stations, industry, transport, nature , oil refineries, combustion	1 value per pollutant, per category of source and per region, plus data for 1400 point sources i.e. ± 200. 000 values in total	2.5	Regional (NUTS III) and location of large emission sources
Water resources	Location of gauging station, drainage basin area, mean and minimum discharge, period : 1970–1985, for the southern regions of the EC	Data recorded for 1061 gauging stations, for 12 variables	3.2	Location of gauging station
Coastal erosion	Morpho–sedimentological characteristics (4 categories), presence of constructions, characteristics of coastal evolution : erosion, accretion, stability	17 500 coastal segments described	25.0	Base file : 1/100 000 Generalisation : 1/1 000 000
Soil erosion risk	Assessment of the potential and actual soil erosion risk by combining 4 sets of factors : soil, climate, slopes, vegetation	180 000 homogeneous areas (southern regions of the Community)	400.0	1/1 000 000
Important land resources	Assessment of land quality by combining 4 sets of factors : soil, climate, slopes, land improvements	170 000 homogeneous areas (southern regions of the Community)	300.0	1/1 000 000
Natural potential vegetation	Mapping of 140 classes of potential vegetation	2288 homogeneous areas	2.0	1/3 000 000

Table 7.1

Theme	Nature of the information	Volume of information / Description	Mbytes	Resolution/scale
Land cover	Inventory of biophysical land cover, using 44 class nomenclature	Vectorised database for Portugal, Luxembourg	51.0	1/100 000
Water pattern	Navigability, categories (river, canals, lake, reservoirs)	49 141 digitised river segments	13.8 / 0.3	1/1 000 000 / 1/3 000 000
Bathing water quality	Annual values for up to 18 parameters, 113 stations, for 1976–1986, supplied in compliance with EEC/76/160 Directive	2650 values	0.2	Location of station
Soil types	320 soil classes mapped	15.498 homogeneous areas	9.8	1/1 000 000
Climate	Precipitation and temperature (other climatic variables : data incomplete)	Mean monthly values for 4773 stations	7.4	Location of station
Slopes	Mean slope per km² (southern regions of the Community)	1 value per km², i.e. 800.000 values	150.0	1/100 000
Administrative units	EC NUTS regions (Nomenclature of Territorial Units for Statistics) 4 hierarchical levels	470 NUTS digitised	0.7	1/3 000 000
Coasts and countries	Coastline and national boundaries (Community and adjacent territories)	62.734 km	0.3 / 3.2	1/3 000 000 / 1/1 009 000
Coasts and countries	Coastline and boundaries (planet)	196 countries	1.5	1/25 000 000
ERDF regions	Eligibility for the Structural Funds	309 regions classified	0.01	Eligible regions
Settlements	Name, location, population of urban centres > 20.000 inhab.	1542 urban centres	0.1	Location of centre
Socio-economic data	Statistical series extracted from the SOEC-REGIO database	Population, transport, agriculture, etc.	40.0	Statistical Units NUTS III
Air Traffic	Name, location of airports, type and volume of trafic (1985–87)	254 airports	0.1	Location of airport
Nuclear power stations	Capacity, type of reactor, energy production	97 stations, update 1985	0.03	Location of station

Table 7.1 (contd)

to obtain a synthesized piece of information: for example, the risk of soil erosion on the basis of climatic, topographic and pedological data. These functions also allow data to be prepared for use in models: for example, the transposition of data for atmospheric emissions into a regular grid.

(v)   Data processing, the production of reports and cartographic documentation. This function was carried out by the data-processing team of the CORINE Secretariat in accordance with the needs of the Commission users and of the CORINE technical working parties (aid to data validation).

The information system set up in DG XI is of a decentralized design. The local network consists of four work stations, a digitizing table, tape streamers and disk drives and printers. The work stations are equipped with GIS software.

The architecture, hardware and software were chosen in accordance with the Commission's data-processing procedure for major projects. When this procedure is completed, specifications will be adapted for Commission guidelines on GIS.

This procedure, initially set up to meet the data processing needs of the CORINE programme, made it possible to identify similar needs in different directorates-general of the Commission and to adopt a consistent and concerted inter-service approach to hardware and GIS software.

The following conclusions can be drawn with regard to setting up and using the CORINE GIS:

(i)   The techniques necessary for constructing the GIS are commercially available. The main problems are related to data processing but to the insufficient availability and comparability of data as well as the existence of a permanent dialogue between data suppliers and information users.

(ii) The existence of geographical reference data of high quality in digital form is a prerequisite for the proper functioning of GIS. Obtaining such data involves a considerable amount of work by specialist bodies, which has to be done ahead of the creation of a GIS.

(iii) The collation of thematic files, their referencing to the geographical framework and cross-referencing among themselves - a key to the proper functioning of the GIS - will be made all the easier if the need for this is taken into account in the data specifications (before any collection starts).

(iv)   To meet the needs of users for CORINE information, the GIS software has been supplemented by software better suited to certain functions: a master system for combinations of large volumes of data, statistical software, graphics presentation software and a low-cost cartographic software.

(v)   In order to be efficient and pay its way, the installation of a GIS and the reference data used have to be designed to serve the whole of an organization - for example the Commission - and not an isolated project or programme. This approach does not rule out decentralized operations such as the decision to set up the European Environment Agency.

(vi)   Awareness of the advantages of GIS is growing. Their potential uses and the value of operational prototypes should be made known. At the same time, the requirements for data and the need to perfect existing systems should be underlined. In this respect, the availability of a GIS in the Commission, close to the user (DG XI and other directorates-general) has proved to be a great advantage from the point of view of the speed of service to the user and also as a way of making better known the potential of this new technique.

7.4   Some generic GIS lessons learned

7.4.1 Data automation: a major challenge

Data automation, which is converting digital geographical data, is a major challenge in setting up a GIS. The costs related to data conversion are in most cases substantially higher than GIS software or hardware. The process of digitizing is labour-intensive and often difficult to automate.

Before digitizing can take place, pre-processing is often needed, since data are outdated, non-uniform or partly lacking. Many cases are reported where available data are not adequate or nonexistent and new data collection procedures have to be set up before data conversion to a GIS can take place.

Data quality, error and uncertainty of data are major elements of importance. These aspects refer to registration accuracy and precision, reliability, parameter variability and measurement error. Data quality in GIS is a major focus of attention. A ranking or grading of datasets in view of data quality should be a point for future discussions. Such meta-information on data is relevant to their use and must be managed as an integral part of an information system.

7.4.2 Data integration: to be identified in the early project phases

A major lesson learnt from CORINE is that integration of data is to be envisaged during data preparation, feature coding and automation. The availability of basic geographical data for geocoding is essential in this context, since it is a common reference for data from different sources. Environmental or health data should refer to a consistent system of spatial references: administrative units, river segments, designated sites, watersheds or human settlements. The various origins of the data sources, with often different scales, projections, measurement units and parameters, renders data integration a major challenge. Identification of these issues in the early phases of a project is a key element for successful implementation of GIS.

7.4.3 Data interchange: essential to GIS

Data transfer, format conversions and interfaces are becoming increasingly important, since data are relevant and common to many kinds of users. GIS acts as an integrative technology, bringing data and people together. The subsequent need for data interchange procedures, data access and interfacing different systems is clear. Data and format standards are of major concern to the GIS community. Data standards are a basis for comparability and harmonization and must be defined where data from different sources are combined. It is clear from past experience within the CORINE Programme that transfer formats can only be successful provided that: (1) transfer format interfaces are implemented in commercial GIS systems or (2) basic interfacing programmes or generic procedures can be delivered together with the format standards. Fortunately, an increasing number of interfaces between different GIS systems have become available, fulfilling many of the current data transfer needs. A continued effort, preferably around a common standard, for providing interfacing modules by software suppliers is needed.

7.4.4 Basis reference data: a topic of common concern

In order to permit data to be converted in a consistent GIS environment, the availability of standard reference data is essential, including standardizing geographical data and coding lists, which is by definition a horizontal, inter-service and inter-organizational task. These reflections are the result of experience within the CORINE Programme, where administrative boundaries have been digitized in the early phase of the Programme. Digital versions of the Community's administrative boundaries were not available at the start of the project, neither within the Commission nor from a commercial source. The automated digital boundary file now serves a large GIS community and has become relevant to and used by many users both inside and outside the Commission. Its distribution and updating has become a major task, which far exceeds the original CORINE Programme objectives but also those of an environmental information system service. After the termination of the CORINE Programme and its integration into the European Environment Agency Task Force, the compilation, distribution and updating of basic geographical data was handed over to EUROSTAT. EUROSTAT has incorporated this into another GIS called GISCO (see Chapter 6 in this book).

7.5   Examples of GIS use in the framework of the CORINE Information System

7.5.1 Modelling atmospheric pollution

For the guidance of Community environmental policy and the monitoring of the application of legislation, the quantities and geographical distribution of pollutants emitted into the air must be known. In collaboration with the OECD and on the basis of international-level experience, the CORINAIR Project led to the establishment of a database on 1985 emissions into the air (Figure 7.1). A new inventory, with 1990 as reference year, is being compiled, for the Community Member States as well as for several central European and EFTA countries. The CORINAIR inventory provides data, per administrative unit, on emissions of pollutants as well as large point sources.

Figure 7.2 shows the distribution of total SO_2 emissions in the Community regions. Models of pollutant dispersion require information in terms of a geometric grid, for which an automated procedure was developed to recalculate the emissions in terms of the ECE-EMEP or OECD grids.

7.5.2 Soil texture

One of the projects of the Community Agricultural Research Programme consisted of the production of a soil map of the Community at a scale of 1:1,000,000 (1986). This map was digitized in the framework of the CORINE Programme, responding to several Commission Programme interests:

(i)   analysis for environmental purposes, particularly of soil erosion risk and land resources;
(ii)   the evaluation of potential agricultural production; and
(iii)   the easy updating and improvement of the original map.

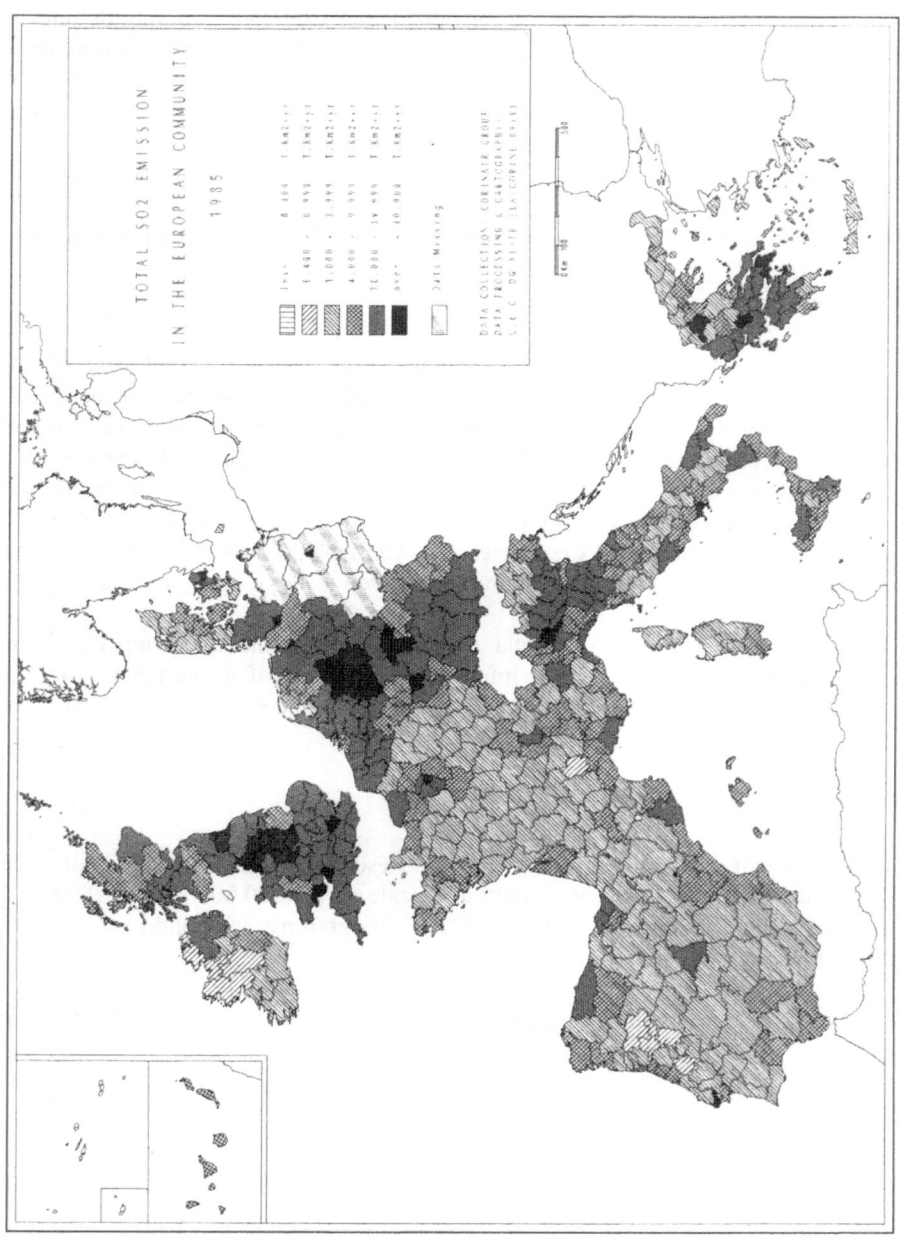

Figure 7.1: Emissions of SO$_2$ in the European Community by administrative area

Figure 7.3 shows a four-level reclassification of the soil map according to soil texture. The 350 original soil classes are regrouped and shaded according to the heaviness of the soil: sandy, medium texture or clayey.

7.5.3 Soil erosion risk

In order to evaluate the soil erosion risk, four types of data were combined in a GIS overlay: climate, soil, slopes and vegetation. Taking into account these basic erosion factors, the model developed enabled the Mediterranean area to be classified according to three soil erosion risk categories (Figure 7.4).

7.5.4 CORINE biotopes

Over 8,000 biotopes (sites of major importance for nature conservation) have been registered and described: geographical location, surface area, type of habitat, type of conservation and animal and vegetation species. Due to the availability of the exact location of each biotope, selections from the database can be mapped. Figure 7.5 illustrates the result of a GIS query on the database, showing the biotopes where the *Lynx pardinia* has been reported.

7.6 Conclusions

Data are key elements in GIS. Data acquisition requires major efforts, due to the high quality standards for digital geographical data and the difficulty in automating the process. Data-sharing is a means to minimize data acquisition and automation costs.

Data-sharing, however, is not only relevant to minimize automation costs, but also, and at least of equal importance, to set up geographical and coding standards. These standards are an essential basis for data consistency across different datasets. The real added value of GIS is achieved when data are not used in isolation but by using GIS as a means to share data between services, departments and institutes on the basis of spatial relationships. Implicit spatial references, on the basis of codes and names, as well as explicit digitization of spatial project data are relevant. Both form a basis for the data transformation that is essential in multi-source data analysis and modelling.

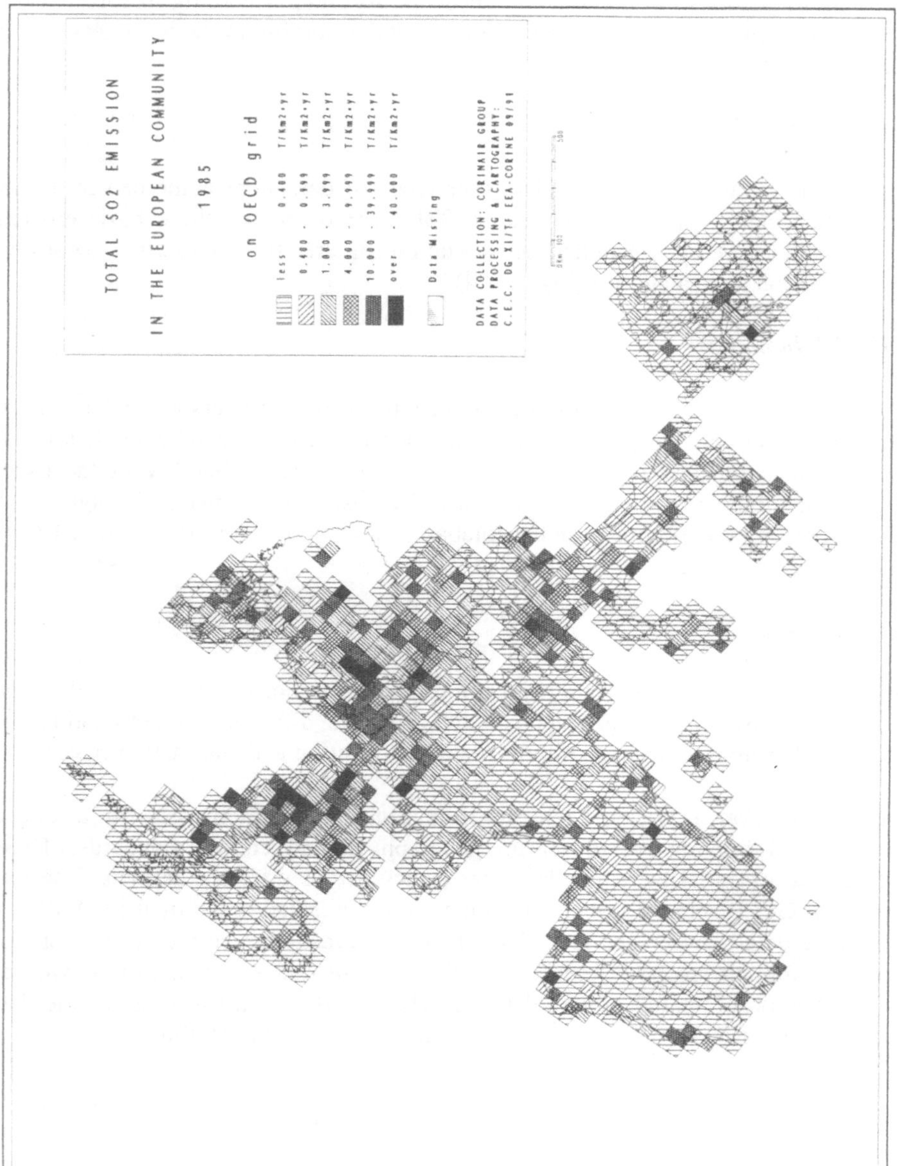

Figure 7.2: Emissions of SO$_2$ in the European Community according to OECD grid cells

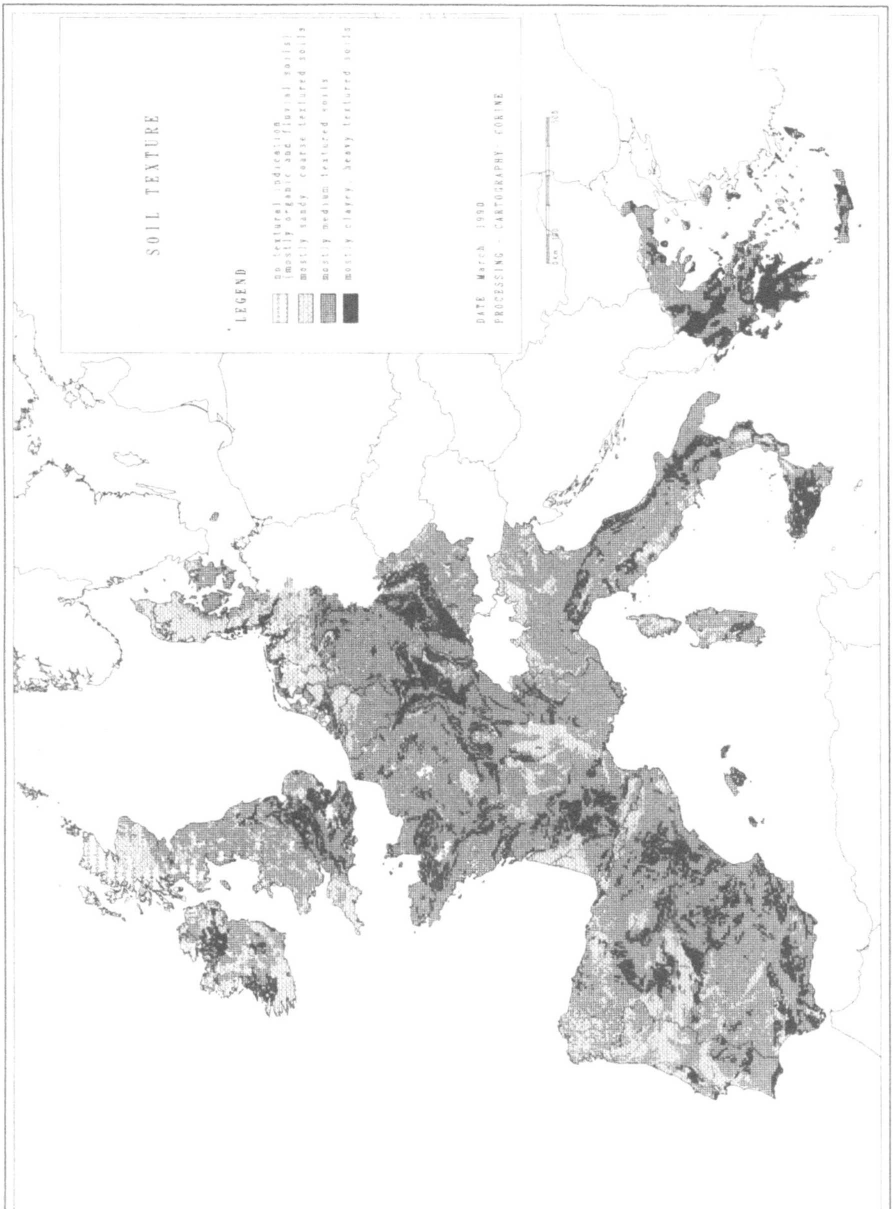

Figure 7.3: Soil texture in the European Community

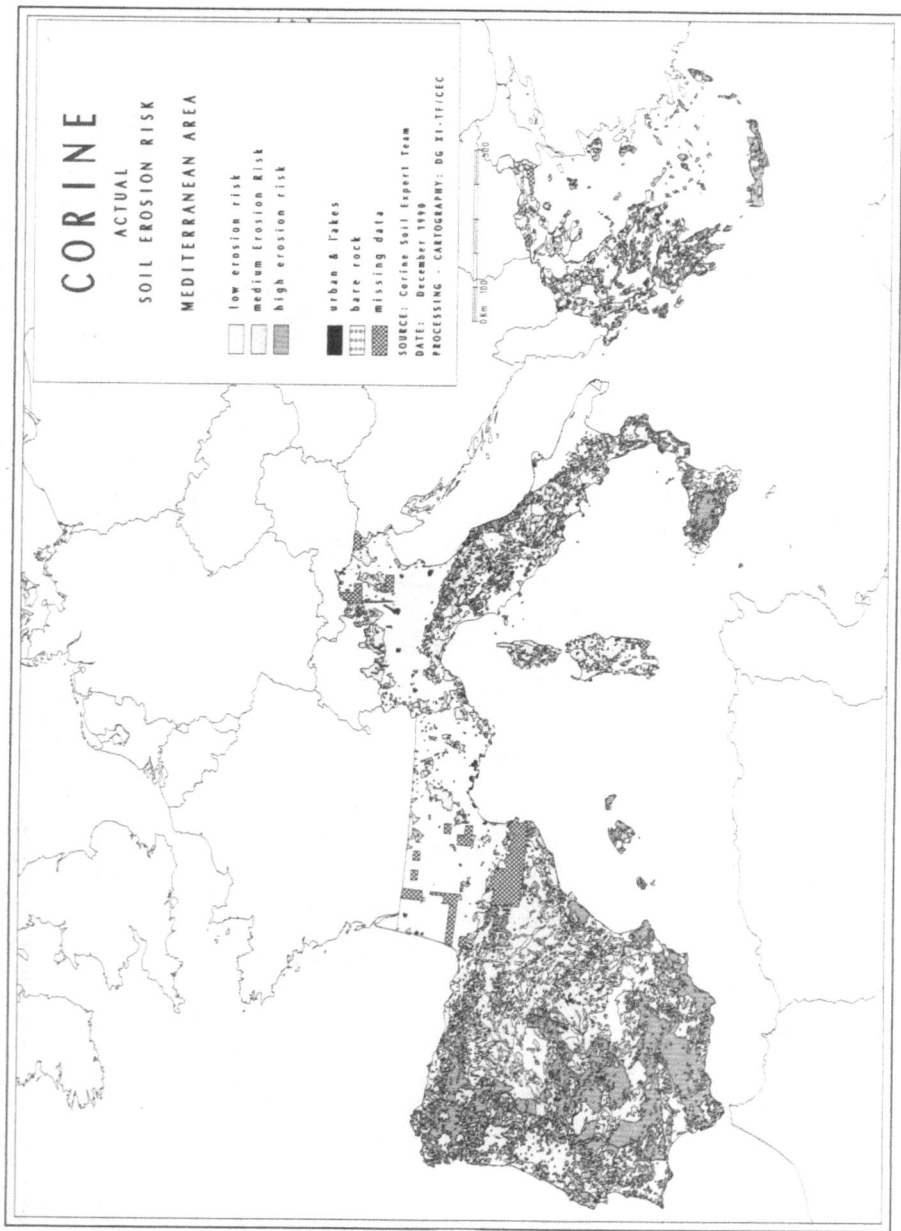

Figure 7.4: Soil erosion risk in the Mediterranean area of the European Community

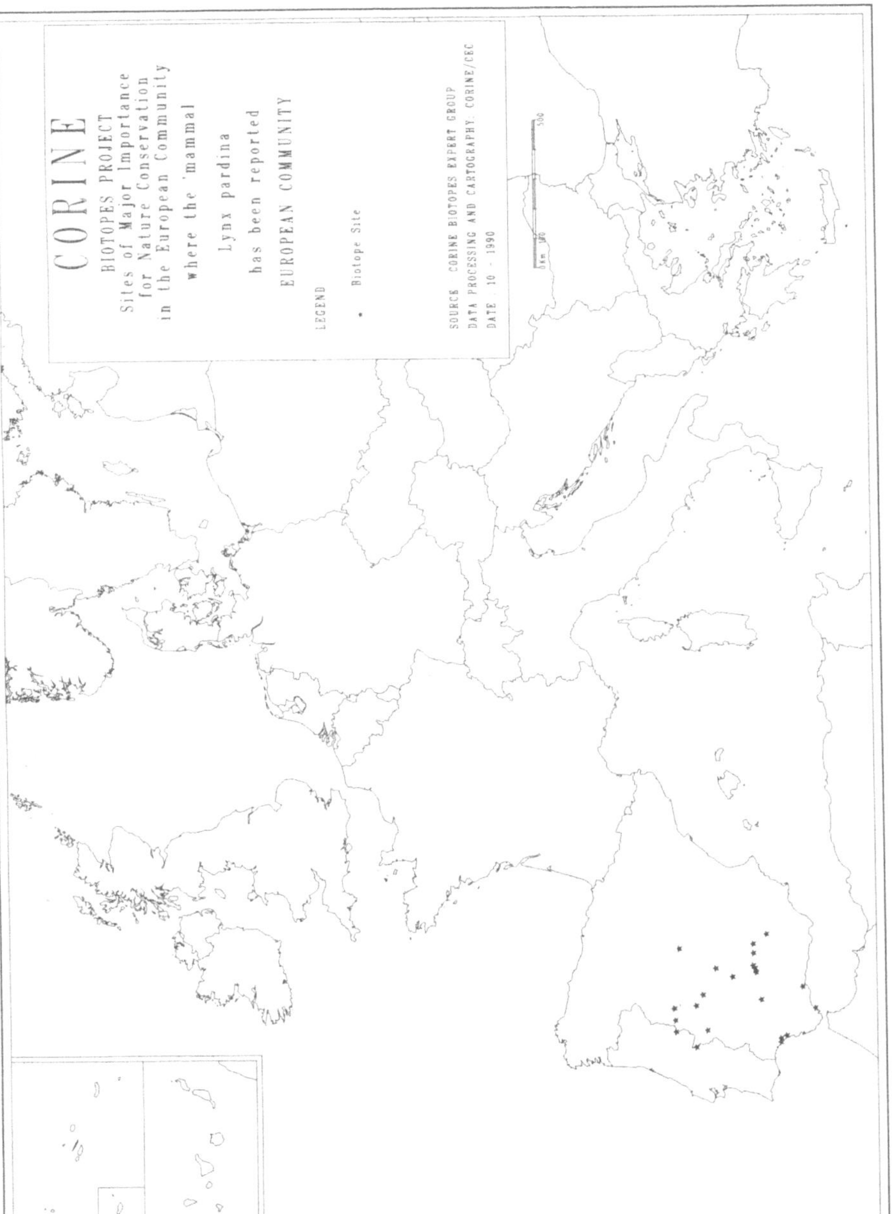

Figure 7.5: CORINE biotopes for *Lynx pardinia* in the European Community

References

Biotopes of major importance for nature conservation in the Community, Final Report, technical manual and cartographic annexes. (EUR 13231, 1992)

Commission of the European Communities (1991). Communication from the Commission to the Council and to the European Parliament concerning results of the CORINE Programme, Com 958, final, 28 May 1991.

CORINE Air Emission inventory, CORINAIR. Final report, in publication

CORINE soil erosion. Final report, in publication.

M.H. Cornaert and J. Maes
European Environment Agency Task Force
Commission of the European Communities
Directorate-General for Environment, Nuclear Safety and Civil Protection XI
Boulevard du Triomphe 174
B–1160 Brussels
Belgium

PART III
ANALYSIS OF SPATIAL INFORMATION

8   SPATIAL ANALYSIS IN HEALTH RESEARCH

Wim Douven and Henk J. Scholten

Abstract
The geographical approach towards the examination of diseases, or spatial health research, primarily focuses on the mapping of diseases and the correlation of spatial distributions by comparing two or more variables. The purpose is to disclose possible spatially determined aspects of disease etiology or even to help in defining hypotheses to be tested in comprehensive epidemiological research. In particular, exploratory spatial analysis could provide valuable instruments for spatial health research, given the fact that disease data are available but that models representing the real world are largely absent. These tools can help solving health research questions from a geographical perspective such as: "What is the spatial distribution of the disease under consideration?", "Can we detect patterns?" and "What are the possible coincidences with disease-causing factors?". In this chapter an attempt will be made to provide an overview of some useful analytical techniques in the context of health research.

8.1   Spatial analysis and geographical information systems

The concept of space plays a central role in spatial research. Very simply, this concept is explained by Tobler's first law of geography, stating that: "everything is related to everything else, but near things are more related than distant things" (Tobler, 1979). Two spatial effects are important in the context of Tobler's law: dependence and heterogeneity. Spatial dependence refers to the relationship between spatially referenced data due to the nature of the variable(s) under study and the size, shape and configuration of the spatial units. Spatial heterogeneity occurs when there is a lack of spatial uniformity of the effects of spatial dependence (Anselin & Getis, 1991). The study of spatially related objects, functions and characteristics can be roughly divided into the description of locational characteristics that differentiate one area from another and the analysis of spatial interrelationships at various scales.

This analysis of spatial objects or spatial analysis consists of a variety of (statistical) techniques. In order to structure the different kinds of techniques and their purposes, Anselin & Getis (1991) make a distinction between selection (of spatial objects at appropriate scale), manipulation (transformation, overlay and interpolation of data), exploration (of the data without a necessary pre-conceived notion) and confirmation (of hypothesis) (see also van Beurden, 1992). Here the focus will be exploratory spatial data analysis because models representing spatial disease characteristics are largely absent in health research. Exploratory and confirmatory spatial analysis are more or less opposite to each other. The principle of explorative data analysis, also called data-driven analysis

M. J. C. de Lepper et al. (eds.), The Added Value of Geographical Information Systems in Public and Environmental Health, 117–133.

techniques without having the aim of testing a pre-defined hypothesis about patterns or relations. As Tukey (1977) states: "It is important to understand what you can do before you measure how well you seem to have done it". This type of analysis might lead to new insights in the characteristics of data and even to the formulation of a hypothesis. An important instrument in the spatial variant of this kind of analysis is a map, as it permits a better, more real-world view of the world and the location of the spatial objects. By viewing maps, information can be acquired about patterns and relations of spatial objects. One of the disadvantages of maps is that its difficult to interpret them in an objective way. As a result, in the 1950s and 60s, an increasing need emerged for statistical techniques that could rationalize the search for patterns and relationships. This need led in the past decades to the testing and improvement of available statistical techniques and the development of new techniques: an important objective was to capture the spatial component of the data in these techniques. Particularly the field of spatial statistics addresses this subject (Ripley, 1981; Unwin, 1981). This development of spatial statistical techniques did not alter the fact that the visual analysis of map information remained an important technique. Confirmatory data analysis, on the other hand, searches for explanations by testing pre-defined hypotheses (model-driven analysis). Summaries are compared with what might be expected from theories of how patterns or relationships could have originated or developed. The two concepts of data analysis are closely linked to each other and often conducted in sequence. Explorative analysis can lead to the formulation of a hypothesis that then can be tested in confirmatory analysis. The distinction between the two concepts is in practice not always clear, as is the case with the various techniques falling in their range. Techniques used in confirmatory analysis, such as regression modelling, can also be valuable in exploratory research.

Geographical information systems (GIS) have proven their usefulness in a wide variety of applications in the past years. The ability to store and combine spatial data from different sources in a structured manner, to perform various manipulations and to present the resulting information in the form of maps feeds this usefulness. Many organizations and institutes working at various scales collect spatially related data for planning and policy purposes and increasingly make use of GIS to store and process this data. The primary fields in which GIS has been successfully implemented are planning, maintenance and monitoring of spatial objects. Also in the scientific field the interest for GIS has been raised for several years as an instrument to support spatial research, in particular, among the regional sciences, where the concept of space plays a central role. Until now, very few exploratory spatial analysis techniques (besides a number of mapping techniques) have been incorporated. In the scientific community, some attempts have been made in the past years to improve this situation. In terms of confirmatory (statistical) analysis within a GIS environment, not much has been achieved. Most of the applications are non-spatial applications of regression analysis and fail to exploit the information on the topology of the observations that is contained in a GIS (Anselin & Getis, 1991). A significant discrepancy between GIS as an instrument for processing spatial data, on the one hand, and the need for exploratory analytical techniques, in some way linked to or implemented in GIS, to analyse spatial data, on the other hand, has come to the fore. This issue is addressed in research literature by, amongst others, Goodchild (1987) and Openshaw (1991).

8.2   Spatial health research: methodological framework

The main areas of spatial health research to be distinguished are disease mapping, associative analysis and spatio-temporal analysis. Related to spatial analysis, these fields of study reflect a change in emphasis from exploratory data analysis to confirmatory data analysis. As we restrict ourselves here to merely exploratory data analysis techniques, disease mapping and associative analysis are of major interest, as spatio-temporal analysis has a more model-driven character. In health research the following stages can be identified in the chain linking aggregate spatial health studies to the identification of possible causal factors:

(i)   the collection and preparation of disease data;
(ii)   the mapping of data to identify spatial disease patterns at a variety of scales;
(iii)   applying objective statistical tests in order to consider whether the variation is significant and, if so, at what spatial scales;
(iv)   measuring the association between disease and other spatially varying factors;
(v)   the interpretation of the results of the previous stages, the indication of areas interesting for further research and eventually the generation of hypotheses;
(vi)   searching for possible causal relationships.

In most cases this process will not be linear, but cyclic, with feedback links to previous stages. From a geographical perspective, stages (i) through (v) are interesting and coincide with disease mapping and associative analysis. An important stage before mapping and analysis is the collection and preparation of the data (stage (i)). Besides other factors, availability, scale, type, level of measurement, reliability and consistency of data are crucial in the whole spatial health process and have to be dealt with before beginning the analysis (Cliff & Haggett, 1988; King, 1979; Matthews, 1990). Stage (vi), which falls under epidemiological analysis, attempts to establish the causality of the merely statistically significant associations detected by associative analysis. This stage requires thorough knowledge of epidemiology and etiology and detailed data coming from case-control studies, which is not the field of study of the regional sciences. As Picheral (1982) stated it: "We act somewhat as a photographer when we bring to light the spatial differences in frequency. Nothing is explained, but a lot is disclosed".

The definition of the scope of spatial health study brings us to two significant questions:

(i)   Which analytical functions and techniques are needed in these fields of spatial health research?
(ii)   Which techniques are already available and used?

Relevant techniques will mainly have an exploratory character, as already discussed, serving disease mapping, the spotting and testing of patterns and the seeking and proving of relationships. Figure 8.1 gives an overview of the areas of spatial health research (disease mapping and associative analysis) with required analytical functions and related analytical techniques. Besides these techniques, a visual analysis of mapped disease data is an important method in exploratory data analysis.

In the following section the spatial analysis techniques will be further outlined, illustrated with applications in spatial health research.

Figure 8.1: Overview purposes, functions and techniques of spatial analysis in disease mapping and associative analysis

Spatial Health Research	Spatial Analysis		
	Purpose	Function	Technique
Disease-mapping	exploratory	- mapping data	- dot maps - choropleth maps - probability maps
		- description patterns	
		point data: . identifying patterns . quantifying patterns	- cluster analysis
		area data: . patterns of value	- autocorrelation - regression modelling
Associative analysis		- description relations . map comparison	- overlay analysis - regression modelling
	confirmative		

The place of spatial analysis and GIS in a broader context of (spatial) health research is portrayed in Figure 8.2. In this framework, GIS (at the right side) plays a central role. GIS, as a system designed to process spatial information, accommodates facilities to store spatial data (spatial database) and to visualize spatial information. Partly due to the fact that the origins of GIS systems lie in the field of cartography, most GIS commercial packages contain a large number of mapping techniques, in contrast to analytical techniques. This holds even more for exploratory spatial statistical analysis techniques. In particular, the scientific community has drawn more attention to the testing and application of these techniques and the relation to GIS. As will be demonstrated in the following section, these developments are interesting from a spatial health research point of view. Some studies show the usefulness of a link of spatial analytical techniques to a GIS for reasons of storage and display of the spatial data. A close linkage or even incorporation would only make sense if the techniques concerned make use of the GIS data structure; otherwise a more loose link between GIS and statistical modules would be more appropriate.

Exploratory spatial health research aims at disclosing possible spatially determined aspects of disease etiology or even to help in defining hypotheses. The indication of areas of extreme disease incidence or prevalence or the indication of possible associative relations might lead to the formulation of hypotheses. In feedback links, this analysis can be refined by selecting other procedures for data preparation or analytical techniques. In the research process, it may also be worthwhile to refine the analysis through a zoom-in process using more detailed data on a lower scale (e.g., from a national to a regional scale) (van der Veen, 1992). The results of this explorative cyclical process can be further examined or confirmed in more comprehensive epidemiological research observing a range of spatial and non-spatial factors in more detail (e.g., case-control studies). The information generated in this research process could be passed on to public health officials in order to support the preparation of health policy.

8.3   Spatial analysis in health research

This section gives an overview of available spatial (statistical) techniques serving disease-mapping and associative analysis, as indicated in Figure 8.2. It is not our intention to provide an exhaustive list of techniques but to indicate interesting techniques. For the general spatial statistical issues of this overview, Unwin (1981), Ripley (1981) and Upton & Fingleton (1985) have been consulted. For for subjects more specific to health, *The atlas of the distribution of diseases* (Cliff & Haggett, 1988) has been a valuable source of information. The techniques will be illustrated by means of spatial health applications.

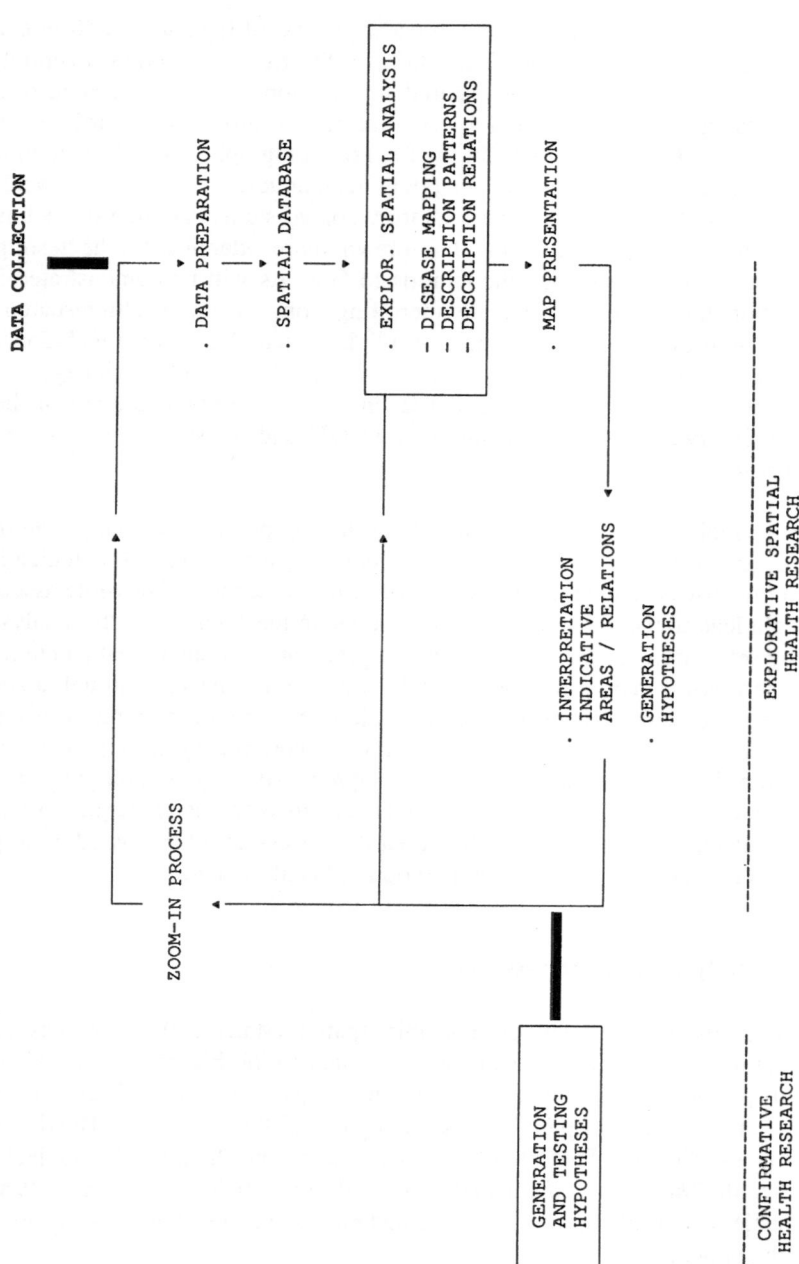

Figure 8.2: Spatial health research framework

The two entries of this overview are disease-mapping and associative analysis. For both, point and areal analytical techniques will be discussed. Points and areas are the most relevant data types from a health research point of view, as most disease data are available either as point information (cases of incidence) or area information (counts of cases, standardized ratio).

8.3.1 Disease-mapping

After the collection of health data, a first stage in spatial health research is the mapping of this data. Maps can provide an efficient method of demonstrating distributions of phenomena in space and can be seen as a useful technique throughout the entire research process. The relevant issues in disease-mapping include:

(i)   cartography;
(ii)   probability maps;
(iii)   map pattern descriptors.

These (sequential) stages in a disease mapping process, might be followed by associative analysis, which will be addressed afterwards.

Cartography

The main aspects in cartography are both the type of data and the level of measurement of data. Both point and area data can be converted to each other: point data can be aggregated to areas (e.g., by using interpolation techniques) and areas can be represented by a point reference. Four levels of measurement of data can be distinguished (nominal, ordinal, interval and ratio); each level determines the type of map to be produced.

Dot maps, dot density, symbol dot maps and proportional symbol dot maps can be generated from point data. The simplest dot map shows a number of dots, one dot representing the point where a specified disease is located. A derived map form is the dot density map, which allows each dot to represent a previously defined number of objects. For these map types, there are no data requirements as far as the level of measurement is concerned. In case the point data is of an ordinal level, symbols can be used to represent the different functions or types of objects. On such symbol dot maps, for instance, different diseases have different symbols. Point data at interval or ratio level allows proportional symbol dot maps to be made. If points have a value besides their georeferenced location, the size of the dot (represented as a circle, box or other symbol) can be varied according to this value. The most common method is to vary the size of the symbol proportionally based upon a linear mathematical function. Other methods make use of exponential functions, especially for practical reasons when the data value range is very broad. A derived map form is the pie chart, where each proportional symbol is segmented in pies showing data that makes up proportions of a whole.

Areas can be seen as spatial objects composed of lines, and in some cases with a point as a reference. Three spatial concepts are important: area, shape and contiguity (the pattern of connections between areas). Area-based maps can broadly be classified in coloured area maps, ordered colour maps and choropleth maps. The simplest map is the coloured area map or chorochromatic map, in which case some symbolism is used to indicate the prevalence of a named attribute (e.g., a certain disease) in a natural or imposed area. Maps showing imposed areas only disclose that the attribute (e.g., a certain disease) is prevalent somewhere in the area and that the area is of the type implied. An important aspect is the selection of the colours to be used. A number of studies are devoted to these cartographic aspects of mapping, emphasizing how people read or perceive colours and combinations of colours. There are no data requirements for chorochromatic maps. Ordered coloured maps are a second type of area-based map that represents ordinal scaled data. In addition to the remarks made about the coloured area maps, here it is of importance that the colours or symbols used do relate to the ordinal graduations. The most complex map type is the choropleth map, where interval or ratio data is represented by using a series of graded colours to indicate increasing data values. One issue regarding the generation of these maps that needs attention is the fact that the selection of the area size can determine the underlying pattern and density of the variable. This problem occurs not only if equally sized areas are used, but even more with irregular and unequally sized zones. This problem has consequences when relating two or more distributions, visually or with statistical techniques. Another issue is that the selection of the number of classes to be mapped and the intervals of these classes determine to a large extent the outcome of the mapping process. Various studies deal with these deficiencies of choropleth maps and give solutions or suggestions to improve the result (Evans, 1977). An important issue in disease-mapping - besides scale and classification - is the way in which data is spatially represented. Standard practice is not to use the spatial pattern of counts (because there is no reference to the population at risk), but, for instance, mortality ratios (e.g., observed count divided by expected number of cases, including standardized mortality rates or standardized registration rates). If the number of cases is small, these measures may lead to misrepresentation of the geographical distribution of any disease incidence (Clayton & Kaldor, 1987). An alternative is the empirical Bayes approach which improves imprecise estimates by estimates from other appropriate areas. Apart from rates, other statistical measures such as the significance of extreme values or residuals can be used. These and other relevant disease-mapping issues are addressed by Cliff & Haggett (1988), Gatrell et al. (1991), Matthews (1990) and King (1979).

GIS software contains a wide variety of tools to generate the kinds of maps outlined above. The functionality of the tools ranges from simple and easy-to-operate mapping to highly sophisticated cartographic modelling, dependent on the GIS software and the quality of the output devices. The availability of simple statistical summarization techniques and statistical graphics in GIS packages is very limited. Alternative approaches are the linking of GIS to existing statistical packages or to add GIS mapping facilities to statistical packages (Bailey, 1992).

Probability maps

After the geographical distribution of a disease is mapped, it is important to be able to determine the areas with extreme data values, which might give more insight in the etiology of the disease under study. A number of techniques are available to define respectively significantly high and significantly low areas of disease incidence. A simple technique to determine which areas have high or low disease rates is to divide the total frequency of the distribution into a number of equal proportions (quantiles). The areas in the upper tail of the frequency distribution may then be defined as significantly high and areas in the lower tail as significantly low. A more detailed method is normal probability distribution, which compares the constructed frequency distribution with one expected under some theoretical assumptions. One of the approaches to identify such a theoretical distribution of a disease is to make use of a normal distribution. From this normal distribution, the probability of obtaining, by chance, values larger than or equal to a certain value can be read, thus determining the areas that are least likely (significantly low) and most likely to have occurred by chance (significantly high). Another approach is to make use of a Poisson distribution, especially when the number of cases is small. It is also interesting to map the differences between observed and expected (as from standard distributions) disease occurrences: the residuals. In order to take into account the number of inhabitants per area, this residual can be standardized or adjusted. Some interesting spatial health studies in a GIS environment have been conducted on mapping and analysing the probability of extreme values (Brown et al., 1990; Hirschfield et al., 1990).

Map pattern descriptors

A map showing the geographical distribution of a disease will generate many questions concerning pattern description. Can we detect patterns in the data distribution? Are these patterns real or more likely a chance occurrence? Probability maps already give indications of the significance of data values but do not address the issue of whether a relationship or interaction with other adjacent areas exists. With the purpose of describing the spatial location of points and areas, a number of spatial analysis techniques have been developed.

A pattern or cluster of points is the characteristic of a spatial arrangement given by the spacing of individuals related to each other. In contrast to dispersion, it is not related to the shape of the study area. The density is the property of dispersal relative to an area but independent of the area shape or the dispersal of the objects therein. For points, there are three ways to objectively characterize these properties of pattern, dispersion and density: these three groups are measures based on density, measures based on distance and measures based on interpoint distance and direction.

Of measures based on density, the simplest method is to calculate the number of points per unit of area (crude density). This method, however, only informs about the property density. Another approach is to count the number of objects falling within a series of regular and equally sized subareas (quadrats), producing a picture of density variation across the map. An advantage of this method is that statistical measures can be used to describe the patterns and compare different patterns (mean, variance, variance/mean ratio). A high variance/mean ratio, for instance, indicates a tending to patterning or clustering. On the other hand, the size and shape of the quadrat influences the outcome to a large extent. In general, two ways of performing the quadrat counts can be distinguished: using a grid so that every point will be counted once (quadrat censusing) or a count with a randomly placed quadrat a number of times (quadrat sampling).

Among the approaches to measure distance between points are measures based on distance, using Euclidean geometry (as the crow flies) or real-world distances (e.g., based on a street network). A first method in this group calculates the map centrality. The mean centre technique, for instance, computes the centre of the map based upon all the locations or a number of points. In case of interval or ratio data, the points can be weighted with their value (weighted mean centre). The standard distance method can then be applied to measure the spatial dispersion of a point, e.g., the mean centre, by the standard variation. A second method in this group, calculates the distance from each point towards its closest neighbour, its closest two neighbours, etc.

A measure based on interpoint distance and direction that tries to tackle the limitations of the previous described measures is based upon Thiessen polygons. The principle is that each point is assigned to the points that are closer to the point in question than any other. These points are then used to create a neighbourhood area. By means of this area with their associated lines, several pattern descriptors can be applied to generate information about patterns.

A number of methods have been developed and outlined in research literature (Alexander et al., 1988) to serve these tests, with only a few, until now, connected to a GIS framework. Most applications are either performed on stand-alone packages or through a link between GIS and statistical packages (Bailey, 1992). One of the GIS applications is the geographical analysis machine (Openshaw et al., 1987). The starting-point is the merging of two sets of point data: the disease cases and a measure of the population at risk. For each point with at least one observed disease case, all other points are ordered by distance and counted by using a quadrat (or circle). The circles with a high significance, tested with a Poisson probability, are drawn. Thus, clusters worthy of further attention appear on the map as sets of densely overlapping circles.

Other methods make more explicit use of the distance between points, as discussed in the second group of measures, such as the method outlined by Cuzick & Edwards (1990). This method looks at the distribution of distances among nearest neighbours and also higher-order neighbours. In addition, Gatrell et al. (1991) worked out a method proposed by Diggle (1990) that also incorporates the character of clusters on a set of motor neurone disease data.

Spatial analysis techniques exist for describing the spatial location of areas. Spatial autocorrelation is a property mapped data possess whenever they exhibit an organized pattern. The emphasis is on the value of the spatial objects rather than on the location, which is a distinct difference with the description of point patterns. The underlying principle of spatial autocorrelation is described in the first law of Tobler. Spatial autocorrelation tests assess whether the observed distribution differs significantly from a random allocation of values. If similar values on the map occur in adjacent areas, this is called (positive) autocorrelation. A good introduction to spatial autocorrelation is given by Goodchild (1985).

Several methods have been developed to capture this self-correlation of spatial data. One of the simplest is the joint counting method. In using this method, data values are split into two classes (e.g., indicated on the map with black and white) and the joints between the possible particular types (black-black, white-white and black-white) are counted. A limitation of the joint-counting method is that the data are reduced to their lowest level, e.g., absent or present. Moran's I statistic (Moran, 1950) accepts data that are not necessarily at their lowest level (Gatrell et al., 1991). Other measures are the generalized proximity values and correlograms. Another approach towards autocorrelation is to extract geographical patterns of disease occurrence by using regression modelling.

In a health research context, some applications can be mentioned: Gatrell & Dunn (1990) have carried out autocorrelation tests in order to find out the spatial autocorrelation on the distribution of the standard registration rates of the cancer of the larynx in north-west England using Moran's I statistic. Kennedy (1988) modified an autoregressive model (ordinary least squares) to take into consideration a particular dependence structure of United States cancer data based on a contiguity measure for neighbouring counties (the first- to fifth-order adjacency counties). The model simultaneously estimates both the regional trend and neighbourhood components of the geographical variation in empirical cancer data. Kehris (1990) programmed autocorrelation tests within a GIS environment.

8.3.2 Associative analysis

So far emphasis has mainly been put on disease-mapping techniques, with the purpose of describing map patterns and testing their significance. After the distribution of single variables is described, it might be interesting to compare this distribution with possible disease-causing variables or maps in order to detect associations. Instead of the description of patterns, attention is given to the description of relations (comparison of two or more maps or variables). Besides visual comparison of maps, which can be very useful in disclosing general features, several (statistical) techniques can be used to perform this map comparison.

One method to isolate the various factors that could have been responsible for the shaping of the distribution of a disease is to superimpose the maps representing the factors (overlay). Its principle is to classify each map into, e.g., two categories (factor present and factor absent) and then to overlay them. The result will indicate for each (new) area the presence and absence of the factors that are studied.

A second method to examine the importance of factors is the use of a Lorenz curve. This curve is used to map two factors. First, the ratio of the two variables is calculated for each area and second, these ratios are ranked, with the lowest given a rank of one. Then each variable is standardized and accumulated, given the previously defined rank. Finally, the two variables are plotted against each other. The result, if both variables are proportionately equal in each area, will be a straight diagonal line. The index of dissimilarity (difference between observed and this diagonal line) gives an indication of the correlation between the factors.

This correlation approach can also be used in regression analys. A regression model involves usually one dependent variable and one or more independent variables. The assumption made is that the value observed for the dependent variable can be simply explained by a linear or non-linear mathematical function of the other variables. For instance, to fit a regression relationship, the ordinary least squares method can be used. The fitted line in itself is not of direct relevance in the search for associative relations. More relevant are the residuals: the differences between this function and the observed data values. The mapping of the spatial distribution of these residuals can give an indication for associative relations: e.g., positive residuals occur in the areas where death rates are underestimated by the regression model.

The associative methods outlined above do not explicitly consider the spatial component of the data; this is especially the case with the last two. It may thus be possible to find a high correlation between two variables, even when the spatial pattern is random. Some attempts have been made in spatial health research to capture this geographical component of the data in the comparison of factors. Some of these applications will be briefly outlined below; only some of them are conducted in a GIS framework.

Cliff & Haggett (1988) use the ordinary least squares regression model and the residuals from this regression to demonstrate the interplay of two factors, defective drainage and source of water supply, in accounting for the geographical distribution of cholera deaths in, mid-nineteenth-century London. Lovett et al. (1986) examine the relationship between mortality and possible explanatory variables over a series of areas where the number of death involved is relatively low. As an example, the analys is carried out on death from ischaemic heart disease in the county boroughs of England and Wales. Instead of the often applied ordinary least squares regression model, use is made of the Poisson regression model because of its advantages when the quantity of data is small. Also in the analysis of cancer of the larynx, use is made of the Poisson regression as an appropriate model for count variables. The dependent variable is the observed count of cases and both the expected variables and the social class are the explanatory variables (Gatrell & Dunn, 1990). Successive variables are added to the null model in an attempt to reduce the deviance. A chi-square test is then applied to test the significance of this reduction.

Another approach to the measuring of associations is the geographical correlation exploration machine (GCEM) (Openshaw et al., 1990), where a GIS is linked to a statistical program. The GCEM simulates an overlay analysis for all the combinations of suspected predictors (in form of maps). After the relevant information is aggregated to the final map, spatial pattern statistics are performed as a measure of an interaction and thus relationship between the predictors and the disease dataset. This method has been applied to the search for geographical correlates of leukaemia.

Methods of testing hypotheses of an elevated incidence of a disease near a set of *a priori* locations (e.g., pollutants) can also be seen as a type of associative analysis and are, in principle, quite similar to the point pattern descriptors. An example is incorporated in the geographical analysis machine (Openshaw et al., 1987). Within a search radius from a location, the cumulative observed cases inside the circle are tested on their Poisson probability. The search radius with the most extreme result is then identified. Finally correction for multiple testing is applied using a Monte Carlo method. Another approach is adopted in additional research on the data on cancer of the larynx (Gatrell & Dunn, 1990). After the suspected locations of pollutants and control sites are buffered, the number of expected disease cases for each circle is examined by overlaying the buffers with the observed cases (Gatrell, 1990; Diggle et al., 1990).

An approach somewhat different from the previous is adopted in SPIDER (Haslett et al., 1990). The main difference is that the focus is visualization and spatial heterogeneity rather than on quantification and dependence. SPIDER consists of some recent developments in the use of statistical graphics. The system enables different views of the data, which can be made active and cross-referenced by linking. A visual impression of the relation between value association (scatter plots) and locational association is offered; the user then has to discover patterns by viewing the dataset.

8.4   Conclusions

The combination of GIS - as an instrument to store, manipulate and display spatial information - and exploratory spatial analysis techniques, with the purpose of describing and summarizing spatial patterns and relations, can serve as a valuable instrument in spatial health research to elicit spatial distributions and relations. This information might lead to the identification of interesting areas and research questions that can be examined further in epidemiological studies. Two issues not explicitly discussed in this chapter, data (e.g., availability at various scales, reliability and consistency) and organization (what information is required?), strongly determine this analytical potential of spatial health research.

A number of applications in this chapter demonstrate this usefulness of exploratory spatial health research. The applications within a GIS environment are very limited but promising and can offer an exciting way to explore data. It is important to keep in mind what the added value of GIS is: the spatial approach to problems. Ideally, this spatial component should not only be prevalent in the data under examination, but - in order to fully exploit the possibilities of spatial analysis - also in the analytical techniques. In this light the developments in (primarily) the scientific community are very interesting; for the purpose of map summarization and description, analytical techniques are being developed and tested and attempts are being made to link these techniques to GIS.

References

Alexander, F., R.A. Cartwright & P.M. McKinney (1988). A comparison of recent statistical techniques of testing for spatial clustering: preliminary results. *In*: P. Elliot, ed. *Methodology of Enquiries into Disease Clustering*. London, Small Area Health Statistics Unit, London School of Hygiene and Tropical Medicine, London

Anselin, L. & A. Getis (1991). *Spatial statistical analysis and geographical information systems*. Paper presented at the 31st European Congress of the Regional Science Association, Lisbon, Portugal.

Bailey, T.C. (1992). Statistical spatial analysis & geographical information systems: a review of the potential & progress in the state of the art. *In: Proceedings of the EGIS'92 Conference*. Utrecht, EGIS Foundation.

Brown, P.J.B., P.W.J. Batey, A. Hirschfield & J. Marsden (1990). *Poisson square mapping, GIS and geodemographic analysis*. Liverpool, Department of Civic Design, University of Liverpool (Working paper 18, URPERRL).

Clayton, D. & J. Kaldor (1987). Empirical Bayes estimates of age-standardised relative risks for use in disease-mapping. *Biometrics*, **43**: 671-681.

Cliff, A.W. & P. Haggett (1988). *Atlas of the distribution of diseases, analytical approaches to epidemiological data*. Oxford, Basil Blackwell Ltd.

Cuzick, J. & R. Edwards (1990). Spatial clustering for inhomogeneous populations. *J R Stat Soc B*, **52**: 73-104.

Diggle, P.J. (1990). *Second-order analysis of spatial clustering for inhomogeneous populations*. Lancaster, Technical Report MA90/35, Lancaster University.

Evans, I.S. (1977). The selection of class intervals. *Trans Inst Brit Geogr n.s.*, **2**: 98-124.

Gatrell, A.C. (1990). *On modelling spatial points in epidemiology: cancer of the larynx in Lancashire*. Lancaster, Lancaster University (North West Regional Research Laboratory, Research Report 9).

Gatrell, A.C. & C.E. Dunn (1990). *GIS in epidemiological research: analyzing cancer of the larynx in north-west England*. Lancaster, Lancaster University (North West Regional Research Laboratory, Research Report 12).

Gatrell, A.C., J.D. Mitchell, H.N. Gibson & P.J. Diggle (1991). *Tests for spatial clustering in epidemiology: with special reference to motor neurone disease* Lancaster, Lancaster University (North West Regional Research Laboratory, Research Report 19).

Goodchild, M.F. (1985). *Spatial autocorrelation*. Norwich, Geo Books (CATMOG No. 47).

Goodchild, M.F. (1987). A spatial analytical perspective on geographical information systems. *Int J Geogr Inf Systems*, **1**: 327-334.

Haslett, J., G. Wills and A. Unwin (1990). SPIDER - an interactive statistical tool for the analysis of spatially distributed data. *Int J Geogr Inf Systems*, **4**: 285-296.

Hirschfield, A., P.W.J. Batey, P.J.B. Brown & J. Marsden (1990). *The spatial epidemiology of food poisoning in Blackpool, Wyre and Pylde: preliminary results*. Liverpool, Department of Civic Design, University of Liverpool (Working paper 14, URPERRL).

Kehris, E. (1990). Spatial correlation statistics in ARC/INFO. Lancaster, Lancaster University. (North West Regional Research Laboratory, Research Report 16).

Kennedy, S. (1988). A geographical regression model for medical statistics. *Soc Sci and Med*, **26**: 119-129.

King, P.E. (1979). Problems of spatial analysis in geographical epidemiology. *Soc Sci and Med*, **13D**: 249-52.

Lovett, A.A., C.G. Bentham & R. Flowerdew (1986). Analyzing geographic variations in mortality using Poisson regression: the example of ischaemic heart disease in England and Wales 1969-1973. *Soc Sci and Med*, **23**: 935-423.

Matthews, S.A. (1990). Epidemiology using a GIS: the need for caution. *Comput environment and urban systems*, **14**: 213-221.

Moran, P.A.P. (1950). Notes on continuous stochastic phenomena. *Biometrika*, **37**: 27-23.

Openshaw, S., M. Charlton, C. Wymer & A. Craft (1987). A Mark I geographical analysis machine for the automated analysis of point data sets. *Int J Geogr Inf Systems*, **4**: 335-358.

Openshaw, S., A. Cross & M. Charlton (1990). Building a prototype geographical correlates exploration machine. *Int J Geogr Inf Systems*, **3**: 297-312.

Openshaw, S. (1991). Spatial analysis and GIS: a review of progress and possibilities. In: J.H. Scholten & J.C.H. Stillwell, ed. *Geographical information systems for urban and regional planning*. Dordrecht, Kluwer Academic Press.

Openshaw, S. (1992). Developing appropriate spatial analysis methods for GIS. *In*: Maquire, D.J, M.F. Goodchild & D.W. Rhind, ed. *Geographical information systems: principles and applications*. London, Longman.

Picheral, H. (1982). Géographie médicale, géographie des maladies, géographie de la sancté. *Espace Géogr*, **11** (3): 161-75.

Pyle, G.F. (1977). International communication and medical geography. *Soc sci med*, **11**: 679-682.

Ripley, B.D. (1981). *Spatial statistics*. London, John Wiley & Sons.

Scholten H.J. & M.J.C. de Lepper (1990). The benefits of the application of GIS in public and environmental health. *World Health Stat Q*, **44:** 160-170.

Tobler, W. (1979). Cellular geography. In: S. Gale & G. Olsson, ed. *Philosophy in geography*. Dordrecht, Reidel pp. 379-386.

Tukey, J.W. (1977). *Explorative data analysis*. Boston, Addison Wesley.

Unwin, D. (1981). *Introductory spatial analysis*. London, Methuen.

Upton, G.J. & B. Fingleton (1985). *Spatial statistics by example*. Vol. 1. *Point patterns and quantative data*. New York, John Wiley & Sons.

van Beurden, A.U.C.J. (1992). Analytical capabilities and possibilities of GIS in environmental research. *In: Proceedings of the EGIS'92 Conference*. Urecht, EGIS Foundation.

van der Veen, A.A. (1992). GIS awareness in epidemiology and environmental studies. *In: Proceedings of the EGIS'92 Conference*. Utrecht, EGIS Foundation.

WHO (1990). *Environment and health: the European Charter and commentary.* Copenhagen, WHO Regional Office for Europe (WHO Regional Publications, European Series No. 35).

Wim Douven and Henk J. Scholten
Free University Amsterdam
Department of Regional Economics
De Boelelaan 1105
NL-1081 HV Amsterdam
The Netherlands

9 STRATEGIES FOR THE USE OF GEOGRAPHY IN EPIDEMIOLOGICAL ANALYSIS

Andrew Westlake

Abstract
The health of populations is affected by a variety of lifestyle and environmental factors, including where people live. Characteristics of locations (including sociodemographic factors and environmental exposure) offer additional information for studies about health. Spatial methods and other geographical procedures can be useful tools in epidemiological investigations, but must serve the analysis and not drive it. Even where accurate environmental data of appropriate resolution are available, our approach should not be "how can we make use of this data?", but rather "what health problems are we trying to solve, and what are the appropriate data and methods to help us find solutions?". This chapter addresses some of the issues in using geography, relating these to the various stages of an investigation.

9.1 Health information and geography

The health of populations is affected by a variety of lifestyle and environmental factors, including place of residence. Characteristics of locations (including sociodemographic factors and environmental exposure) offer additional information for studies about health, though there are considerable problems related to the relevance, specificity and accuracy of such data. Even where accurate environmental data of appropriate resolution are available our approach should not be "how can we make use of this data?", but rather "what health problems are we trying to solve, and what are the appropriate data and methods to help us find solutions?".

Geography is an important source of information for epidemiological studies and provides a way of classifying data records into groups (e.g., membership of administrative areas or distance from a point source) separately from the personal characteristics of the individuals. This geographical information can be used to investigate associations at the group (area) level, to look for associations with nearness to a point source of interest or to examine any other aspects of geographical location that are not captured by variables observed directly for the individuals. Groups defined by area can provide a link to other sources of information based on the same groups. The work of Barker and colleagues (Barker & Osmond, 1987) provides an important example of epidemiological analysis using ecological data. Where detailed location information is available for individuals it is possible to fit models that include spatial correlation components.

M. J. C. de Lepper et al. (eds.), The Added Value of Geographical Information Systems in Public and Environmental Health, 135–144
© 1995 Kluwer Academic Publishers.

Location may be sufficient for the purposes of exploring variations in health data. The identification of an excess of a particular condition at a particular location is an important discovery in its own right. The availability of geographically grouped health and population data (of appropriate quality and resolution) is a prerequisite for any such analysis. The explanation for such an excess can then be sought, and may well lie in the characteristics of the groups induced by the geography, rather than in any direct aspect of the geography. For example, variations in the demographic or social structure of the population at risk may provide a complete explanation. A specific study of the individuals may be indicated, and other information about the location may provide clues towards an explanation. Environmental data (of appropriate quality and resolution) have a place in such explanatory processes, but are by no means the only input that should be examined by the thoughtful investigator.

Many statisticians and epidemiologists express considerable concern about the use of mapping as an analytical step in an investigation. However, maps provide an important mechanism for presenting the geographical component of the conclusions from an analysis.

9.2   The process of an epidemiological investigation

We can identify five stages in an epidemiological investigation of a concern over a health issue:

(i)   A specific concern is identified.

(ii)   The existence of an excess is confirmed.
        This may be an excess compared with an established background rate or variability which is in excess of that expected from the underlying natural variation. A specific outcome (or more than one) must be identified, together with variables which reveal the excess or variability. Statistical methods are needed to confirm that the excess is not simply a chance outcome consistent with the underlying statistical model.

(iii)   An explanation for the excess is sought.
        Having established that distinct outcome groups can be differentiated we use statistical and other methods to identify possible explanatory variables or groupings, being cautious not to confuse association and causation.

(iv) Confirmation of the explanation is sought.
This must be found in new information, whether in new data related to the original study, new studies or trials or even in experiments to demonstrate the mechanism of the proposed explanation.

(v) The conclusions of the investigation are presented.

Various types of data are required for the different stages and vary for different types of investigation. For a concern about a localized excess for some health outcome at stage (ii) we need information on the health events concerned and the population at risk. In addition, we need to know about the location of the events and the population, so that we can confirm that there are indeed different risks at different locations. We may wish to include information about factors known to be associated with the events, such as age and sex.

At stage (iii) we need data about additional variables to pursue a particular explanation, which may be found through the population structure (demography, socioeconomic factors, social or occupational groupings or even history) or through factors more directly related to the physical geography (soil types or geology, point sources of pollutants or more general environmental factors). To confirm an explanation (stage (iv)), we need more outcome data from different places and/or different time periods or, alternatively, a more direct method to demonstrate the proposed link. Maps can be an important tool for presenting the conclusions of the investigation, at stage (v).

9.3   Uses of geographical information

9.3.1 Confirming a geographically related excess

Location can be associated with epidemiological information in several ways, but most usually this is done either through attaching point locations to individual cases or through attaching membership of some geographical group (area) to each case. Location can then be used to investigate whether it makes any difference where the cases are. Two different approaches to confirmation can be used, depending on whether there is a focus on a particular source of risk or not. For the moment we concentrate on the latter situation, with no proposed focus.

Location within some area grouping (often the existing administrative areas) can be used to allocate populations into near (and hence similar?) groups. Membership of these groups in itself is not important, but if the groups exhibit (significant) differences then it is possible to suggest that there may be some (so far) unobserved factor associated with the outcome under study, the effect of which varies more between groups than within them.

Using standard analysis of variance techniques (or an appropriate generalization thereof), one can investigate whether the average level (probability or rate) of the outcome is significantly different in different areas (health districts, for example). This technique can take account of other (observed) characteristics of groups, such as the age and sex structure. Frequently we choose to aggregate data within areas and use the areas (rather than the cases) as analysis units, usually with rates as the response variable. Sometimes we have no choice, since the area aggregates may be the only outcome data available.

Where existing groups or administrative boundaries are not appropriate, the spatial processing component of a system for handling geographical structures (a geographical information system) can be useful in defining new membership areas based on the characteristics of structures (such as roads, estuaries, industrial areas) that may be related to the outcome under study. This is most accurately done when outcome and population data are available at very fine resolution. To work well, it is sufficient that the spatial resolution of the outcome and population data grouping is well below that of the resulting areas, so that allocation problems at the boundary are relatively insignificant. As with all statistical investigations, it is important that the rules for grouping and analysis be defined in advance of examination of the data. Systems that search for areas with high response will always find some, but the process of searching invalidates the usual statistical interpretations.

Where there is some predefined focus for attention, the distance from the focus can be used to look for gradients in risk that peak at (or near to) the focus. In particular, this would be used in testing hypotheses about the association of outcome rates with distance from a point source. The word 'distance' is used here in its most general sense, to refer to any single-valued function computed from the location of two points. This includes measures such as Euclidean distance, elliptically weighted distance or even travel time.

In most situations it is essential to make allowance for variables (confounders) that are known to be associated with the outcome under investigation. Age and sex have already been mentioned and there are many other well-established links between health outcomes and individual characteristics or behaviour. These should be examined carefully to determine which should be allowed for when deciding whether a real excess exists and which should be treated as potential explanatory variables for interpreting the excess. The borderline is not always clear.

9.3.2 *Explaining a geographically related excess*

To explain variation between geographical areas, we usually look for some measurement about the groups (explanatory variable) that is associated with the differing levels of outcome. This may be some property of the individuals in each group or a property of the group areas themselves. Frequently the explanation will be found in the aggregate socioeconomic, demographic or occupational characteristics of the groups and has nothing to do with the actual geographical groups for which the excess is observed. Rather, there is something about an area which has drawn together people with similar social or other characteristics and it are these characteristics of the population which are related to the outcome. It should then be possible to demonstrate the association much more strongly by looking at individual data (through specific studies) instead of aggregates of outcome and explanatory variables for the groups. In other situations the explanation may indeed be found through the areas themselves, for example through the water supply, or specific sources of pollution, factors truly shared by all the individuals situated in the area.

Notice that membership can be used to provide a link to other information about the areas even when this is neither collected directly for the individuals under study nor even directly related to them. In Great Britain, the census small area summary statistics are the most obvious examples of this, providing demographic and social data of high quality for about 140,000 enumeration districts with an average population of about 400. Of course, summary data about an area is (statistically) much less efficient than the actual data about cases and there is the usual danger from the ecological fallacy of assuming that the properties of the area relate to the cases. None the less, provided the cases are (at least) an appropriate sample from the area and that the usual warnings against inferring causation from association are observed, this linking of data through area membership has a useful role in epidemiological investigations. For example, Barker & Osmond (1987) show that (at the level of local authorities) death rates from ischaemic heart disease are strongly correlated with the rates of both maternal and neonatal mortality (taken as surrogates for maternal stress on infants) 60 years previously.

9.3.3 *Geographical analysis*

The procedures mentioned above make use of geography, but none of them can truly be described as geographical analysis techniques. Geography is used as a source of data, but the methods then used are not affected by the nature of that source.

Many statistical methods make the basic assumption that cases are independent after adjustment for explanatory variables. This is equivalent to assuming that knowing that a case is a member of an area group captures all aspects of the similarity of cases in the group, whereas in fact we would usually assume that, even within a group, nearby cases are more likely to be similar than those further apart. Thus, the methods above, although they may give reasonable results in many situations, do not make full use of geography.

A more realistic statistical geographical model will include as a component an unobserved geographical factor, varying slowly compared with the distance between cases, which causes close cases to be more alike than distant ones. This approach leads to analysis using:

(i)   spatial correlation models based on adjacency or nearness.
(ii)   smoothing techniques based on distance; and
(iii) clustering techniques based on cases rather than areas.

All these techniques are complex and computationally demanding and in particular require efficient methods for the retrieval of nearby cases.

9.3.4 Spatial analysis

Geographers use this term to refer to the manipulations that take place between spatial structures, such as placing points in polygons or computing the overlap between two areas. Epidemiologists and statisticians reserve the word analysis to refer to investigations of data or structure and not to their manipulation. It would be useful if the term spatial analysis could be dropped in favour of reference to spatial operations, spatial manipulation or spatial exploration.

9.4   Use of maps

9.4.1 General considerations

Maps have an important role to play alongside the epidemiological use of geographical data, in the presentation and emphasis of conclusions drawn from the analysis. They have great potential as a method for communicating results, but great care is needed to ensure that the correct message is presented.

The value of a graphic image in communicating information is widely accepted, because it allows attention to be drawn to the particular characteristics of the data we wish to emphasize. Whereas numbers are neutral (and, as such, often indigestible), images and colours are not. If we understand the nature of perception of image and colour, we can use it to reinforce the selected message.

The use of bar and pie charts is generally felt to be appropriate, since it allows information about the relative magnitude of different categories to be perceived very quickly. Scatter diagrams are useful for displaying the general pattern of configuration of data and for detecting outliers (in two dimensions).

Maps show well the spatial configuration of information, just as traditional maps show the spatial configuration of physical structures. When we overlay statistical data (often rates) on a map we retain the underlying physical structure (and thus the relative configuration of areas) but have fewer choices for representing the statistical numbers. There is little scope for representing values through size (since the size attribute has already been used to represent areas), so colour or shading are usually adopted, for which judgements about relative values are much more difficult. The association of colours with bands of values can be used to draw attention to the relative location of areas in the same bands.

The representation of more than one value for an area (whether an associated variable or the precision of the first value) is particularly difficult. In effect, some of the dimensions of presentation that might be used to represent the inner details of statistical information have already been used up in showing the spatial component. This is often an acceptable trade-off, but it is important to recognize the limitations of the representation of statistical information in a map. None the less, where a simple message is to be conveyed, a map can be a powerful means to carry that message.

9.4.2 Presentation choices

Different considerations arise in different contexts. When results are presented for relatively few areas (such as the counties in Great Britain), a map gives the reader a context in which to view the data and allows the comparison of results for adjacent areas. The numerical values of results and their precision (standard errors), can readily be shown, either on the map or in an adjacent table.

With large numbers of areas, considerable caution should be exercised. In this situation there is too much information for explicit comparisons to be made, so the eye is drawn to looking for patterns, which it will invariably find. This can be acceptable when the information presented is directly related to the geography the eye sees, but there is a great danger that characteristics of the statistical measure will be confused with characteristics of the underlying geographical distributions. This is particularly a problem with the effects of variation, that is, where different areas have different statistical precision in their measures. We can identify a number of potential problem areas.

Effects of population density: with maps of cases over wide areas, there is a danger that only the underlying distribution of population will be shown. It is obvious that a map showing the location of lung cancer cases over a wide area of varying population density would show little more than the population density, since the cases (approximately a random sample of the population) will be clustered in accordance with the population density. Where population is more uniform this will not be a problem. For example, John Snow was able to identify the Broad Street pump by looking at the distribution of cases of cholera over the relatively uniform population distribution in a central area of London. Similarly, maps showing land use allow the eye to draw proper conclusions about clustering of usage because this is directly related to the area shown for the use.

Rates versus significance: changing from maps of cases to maps of rates for areas changes the effect of this problem but does not provide a complete solution. If areas are chosen with (approximately) equal population, then areas of low population density will be physically large and thus will catch the eye if solidly filled on a map. If areas are more equal in size, then those with lower total populations will have rate estimates with higher variance, and so the highest rates are likely to be observed in areas of low population density. This is a particular problem with rare conditions, where the highest observed rates often represent single cases in areas with low population. This is an example of the problem mentioned above, that it is difficult to show both level and precision (average and variability) in a single map.

Another solution is to display not the rates but their significance levels (or something equivalent such as chi-square values). Here we suffer from the reverse problem, that only areas with large populations will have the statistical power to stand out. There are situations when this is appropriate, but it would not be so when looking for small localized excesses. The empirical Bayes procedures proposed by Clayton & Kaldor (1987) offers a promising compromise between the rate and significance approaches.

Brown et al. (1991) have suggested that shading or colouring on a map of population-based characteristics should be restricted to built-up areas (as shown in Great Britain on high-resolution Ordnance Survey maps) with the remainder left blank. This seems a big step towards the true underlying geography but has significant implications for the amount of geographical information needed.

Confounding: many common diseases are known to be associated with socioeconomic factors that themselves exhibit geographical clustering. The perception of a map of rates for a disease related to smoking may be heavily influenced by patterns in socioeconomic status. Standardization can help to avoid this problem but is not without its own (statistical) pitfalls.

Such sociodemographic interpretations show the danger of relying on mere associations. Lung cancer rates may well be higher in built-up areas and various air pollution measures may also be high. A definite association exists, but this does not show that air pollution causes lung cancer. Rather, populations in built-up areas tend to be of lower socioeconomic status, such populations are heavier smokers and smoking causes lung cancer. Thus, built-up areas are correctly identified as having higher rates, but the cause of this is not a property of the areas, but rather of the individuals within the areas. Seaside towns in the south of England are likely to have high crude mortality rates, not because the seaside is intrinsically unhealthy, but because elderly people choose to retire there, producing an unusual age distribution in the local population.

9.4.3 Mapping as an investigative method

The eye is highly trained and sophisticated at detecting patterns in noisy and seemingly random information. Maps are not good at representing the complex interrelationships that can exist between statistical response and explanatory variables. Herein lies a serious problem, which arises when we try to use a map or other image for data exploration (rather than to put across a selected message). It is difficult, even for the trained investigator, to avoid the pitfall of seeing patterns where no real association exists. There is thus considerable danger in using maps as a central component of data exploration, and a special danger in presenting uninterpreted data (as maps) to untrained users. It is worthy to note that few geographers promote the use of maps in this way, but it is seen as a major attraction by end-users, who have difficulty with the interpretation of numerical statistics. Ultimately, there is no alternative to careful interpretation, and to avoid it is to abrogate scientific responsibility.

9.5 Conclusions

The potential uses of geographical inputs in epidemiological investigations are many. Location is an important additional input to an analysis since it gives us a way to look for variation associated with nearness but not captured by the variables observed in a study. Explanations for such variation may be found in the properties of areas (environmental data) but are frequently strongly influenced by the nature of the population grouped in the areas.

A geographical information system, seen as an implementation of methods for geographical manipulation and presentation, can provide useful tools when working towards an understanding of the underlying process, alongside database and statistical analysis tools. However, geography is only a part of the overall tool-kit and must always be subservient to the overall epidemiological objective.

There is considerable danger in using a map as a tool for data exploration. It is easy to see spurious patterns, and inappropriate prior examination of the data can invalidate subsequent statistical analysis.

We should return to the fundamental questions for health researchers: what problem are we trying to solve and what message are we trying to communicate? This is what must drive our methodology and not just the availability of powerful mapping systems.

References

Barker, D.J.P. & C. Osmond (1987). Death rates from stroke in England and Wales predicted from past maternal mortality. *Br Med J*, **295**: 83-86.

Brown, P.J.B., W.J. Batey & A. Hirschfield (1991). Prerequisites for the Poisson chi square mapping and geodemographic analysis of relatively rare conditions. *In: Proceedings of a Workshop on Geographical Data Structures for Small Area Statistical Analysis*. London, London School of Hygiene and Tropical Medicine.

Clayton, D. & J. Kaldor (1987). Empirical Bayes estimates of age-standardised relative risks for use in disease mapping. *Biometrics*, **43**: 671.

Andrew Westlake
London School of Hygiene and Tropical Medicine
University of London
Keppel Street
London WC1E 7HT
United Kingdom

10 ANALYSING SPATIAL PATTERNS OF DISEASE: SOME ISSUES IN THE MAPPING OF INCIDENCE DATA FOR RELATIVELY RARE CONDITIONS

Peter Brown, Alexander Hirschfield and John Marsden

Abstract
This chapter examines some of the methodological issues encountered when attempting to interpret patterns in data recording the occurrence of relatively rare events or conditions. Such issues are of particular concern to those working in the field of spatial epidemiology, where understanding the significance of patterns in the incidence of individual morbidity and mortality conditions can contribute to reducing the risk of exposure to these conditions and to the alleviation of suffering. This chapter outlines some of the advantages and disadvantages of alternative ways of representing aspects of the occurrence of relatively rare conditions in graphic form. Discussion focuses on the benefits and limitations of using a Poisson chi-square procedure in the production of choropleth maps. Consideration is also given to ways in which geographical information system features can be exploited in an effort to improve the representation of rare condition information in areas of relatively low population density. Use of the methods discussed is illustrated by drawing upon examples from applications that have been undertaken in a series of health-related projects carried out by the Urban Research and Policy Evaluation Regional Research Laboratory in the north-west of England.

10.1 Introduction

Interest in the issues raised in this chapter has been stimulated by experience gained in a series of studies undertaken by the Urban Research and Policy Evaluation Regional Research Laboratory (URPERRL) staff in collaboration with district health authorities in the north-west of England. These studies have shared the theme of a concern with the spatial distribution of a range of morbidity or mortality conditions and attempts to facilitate the interpretation of maps depicting these distributions, primarily for the benefit of district health authority staff.

Much of this work has been exploratory and experimental, with an emphasis on spatial analysis and hypothesis generation. It has featured attempts to exploit some of the more straightforward features of widely available geographical information system (GIS) software to enable those unfamiliar with GIS technology to achieve a better understanding of the patterns that may be present in the incidence of a variety of conditions.

M. J. C. de Lepper et al. (eds.), The Added Value of Geographical Information Systems in Public and Environmental Health, 145–163.
© 1995 Kluwer Academic Publishers.

incidence of morbidity and mortality to socioeconomic conditions. This work was undertaken as part of a project concerned with the development and testing of modifications of a formula-based method, employed by Mersey Regional Health Authority, for allocating resources between health service units (Batey et al., 1985). Choropleth maps were produced to depict the variation present in a range of census-derived indicators of deprivation at census ward level. These were compared with corresponding maps of simple annual rates of incidence per 10,000 population, standardized mortality ratios (SMRs) and standardized incidence ratios (SIRs) for a variety of conditions, presented using similar methods to those employed in the production of atlases of diseases (Gardner et al., 1983; Kemp et al., 1985).

Some aspects of the experience gained in the course of this exercise are relevant to the theme of this chapter. These relate to the following three topics:

(i)   the principles adopted in defining shading conventions in choropleth maps;
(ii)   the specification of spatial units for similar mapping purposes; and
(iii)   the treatment of extreme values, including the choice of method for assessing their
        significance.

The first of these will be discussed very briefly before more detailed consideration is given to the specification of spatial units and treatment of extreme values.

The above work underlined the inadequacy of shading conventions based on simple quantiles as a means of representing variation in census-derived and other indicators that display a wide range of degrees of skewness and spatial concentration. For this purpose, a nested means approach proved more satisfactory, as the resulting intervals reflect the distributional characteristics of the data, unlike arbitrary classification schemes, such as quantiles. The approach is based upon the idea of initially distinguishing between the cases that are greater than and less than the overall mean value. The means of the cases that are, respectively, greater than and less than the overall mean then serve as further critical values with which to distinguish four subsets of cases. The method thus isolates relatively high and low values, which are readily interpretable, in a consistent manner, as long as the distribution is not too strongly skewed. With the same proviso, it is possible to apply the same principle in splitting each of the four subgroups of cases at its mean value to produce eight levels of shading.

Similarly, SMR and SIR plots (such as the one featured in Map 10.2) achieve a degree of consistency because they take into account local variations in the age and sex distribution of the population. Intervals, such as < 50, 50-100, 100-150, etc. are fairly readily interpreted as corresponding to values that are less than half the mean, lying in the interval between half the mean and the mean, and so on. However, it can remain difficult to judge whether the differences between sub-units lie within a range that might be expected or is large enough to be regarded as in some way significant.

None of these methods addresses another fundamental criticism of conventional choropleth mapping - the failure of spatial units defined solely for administrative purposes, such as the enumeration districts (EDs) employed in carrying out the British Census of Population, to reflect the true distribution of this population.

It can also be difficult to assess the significance of extreme values - whether the high (or low) values are higher (or lower) than might be expected, based on randomness assumptions and the consequent inevitability of chance clustering. This becomes all the more important in the analysis of relatively rare conditions, which are characterized by small numbers of occurrences in individual spatial units and highly skewed distributions, when these counts are related to the appropriate populations at risk.

Before we elaborate upon these points, we should note that the account presented here only addresses a subset of the issues raised by the recent debate and discussion (largely prompted originally in the United Kingdom by the Black (1984) report) of statistical and related methods for the detection of clusters and/or the clustering of relatively rare diseases in space or over space and time (Clayton & Caldor, 1987; Gardner, 1989; Hills & Alexander, 1989; Openshaw, 1990; Cuzick & Edwards, 1990; Wartenberg & Greenberg, 1990).

In this respect, Besag & Newell (1991) make a useful distinction in their discussion of tests of clustering in which the usual question to be answered is whether the observed pattern of disease could reasonably have arisen by chance alone. They distinguish two types of test: general and focused. "The former are concerned with the overall pattern of disease over a large region, rather than with any specific cause; such tests are often to be found accompanying maps of disease incidence in cancer atlases, for example. In contrast, focused tests concentrate on one or more smaller regions selected ostensibly because of some factor (e.g. the location of a nuclear installation) that has been previously hypothesised to be associated with the disease; comparisons are usually made with the incidence rates outside the special region or regions)."

Although some reference is made to the latter form of focused testing, the main emphasis here will be upon methods relevant to the former, general case of the examination of overall patterns of disease across a region.

10.2 Treatment of spatial units

Analysis of the type described above is influenced fundamentally by the specification of the spatial units employed. This applies to both the units used to record the location of sufferers or victims of the conditions of interest and the population at risk.

If the underlying population were to be uniformly distributed, then it would be appropriate to use point plots to display the location of the victims or occurrences of the condition and to employ nearest-neighbour methods (Cliff & Ord, 1981) as a means of establishing the significance of any pattern that may be observed. In the more typical circumstances represented by a heterogeneous population distribution, Schulman et al., (1988) employed a density-equalizing map projection to create areas of a size proportional to their population to which similar methods can be applied. However, among the disadvantages of this approach is that the resulting display bears no relationship to geographical space and cannot be used in conjunction with, for example, maps of land use (Alexander et al., 1991). As an alternative, Cuzick & Edwards (1990) have developed an extension of the traditional nearest neighbour approach that can accommodate a heterogeneous population distribution.

Besag (1989) put forward a different method of extending nearest-neighbour approaches that has been taken further both by Alexanderet al. (1991) in the form of the nearest neighbour area (NNA) method and by Besag & Newell (1991). These innovations involve consideration of disease incidence in a number of overlapping circles in a similar manner to that employed in the geographical analysis machine developed by Openshawet al.,(1987, 1988) in seeking to identify the localized clustering of cases. Various aspects of the evolution of this approach are reviewed by Openshaw (1990).

The development of these approaches has been prompted by attempts to overcome the problems associated with the use of a set of irregularly shaped spatial units to record the distribution of a population that is itself unevenly distributed over space. Where the population data are derived from a census, the choice of unit of analysis is restricted to the areal units which, typically, are defined primarily to minimize effort in data collection. Other considerations in their specification include the need for units to be consistent with, or aggregate to, irregularly shaped administrative units, such as the ward or district, and to exhaust the geography of the country.

Aggregation compounds these underlying weaknesses, as characterized by the modifiable areal unit problem (Openshaw, 1984). This is encountered when the results of an analysis are shown to vary significantly with the spatial scale. Bracken (1989) summarizes the problem well, in the following terms: "in general, as the data values associated with each tract are area aggregates, the actual values may be as much a product of the zonal boundaries and their locations, as a representation of an underlying geographical distribution" (p. 307).

Bracken goes on to illustrate how the use of such areal tracts as census EDs or wards for the presentation of information in choropleth map form can give rise to extremely misleading impressions of the spatial distribution of phenomena. Two rural examples are reproduced here (see Map 10.1). These also serve to convey a clear impression of how the potential for misrepresentation is greater when the population is more clustered. The first shows how to ensure geographical coverage: a typical arrangement of census units in a valley-type community will stretch from the valley floor to the watershed of the next valley, despite the fact that the population is concentrated in the valley floor. The second example relates to a typical large village in which it is suggested that four enumerators have been used. In this case, the community has been split into four EDs, each of which includes a substantial area of unpopulated farmland.

Map 10.1: Typical census enumeration district (ED) arrangements in (left) a valley-type community and (right) a large village in the United Kingdom

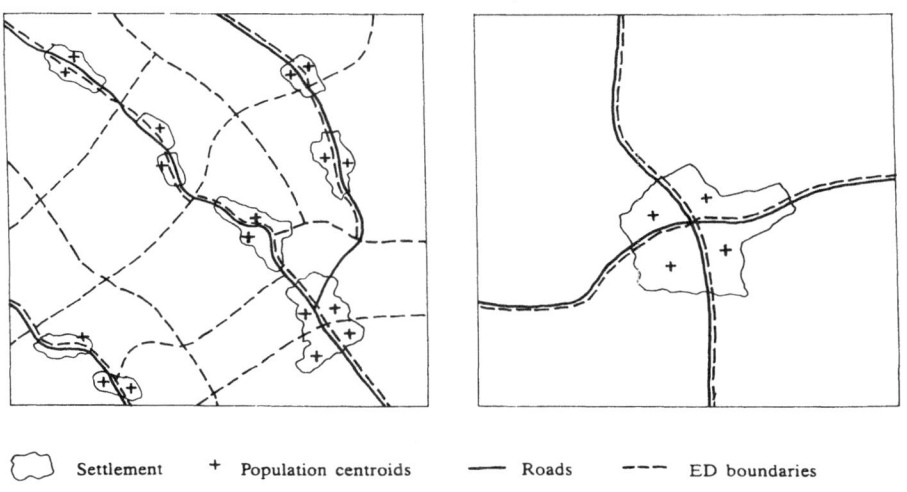

�⌣ Settlement   + Population centroids   —— Roads   - - - ED boundaries

Source : Bracken (1989)

In both cases, it is inevitable that the EDs contain land with very different characteristics. Any effort to relate data at the disaggregated or aggregated level to the actual extent of the residential area is impossible.

In addition, the drawing of census unit boundaries in this way creates further problems for the epidemiologist. By splitting up the populated area in an effort to even out enumerator workloads, the detection of clusters of cases based on rates of incidence per head of population is rendered more difficult. For example, the occurrence of a case in each of the four EDs illustrated in the village depicted in Map 10.1 could yield unremarkable incidence rate figures for the individual EDs and thus fail to attract the further attention this spatial clustering would be likely to warrant.

Similar shortcomings of census geography as a means of representing the distribution of population are evident in urban areas. What appear, on an Ordnance Survey map, to be continuously built-up areas are often found to comprise a mix of residential and other forms of development, interspersed with large areas of open space, such as playing fields, parks, commercial premises and derelict sites.

A more realistic description of the distribution of population will require the separate representation of the extent of residential areas. Furthermore, these more tightly defined boundaries should be related to the spatial units to which the census-derived measures or indicators refer. With the residential area and census unit boundaries held in digital form, this link can be achieved relatively easily within a GIS framework.

In Maps 10.2 and 10.3 we illustrate some of the benefits of capturing the boundaries of residential areas in digital form and their use as a filter through which information relating to the corresponding set of census units (wards) is passed. However, before noting some features of these maps, it is appropriate to describe the context in which the maps were produced.

Map 10.2: Standardized incidence ratio (SIR) per 100,000 population for
           Campylobacter: Blackpool, Wyre and Fylde Health Authority - choropleth
           map of 1981 census wards

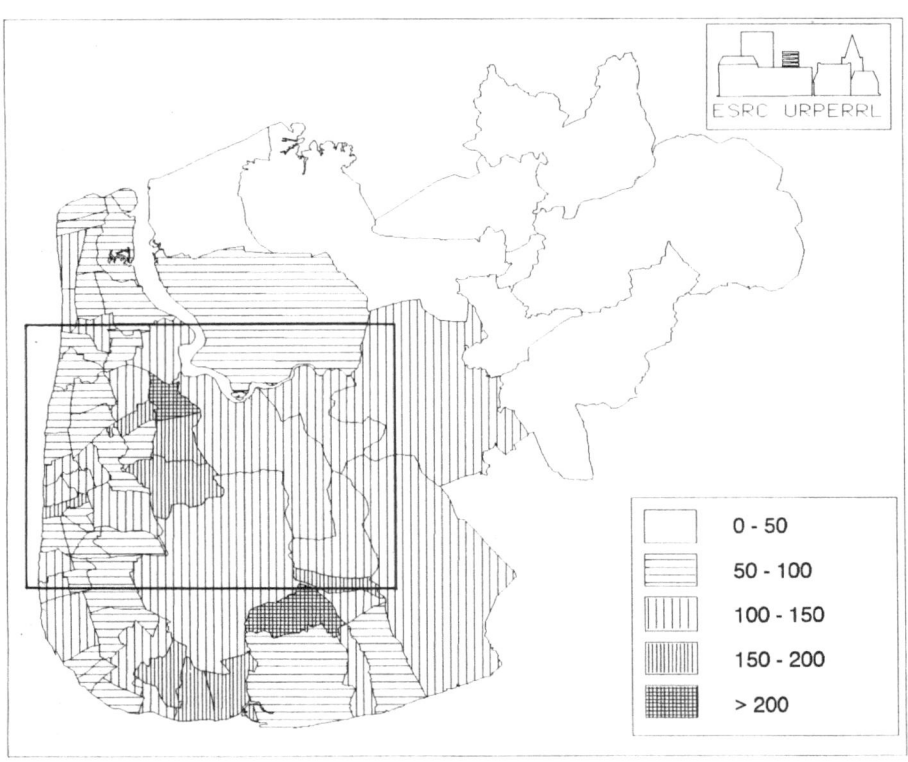

The data used in producing Maps 10.2 and 10.3 were drawn from an investigation of
the incidence of food poisoning in the Blackpool, Wyre and Fylde Health Authority
area (Hirschfieldet al.,1990). For the benefit of those unfamiliar with the area,
Blackpool is an important tourist centre and a major seaside resort located in north-
west England. The area is noted for having a large number of small bed-and-breakfast
hotels, a sizeable stock of privately rented housing and an abundance of fast-food
outlets.

Map 10.3: Standardized   incidence   ratio   (SIR)   per   100,000   population   for
           Campylobacter: Blackpool, Wyre and Fylde Health Authority - windowed
           plot of choropleth map with shading restricted to built-up areas within wards

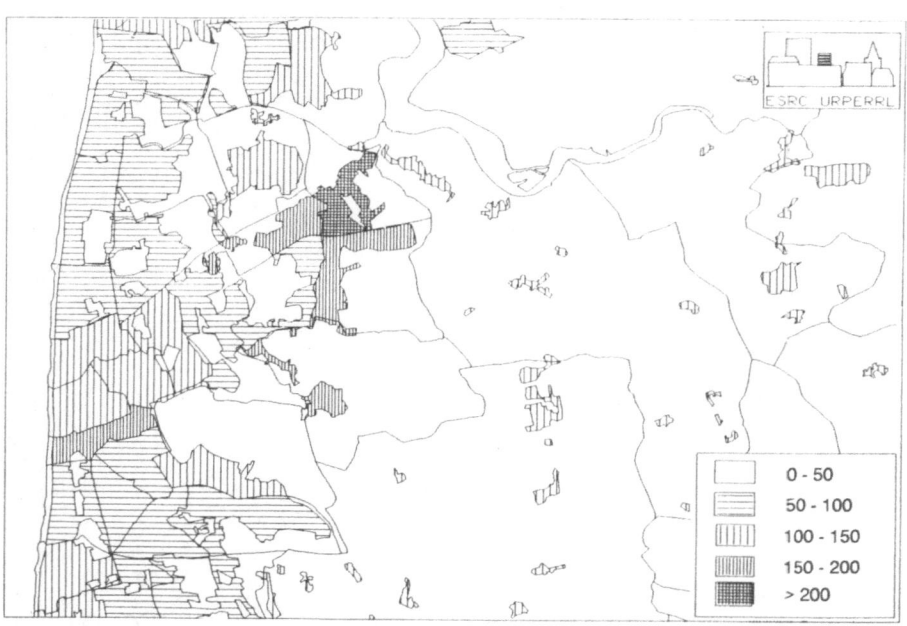

The study focused upon the analysis of data relating to 1300 notifications of three forms
of food poisoning condition that were recorded over the period 1982-1989. The three
conditions were *Campylobacter, Salmonella enteritidis* and other species of *Salmonella*.
The individual records relating to each case included only the date of birth, date of
specimen, sex and postcode of residence of the sufferer. This has placed a limit on the
forms of analysis that are possible. Efforts to seek an explanation for, or a cause of, the
outbreaks recorded are handicapped by the absence of any indication as to where the
condition might have been contracted. Nevertheless, a number of useful analyses have
been performed that help to throw light on possible associations between high incidence
and features of the population structure and the areas in which they live.

Map 10.3 provides an example of the way in which the isolation of the residential area
can be expected to influence the appearance and interpretation of choropleth maps. This
serves to illustrate the use of GIS software to both window in on an area of interest and

to overlay different sets of map information - in this case, ward boundaries and built-up area boundaries. For those unfamiliar with the geography of the area, it should be clear that most of the population is concentrated in the coastal strip, running from Fleetwood in the north, through Blackpool and down to Lytham St. Annes in the south. The map indicates that much of the remainder of the area is rural and characterized by very small settlements.

The measure that is plotted in Maps 10.2 and 10.3 is the SIR relating to the food poisoning condition of *Campylobacter* estimated for the 71 wards in the area (based on 838 cases). The two maps appear very different. It is apparent that the initial choropleth map conveys a potentially misleading impression of the incidence of *Campylobacter*. The exclusion of rural areas, airports, parks and other types of open space reveals that, in many wards, the area to which the indicator relates is confined to a comparatively small tract of residential development and, in some cases, only a tiny proportion of the total ward area.

There is an evident improvement in the representation of the extent of the areas to which the measure of interest relates. Interpretation may also be facilitated by the further superimposition of some of the main features of the road network. However, interpretation of the pattern displayed by the SIR should still recognize that this depiction of the measure cannot be claimed to reflect variation in the density of residential development. Nevertheless, the technique is reasonably successful in filtering out the more obvious non-residential areas to produce a more refined and, arguably, less misleading form of map.

Before leaving this topic, we can note that Bracken (1989) and Bracken & Martin (1989) describe a promising alternative means of achieving a similar effect in representing data relating to residential areas that is based upon raster (cell or pixel based) as opposed to vector graphics principles. This involves generating a raster surface from ED locations and their populations using a form of spatial moving average. An average is estimated for a point based on values for all points that fall within a window or kernel, a procedure described as a moving-kernel density-estimation approach. The method requires the analysis of inter-ED centroid distances "to estimate a plausible distribution of residential areas that is independent of zonal boundaries" (Bracken, 1989, p. 311). The approach has proved to be particularly effective in rural or less densely populated urban areas, as illustrated in Map 10.4 in a case involving the derivation of a surface to represent unemployment rates based on a 200 m × 200 m window.

Map 10.4: 200 m × 200 m cell raster surface of modelled unemployment (1981 data) distribution for the Cardiff Region, plotted in three classes

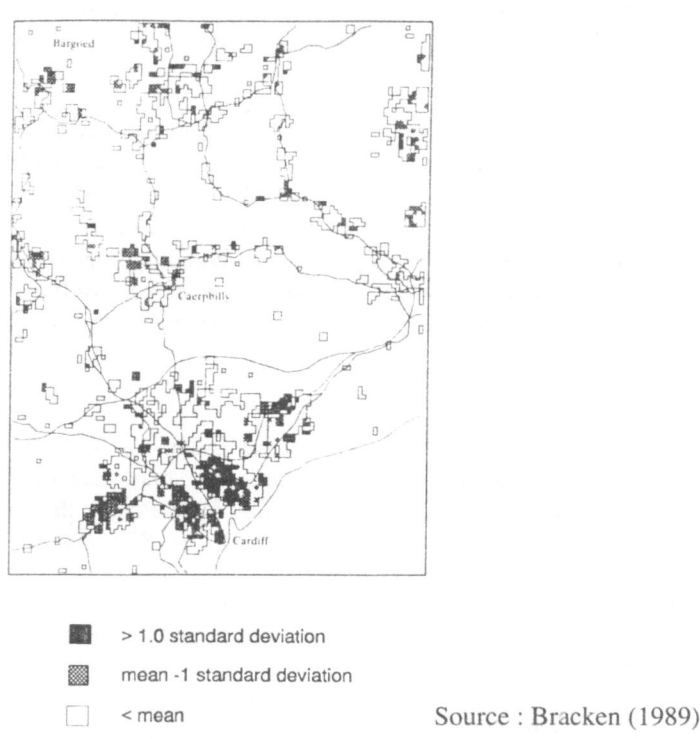

■   > 1.0 standard deviation

▓   mean -1 standard deviation

□   < mean   Source : Bracken (1989)

10.3 The treatment of extreme values

In the introduction, we noted the widespread use that has been made of choropleth maps to depict the variation that is present in indicators relating to a population. We have discussed some of the problems that can be associated with the configuration of the spatial units for which the indicators are derived. A further critical issue relates to the treatment of relatively high and low indicator values. If any map of this form is to be useful, the display should be restricted to values or patterns that are in some way significant or unusual. At the very least, some means of judging the likely significance of different levels of shading should be provided. For more common conditions, the standardized ratio or nested mean approaches noted above can provide a reasonable basis for such a scheme. However, in the case of rare conditions, account must be taken of the fact that fluctuations in values may merely represent random vari ation.

One approach to this problem has been based upon the use of Bayesian estimates of the standardized ratios (Clayton & Kaldor, 1987). Another variant involves the use of Poisson regression modelling to provide a basis for testing for heterogeneity in the population (Frome, 1983). Here we shall focus upon a Poisson chi-square-based approach that has been adopted to the specification of the intervals that are used to represent spatial variation in conditions of interest in several projects carried out by URPERRL (Brownet al., 1990). It is appropriate to outline how this approach came to be adopted and to note some of its virtues and shortcomings.

The origin if this approach can be found in the work of the Census Research Unit at the University of Durham. The Durham group examined the problems encountered in the mapping of ratios or measures derived with respect to a population or denominator that varies widely in size (Visvalingam, 1978; Clarkeet al., 1980; Rhind et al., 1980). It was observed that, the larger the base population, the more reliable is the resulting ratio, yet the less likely it is to fall in a non-average or exceptional class - where deviation from national or local average values is the primary concern. This effect often results in extreme values being associated with thinly populated census units, typically those in rural areas. An alternative approach, involving the mapping of absolute counts in particular categories, and leaving interpretation to the user, tends to give too much emphasis to urban areas.

The signed chi-square statistic was put forward in these circumstances as a compromise measure that takes into account both the absolute deviation from an average (or expected) value and the relative deviation (effectively the proportionate difference from the expectation) (Visvalingam, 1978).

The chi-square is computed as: $\chi^2 = \Sigma[(O - E)^2/E]$, where O is the observed number of individuals or cases, E is the expected number (based on the national average), and χ^2 is the chi-square value, summed over the appropriate number of classes. It is then given a negative sign if the expected number is greater than the observed number. This is a departure from conventional usage, whereby the sign of any difference between observed and expected is eliminated by squaring and is of no subsequent concern.

This measure results in units with a high population obtaining a larger chi-square value than a low population unit with the same percentage difference from the average or expected value. In other words, a weighting is applied such that small population units must be very unusual indeed before they are given a high chi-square value.

The signed chi-square mapping technique was employed in the production of a national atlas of census-derived indicators (Clarke et al., 1980). Among the problems faced by the developers of the atlas was the specification of the extreme values to be highlighted and an appropriate method for labelling the maps. After choosing to identify only four categories of deviation from the national mean, the solution to the first problem was to adopt the -3.84 and 3.84 values of the signed chi-square statistic as the inner limits of the two extremes. These are the tabulated 95 per cent significance levels for chi-square with one degree of freedom. However, "with spatial data of this non-binomial nature we do not wish to press an interpretation based on statistical significance" (Rhindet al., 1980, p. 8). The associated labelling problem was overcome by applying a qualitative, semantic scale to the ranges described by these values, to include very low/low/high/very high.

This approach provided the starting-point for the mapping work described here. Although the signed chi-square approach was adopted in circumstances in which interest focused on differences from the national mean, Visvalingam (1978) notes that the observed value or count may be compared with an expected value derived by some other means than applying the national average to the local base population. A cumulative Poisson model is a widely used method of deriving an estimate of the likely probability of occurrence of a particular number of cases in a spatial unit, especially where the mean incidence rate is relatively small (Choynowski, 1959). This takes the following form, where $P(x)$ is the probability of the observed value or count in a given area being x, given knowledge of the mean incidence rate μ:

$$P(x) = \frac{e^{-\mu} \cdot \mu^{x}}{x!}$$

where e is the constant 2.7183

In order to derive the probability of observing, say, two or fewer cases in an area, we need to estimate the probability of none, one or two cases, i.e., to obtain a measure of the cumulative probability. This probability distribution is then used as a basis for deriving the expected number of cases in an area. This is achieved by successively applying the probability of observing one case, two cases, etc. to the population of the area in order to generate the expected number of occurrences, which can then be aggregated to produce a total count.

Initially, a very crude and simple form of Poisson model was employed as the basis for deriving an expected value for the number of cases of a condition in a ward, using the mean rate of incidence and the overall population. The mean rate was most appropriately specified as the national average or mean annual rate, although a local mean was adopted in the absence of a national figure. This was then modified to reflect the effect of age-

and sex-specific incidence rates. (The data requirements of this approach are discussed in more detail in Brown et al. (1991).)

The annual expected value derived in this way is then used, together with the corresponding observed annual count of cases, to derive a chi-square statistic for each ward. A shading convention is adopted in the production of maps of the chi-square values in which critical values are specified with respect to the 20, 5 and 1 per cent significance levels: 1.64, 3.84 and 6.64 respectively.

Map 10.5: Blackpool, Wyre and Fylde Health Authority: Poisson chi-square map for Crohn's Disease (1976-1986)

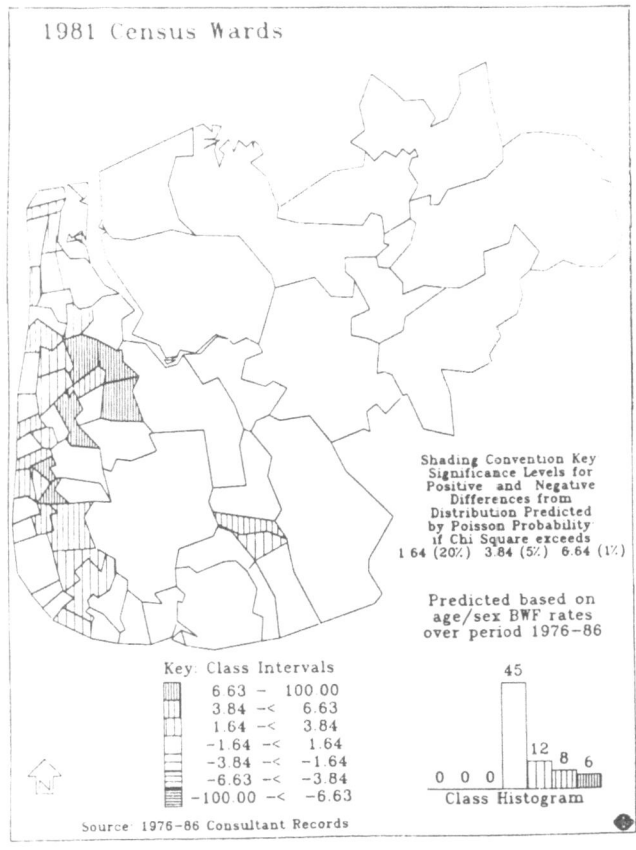

Map 10.5 illustrates the type of map that has been produced in this manner in which data relating to the incidence of 211 cases of Crohn's disease, drawn from an individual consultant's postcoded records of cases in the Blackpool, Wyre and Fylde area over the period 1976-1986, have been the subject of detailed investigation (Bowman et al., 1990). In this case, the Blackpool, Wyre and Fylde mean rates for 12 age bands by sex (using the 1981 population - the midpoint in the interval 1976-1986 - as the denominator) were used as the basis for deriving Poisson estimates of expected numbers of cases by ward.

The histogram indicates the number of wards (of the total of 71) falling in each class interval that correspond to the above critical values. In this case, it can be seen that six wards are isolated as having an overrepresentation of Crohn's disease cases that appears to be significant at the 1 per cent level. A further 8 and 12 wards are identified as having greater counts than expected at the 5 and 20 per cent levels respectively.

Map 10.6: Blackpool, Wyre and Fylde Health Authority: point plot of postcode locations of people with Crohn's disease (1976-1986)

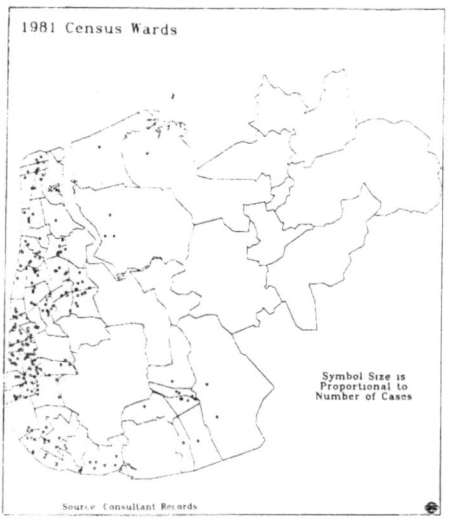

Interpretation of this pattern is assisted by reference to Map 10.6, in which the location of the postcode of residence of each of the 211 cases is plotted, based upon the crude 100-m grid square resolution of the postcode grid reference. This provides a visual impression of the number of cases in each ward that have given rise to the significant ward values noted above. Indeed, we would argue that such a map is an essential complement to the Poisson chi-square map featured in Map 10.5, as it can contribute to alerting the reader to the very limited number of cases that are often responsible for pushing the resulting incidence rate value into a critical region.

For the present purpose, the level of resolution of the information used to relate the postal address to the postcode, and thus to the ward geography, can be described as adequate. However, for many other purposes, the link between postcode and census geography is acknowledged to have a number of shortcomings (Gatrell, 1989; Openshaw, 1989). Some of these stem from the fact that the grid reference of the postcode is accurate, at best, to a resolution of 100 m. This is due to the fact that the Ordnance Survey grid reference of a postcode is defined as "the grid reference of the bottom left hand corner of the 100-m grid square within which the lowest numbered address is located".

This relatively crude specification is apparent, even at the scale of the map depicted in Map 10.6, from the regular, grid-like pattern of the plotted symbols. In more localized studies, the effect is still more pronounced and places severe limits on the extent to which it is possible make claims for the existence of spatial clustering. The crudeness of the spatial reference also gives rise to much scope for error in assigning individual postcodes to census enumeration districts, and thus to an area typology code such as in geodemographic applications (Brown, 1990, 1991; Brown et al., 1991). However, it is possible to claim that, in some circumstances, this crudeness brings with it the advantage of preserving the anonymity of the individuals represented by the postcoded incidence records, such as, for example, in the case of registered users of a drug-dependence clinic in Liverpool (Fazey et al., 1990).

10.4 Conclusions

This chapter has sought to examine some of the methodological issues encountered when attempting to interpret patterns in data that record the occurrence of relatively rare medical events or conditions. It has also served to illustrate how some rudimentary GIS features have been used to enhance the presentation of information that has been analysed in the course of a series of health-related projects undertaken by URPERRL staff. Discussion of these issues has been set in the context of a recent growth in interest in the application of statistical methods in the analysis of patterns of disease incidence across both time and space among a heterogeneous population.

Some of the problems that are associated with the use of conventional choropleth maps have been highlighted and some comments offered on the relative merits of alternative shading conventions. It has also been demonstrated how, with the aid of relatively straightforward GIS functions, the capturing of built-up area boundary data in digital form can help to reduce the scope for misinterpretation of maps of this type.

This was followed by discussion of some of the issues raised in the specification of methods that can be used to identify significant values to which attention is to be drawn in such maps. The origins of a Poisson chi-square based approach have been outlined and a brief discussion presented of the results obtained when it is applied to the distribution of a relatively rare condition.

Although the Poisson chi-square statistics appear to perform adequately as a means of isolating areas in which unusually high (or low) rates of occurrence are observed, it would be desirable to undertake a systematic comparison with alternative methods of isolating significant values or patterns, such as the Bayesian methods advocated by Clayton & Kaldor (1987). This could be carried out by applying the same tests either to a variety of different real-world datasets or to artificial or simulated datasets of known distributional or statistical properties. The results could then be compared in terms of areas identified and cases isolated at different levels of significance. In pursuit of the former approach, URPERRL is embarking on a further project which is to focus upon the study of the temporal and spatial distribution of cases of another relatively rare condition, cerebral palsy, and alternative methods of isolating extreme values will be employed.

Acknowledgements

Some of this research has been made possible by the grant (No. A504 28 5003) awarded by the Economic and Social Research Council (ESRC), which led to the establishment of URPERRL, as part of the Regional Research Laboratory Initiative. The authors gratefully acknowledge this support, together with the further funding provided by Blackpool, Wyre and Fylde Health Authority for aspects of the work reported here.

References

Alexander, F.E., T.J. Ricketts, J. Williams & R.A. Cartwright (1991). Methods of mapping and identifying small clusters of rare diseases and applications in geographical epidemiology. *Geogr Anal*, **23**: 158-173.

Batey, P.W.J., P.J.B. Brown & M. Madden (1985). *Social deprivation measures and the calculation of revenue targets*. Report commissioned by Liverpool District Health Authority. Liverpool, Department of Civic Design, University of Liverpool.

Besag, J. (1989). Contribution to the discussion at the Royal Statistical Society Meeting on Cancer Near Nuclear Installations. *J R Stat Soc A*, **152**: 367-368.

Besag, J. & J. Newell (1991). The detection of clusters in rare diseases. *J R Stat Soc A*, **154**: 143-155.

Black, D. (1984). *Investigation of the possible increased incidence of cancer in West Cumbria*. London, HMSO.

Bowman, S., P.J.B. Brown & P.W.J. Batey (1990). *Mortality and morbidity in Blackpool*, Wyre & Fylde: a statistical and spatial analysis. Liverpool, Department of Civic Design, University of Liverpool. (Working paper 11, Urban Research and Policy Evaluation Regional Research Laboratory).

Bracken, I. (1989). The generation of socioeconomic surfaces for public policy making. *Environ Planning B*, **16**: 307-326.

Bracken, I. & D.J. Martin (1989). The generation of spatial population distributions from census centroid data, *Environ Planning A*, **21**: 537-543.

Brown, P.J.B. (1990). *Geodemographics: a review of recent developments and emerging issues - towards an RRL research agenda*. London, Economic and Social Research Council (Regional Research Laboratory Initiative Discussion Paper 5.)

Brown, P.J.B. (1991). Exploring geodemographics. *In*: I. Masser & M. Blakemore, ed. *Geographic information management: methodology and potential applications*. London, Longman, pp 221-258.

Brown, P.J.B., P.W.J. Batey, A.F.G. Hirschfield & J. Marsden (1990). *Poisson chi-square mapping, GIS and geodemographic analysis: the spatial and a spatial analysis of relatively rare conditions*. Liverpool, Department of Civic Design, University of Liverpool. (Working Paper 18, Urban Research and Policy Evaluation Regional Research Laboratory).

Brown, P.J.B., P.W.J. Batey, A.F.G. Hirschfield & J. Marsden (1991). Prerequisites for Poisson chi-square mapping and geodemographic analysis of relatively rare conditions. *In*: A. Westlake, ed. *Proceedings of the SAHSU Workshop on Geographical Database Structures for Small Area Statistical Enquiries*. London School of Hygiene and Tropical Medicine, 87-108.

Brown, P.J.B., A.F.G. Hirschfield & P.W.J. Batey (1991). Applications of geodemographic methods in the analysis of health condition incidence data. *Papers Regional Sci J Regional Sci Associ Int*, **70**(3): 329-344.

Choynowski, M. (1959). Maps based on probabilities. *J Am Statistical Assoc*, **54**: 385-388.

Clarke, J.I., J.C. Dewdney, I.S. Evans, D.W. Rhind & M. Visvalingam (1980). *People in Britain: a census atlas*. London, HMSO.

Clayton, D. & J. Kaldor (1987). Empirical Bayes estimates of age-standardized relative risks for use in disease mapping. *Biometrics*, **43**: 671-681.

Cliff, A.D. & J.K. Ord (1981). *Spatial processes: models and applications*. London, Pion.

Cuzick, J. & R. Edwards (1990). Tests for spatial clustering of events in heterogeneous populations. *J R Stat Soc B*, **52**: 73-104.

Diggle, P.J. (1990). A point process modelling approach to raised incidence of a rare phenomenon in the vicinity of a prespecified point. *J R Stat Soc A*, **153**: 349-362.

Diggle, P.J., A.C. Gatrell & A.A. Lovett (1990). Modelling the prevalence of cancer of the larynx in part of Lancashire: a new methodology for spatial epidemiology. *In*: R.W. Thomas ed. *Spatial epidemiology*. London, Pion, pp 35-47.

Fazey, C., P.J.B. Brown & P.W.J. Batey (1990). *The epidemiology of patients attending a drug dependency clinic covering Liverpool and South Sefton Health Districts: an exploratory analysis*. Liverpool, Department of Civic Design, University of Liverpool. (Working Paper 10, Urban Research and Policy Evaluation Regional Research Laboratory).

Frome, E.L. (1983). The analysis of rates using Poisson regression models. *Biometrics*, **39**: 665-674.

Gardner, H.J., P.D. Winter, C.P. Taylor & E.D. Acheson (1983). *Atlas of cancer mortality in England and Wales 1968-1978*. Chichester, Wiley.

Gardner, M.J. (1989). Review of reported increases of childhood cancer rates in the vicinity of nuclear installations in the UK. *J R Stat Soc A*, **152**: 307-325.

Gatrell, A.C. (1989). On the spatial representation and accuracy of address-based data in the United Kingdon. *Int J Geogr Inf Systems*, **3**: 335-348.

Hills, M. & F.E. Alexander (1989). Statistical methods used in assessing the risk of disease near a source of possible environmental pollution: a review. *J R Stat Soc A*, 152: 353-363.

Hirschfield, A.F.G., R. Barr, P.W.J. Batey & P.J.B. Brown (1989). The Urban Research and Policy Evaluation Regional Research Laboratory. *Mapping Awareness*, **3**: 34-37.

Hirschfield, A.F.G., P.W.J. Batey, P.J.B. Brown & J. Marsden (1990). *The spatial epidemiology of food poisoning in Blackpool, Wyre & Fylde: preliminary results*. Liverpool, Department of Civic Design, University of Liverpool (Working Paper 14, Urban Research and Policy Evaluation Regional Research Laboratory).

Kemp, I., P. Boyle, M. Smans & C.S. Muir, ed. (1985). *Atlas of cancer in Scotland, 1975-1980: incidence and epidemiological perspective*. Lyon, IARC (Publication No. 72).

Martin, D.J. (1989). Mapping population data from zone centroid locations. *Trans Inst Br Geogr*, 14: 90-97.

Openshaw, S. (1984). *The modifiable areal unit problem*. Norwich, Geo Books (Concepts and techniques in modern geography 38).

Openshaw, S. (1990). Automating the search for cancer clusters: a review of problems, progress and opportunities. *In*: R.W. Thomas, ed. *Spatial epidemiology*, London, Pion, pp 48-78.

Openshaw, S., M. Charlton, C. Wymer & A.W. Craft (1987). A mark 1 geographical analysis machine for the automated analysis of point data sets. *Int J Geogr Inf Systems*, **1**: 335-358.

Openshaw, S., A.W. Craft, M. Charlton & J.M. Birch (1988). Investigation of leukaemia clusters by use of a geographical analysis machine. *Lancet*, **i**: 272-273.

Rhind, D.W., I.S. Evans & M. Visvalingam (1980). Making a national atlas of population by computer. *Cartogr J*, **17**: 3-11.

Schulman, J., S. Selvin & D.W. Merrill (1988). Density equalized map projections: a method for analyzing clustering around a fixed point. *Stat Med*, **7**: 495-505.

Visvalingam, M. (1978). The signed chi-square measure for mapping. *Cartogr J*, **15**: 93-98.

Wartenberg, D. & M. Greenberg (1990). Space-time models for the detection of clusters of disease. In: R.W. Thomas, ed. *Spatial epidemiology*. London, Pion, pp 17-34.

Peter Brown, Alexander Hirschfield and John Marsden
Urban Research and Policy Evaluation Regional Research Laboratory
Department of Civic Design
University of Liverpool
P.O. Box 147
Liverpool L69 3BX
United Kingdom

APPLICATIONS OF GEOGRPAPHICAL INFORMATION SYSTEMS IN PUBLIC AND ENVIRONMENTAL HEALTH

PART IV

APPLICATIONS OF GEOGRAPHICAL INFORMATION SYSTEMS IN PUBLIC AND ENVIRONMENTAL HEALTH

11 THE EXPLORATION OF THE POSSIBLE RELATIONSHIP BETWEEN DEATHS, BIRTHS AND AIR POLLUTION IN SCOTTISH TOWNS

Owen L. Lloyd

Abstract

Large-scale geographical patterns of diseases can be seen in most atlases of mortality or morbidity. These patterns reflect ill-defined regional differences of environments composed of intermingled and complex socioeconomic factors. Analysis of these broad patterns, however, may show individual communities that stand out as exceptions. These exceptions provide opportunities for more focused epidemiological and environmental studies to explore the possible environmental causes of these epidemiological abnormalities. In *Atlas of mortality in Scotland* from 1987, exceptional rates of various diseases were found in several towns. For the investigation of the high rate of lung cancer in one town and for the replication studies of other towns, the communities were subdivided into small areas for the combined epidemiological and environmental investigations and the results were subjected to geographical analysis. The detailed scale of these geographical frameworks suggests some possible mechanisms of environmental toxicity. These studies also demonstrated that two types of data resource might have substantial value for environmental epidemiology when geographical analysis by small area is required. Firstly, to interpret epidemiological studies related to environmental air quality in a community, detailed patterns of airborne pollution should be known. These patterns should be based on measurements of pollutants in samples collected simultaneously at many sampling sites; e.g., low-technology samplers, such as mosses, lichens and soils, collect metallic pollutants in the ambient air. Secondly, for demonstrating the effects of some toxic environments with minimal delay, obstetric data might perform a useful screening function, a view supported by some of our investigations in Scotland.

11.1 Introduction

From the earliest years of public health, the knowledge that many diseases do not have uniform geographical distributions has stimulated the search for environmental causes of those diseases. More recently, this awareness has been joined by a growing knowledge of the actions of environmental toxins, many generated by industrial activities. With its relatively high rates for diseases such as lung cancer and heart disease, Scotland is a useful subject for the geographical analysis of abnormal patterns of mortality. An *Atlas of mortality in Scotland* (Lloyd et al., 1987) based on data published annually by the Registrar General, showed the geographical patterns of major categories of disease in communities during periods between 1959 and 1983. The consistency of the large-scale patterns of several diseases in this atlas supported the view that toxic factors (distinct from the non-specific influences of social class) were, or had been, in the environment of many communities: a notable example was the influence of urbanization and industrialization on the patterns of bronchitis and lung cancer (Williams et al., 1987). However, within this pattern for lung cancer in 1969-1973, a spectacular non-urban

167

M. J. C. de Lepper et al. (eds.), The Added Value of Geographical Information Systems in Public and Environmental Health, 167–180.

exception was provided by a small community situated in the countryside midway between Edinburgh and Glasgow: the town of Armadale (Lloyd & MacDonald, 1984).

Founded around the start of the nineteenth century as a staging post on the Great Road between Edinburgh and Glasgow, Armadale grew steadily during the early decades of the Industrial Revolution, mainly due to the exploitation of local deposits of coal and iron (Lloyd et al., 1985a). That source of prosperity eventually passed, however and by the late 1960s the remaining heavy industry consisted of one small steel foundry in the southeast quadrant of the town and a brickwork south of the town. With a stable population of about 7000 inhabitants, Armadale resembled neighbouring communities in its general socioeconomic characteristics. Yet the routinely published mortality data for the town in 1969-1973 were unique: the highest standardized rates for total mortality and for lung cancer of any community in Scotland; and high standardized mortality ratios (SMRs) also for other major diseases including ischaemic heart disease, cerebrovascular disease, bronchitis and other cancers. The time-trend for lung cancer was unusual: between 1951 and 1967, the annual SMRs exceeded 100 in only three years (1954, 1958 and 1963), whereas between 1968 and 1977 all the SMRs were high; only 16 deaths were registered during the 8 years of 1961-1967, compared with 55 deaths during the following eight years; most cases were later confirmed by data from hospital records and the cancer registry. At the early stages of the investigations, however, no etiological clues explaining the high rate of lung cancer could be found: mortality rates were high in both males and females; the male age groups most affected were those between 55 and 84 years; no unusual histological features of the tumours; and no strong links with the causal circumstances commonly linked with lung cancer (i.e., occupation and tobacco usage). However, the routinely published data did not include the residential locations of the deaths within the town; so we decided to obtain this information and map it, in case the geographical distribution of the lung cancer might provide a clue about its possible causation.

The metallurgical process in the steel foundry had been changed in the mid-1960s, so that the nature of the associated air pollution in the town would have altered (probably including smaller particles). The geography of air pollution by metals in the town therefore had to be ascertained. Although the prevailing winds in central Scotland blow from the south-west, these are generally vigorous winds that disperse air pollutants effectively when the topography is relatively open - the case in the vicinity of Armadale. The winds most frequently associated with the accumulation of air pollution are those from the north-east (the second most frequent wind direction) (Lloyd et al., 1985a) and so it was predicted that air pollution from the low fume-stacks and roof vents of Armadale's foundry would most affect the residential area directly south-west of that foundry. Nevertheless, this prediction had to be supported by objective evidence of detailed patterns of pollution for two reasons: the foundry's air pollution was emitted close to the ground and therefore the concentrations of pollutants were likely to change appreciably over short distances; and the distribution of the pollution plume was likely to have been altered by the uneven nature of the surrounding topography.

To obtain detailed patterns of pollution, a high density of sampling sites was required, even where high values of pollution were not predicted. Conventional high technology samplers are costly, tend to suffer breakdown or vandalism when exposed over long periods in the general environment and are restricted by the availability of power supplies. These drawbacks are avoided by low-technology samplers (LTSs) utilizing materials such as mosses and soils (Martin & Coughtrey, 1982). Methodological studies have clarified the influences of factors such as the duration of exposure and the size and shape of the samplers (Gailey & Lloyd, 1986). To confirm the predicted geography of air pollution in Armadale, therefore, we decided to use several techniques of low-technology sampling (Gailey & Lloyd, 1986; Lloyd & Gailey, 1987; Yule & Lloyd, 1984; Smith et al., 1986).

Environmental pollution can have adverse effects on reproductive processes. Disturbances of one obstetric parameter, the sex ratio of births (males:females), which is around 1.06 in most countries, have been associated with exposure to occupational and environmental toxins and stresses (Hytten, 1982; Lloyd et al., 1988a). Hence, we investigated the obstetric biology of Armadale's population during the epidemic of lung cancer by reviewing time trends of the sex ratios of births and fertility rates for the town and by mapping the distribution of the sex ratios within the town (Lloyd et al., 1985a).

Replication studies were required to ascertain the consistency of any geographical associations found between air pollution and lung cancer in Armadale. In the nearby town of Bathgate, a small steel foundry had introduced during the early 1960s a process similar to that introduced by Armadale's foundry a few years later; and subsequently the annual SMRs for lung cancer for the whole town had shown a transient increase (Lloyd et al. 1985b). Because of the industrial and epidemiological similarities with Armadale and despite its topography differing markedly from that of Armadale in being considerably more hilly, Bathgate was chosen for the replication study. This study was to test the links between three types of pattern: the previous pollution by metals from that foundry and the patterns of lung cancer and of abnormal sex ratios of births within the town.

The old town of Kirkintilloch borders a small valley containing a stretch of the Forth-Clyde Cana (Smith et al., 1987). Earlier this century, the town had contained several iron foundries, but these had closed in turn, the last about five years before the start of the present study in 1984. In the 1920s, a nickel refinery had closed. Although the town as a whole had not shown a high mortality for lung cancer and did not contain a steel foundry, a combined epidemiological and environmental study was undertaken to examine the geographical relationship between pollution from the iron-founding industry and the patterns of lung cancer within the town.

Thus, the investigation of the etiology of the epidemic in Armadale ultimately involved four interconnected studies, all depending on the assessment of geographical distributions:

(i)   the analysis of the spatial patterns of mortality of lung cancer in the town;
(ii)   the analysis of the spatial patterns of metallic air pollution there shown by low-technology sampling;
(iii)   the follow-up studies of lung cancer and air pollution in two other towns; and
(iv)   the assessment of obstetric epidemiological data to indicate abnormal or potentially toxic environments.

11.2 Methods

The addresses of all residents in Armadale who died from lung cancer during appropriate groups of years between 1961 and 1986 were plotted on maps of the town. The denominators were the populations in residential zones formed from aggregates of census enumeration districts of the censuses of 1966, 1971 and 1981. The zones' SMRs were calculated by indirect standardization (Williams & Lloyd, 1988).

To ascertain the detailed patterns of air pollution in Armadale by means of low-technology sampling, the town was first divided into 65 grid squares, each with an area of 400 m^2; the numbers and varieties of lichens at one site within each grid square were counted to give an index of atmospheric pollution within the town (Yule & Lloyd, 1984). Secondly, to investigate the pollution in more detail, other low technology sampling techniques were used. The materials were both indigenous and transplanted and, included species of *Sphagnum* and *Hypnum* mosses, *Lecanora* and *Hypogymnia* lichens, surface and core soils, grasses and the fabric ('tak') (Gailey & Lloyd, 1986; Lloyd & Gailey, 1987; Yule & Lloyd, 1984; Smith et al., 1986). The indigenous LTSs were used to show the patterns of pollution in previous years, whereas the transplanted LTSs show current pollution. For all samplers except the soil cores, 47 sites within the town were selected. The transplanted LTSs were deployed there during eight consecutive two-month periods. The values of a range of metals (iron, manganese, lead, zinc, chromium, nickel, cobalt and arsenic (in soil cores only)) in the LTSs were measured, using atomic absorption spectrophotometry and (for part of the study) X-ray fluorescence. For the geographical analyses, several types of map were constructed, including computer-generated maps of expected concentrations.

To assess further the flow of air pollution from the foundry, a polystyrene model of Armadale's topographic environment was placed in a wind-tunnel. The distribution of smoke emitted from a nozzle at a site on the model corresponding to the position of the foundry's fume-stack was studied under various conditions of airflow (Lloyd et al., 1985a).

In the replication study of Bathgate (Lloyd et al., 1985b), the historical nature of the pollution indicated that the appropriate type of LTS for assessing the pattern of pollution was soil cores; 19 sampling sites were chosen. Because the local geography differed completely from that of Armadale, the probable flow of pollution from the relevant steel foundry was also investigated by means of a wind-tunnel study. To ascertain the spatial patterns of lung cancer within the town, death certificates for the periods 1960-1965 and 1966-1976 were assigned by residential locality to census enumeration districts for 1966 and 1971; these districts were aggregated into three zones according to the degree of metal pollution shown by the environmental study and the zones' SMRs were calculated.

In the study of Kirkintilloch (Smith et al., 1987), where the pollution was entirely historical, only soil cores were used for the environmental sampling; the concentrations of several metals at 51 sites within the town were measured. The geographical distribution of lung cancer for 1966-1976 was studied; five residential zones, based on enumeration districts of the 1971 census, were arranged a priori according to probable exposure to fumes from the last working iron foundry.

For the assessment of the association of obstetric parameters with environmental pollution from steel foundries and with lung cancer, the sex ratios of births and fertility rates of Armadale's population were ascertained for the town and its subdivisions for periods of several years on either side of 1968, which marked the first year of the epidemic of deaths from lung cancer. The same associations were also tested in Bathgate.

11.3 Results

Within Armadale, the geographical study of the residences of the victims of lung cancer during 1968-1974 (later extended to 1968-1976) showed a statistically significant cluster of deaths in the residential zone directly to the south-west of the local steel foundry (Map 11.1) (Lloyd et al., 1985a; Lloyd, 1978).

Map 11.1: Geography of lung cancer in Armadale, 1968-1976; dots = residences of
           deaths; the ground slopes from the foundry northwards down to a valley.

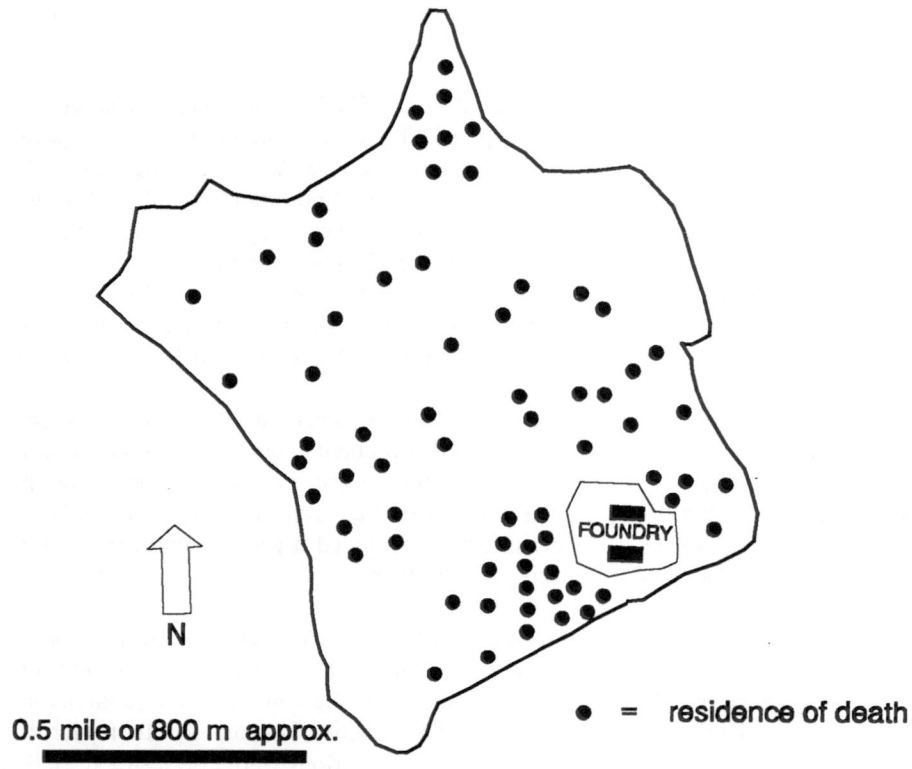

Routine meteorological data had previously indicated this area as probably the most
vulnerable to air pollution from the foundry (Lloyd et al., 1985a). This cluster was also
demonstrable using spatial analytical techniques (such as radial clustering and
computer-based approaches), which avoided reliance on the boundaries of the
enumeration districts (Lloyd, 1981; Lawson, in press). The small cluster in the north of
the town was not statistically significant and its probable biological significance was
unclear. Before 1968, neither area showed any clustering of deaths (Lloyd et al., 1985a)
and the reduction of industrial air pollution later in the 1970s was accompanied by the
decline of the relevant SMRs for those areas (Williams & Lloyd, 1988).

The index of atmospheric pollution study for Armadale indicated that most pollution was
near the foundry. In the LTSs, high values of most metals were found generally on either

side of the foundry, along an axis running south-west to north-east, which corresponded with the common wind directions; signs of a northward drift of the pollution were also seen. (The only consistent exception to this pattern was seen for lead, which had high values near the central crossroads - the effect of traffic pollution.) In soil cores, where the metal values represent the effects of more historically distant pollution, the highest concentrations were to the south-west of the foundry, but high concentrations were also to the north of the foundry (Map 11.2). The evidence, therefore, was that the pollution first moved from the foundry south-west to the area with the large cluster of lung cancer and then north down a gentle slope into the small valley in the north of the town (i.e., the area with the small cluster). This spatial pattern was confirmed by the wind-tunnel technique: irrespective of direction, winds with low speeds carried pollution from the foundry site first south-west and then north (Lloyd et al., 1985a).

Map 11.2: The pattern of arsenic pollution in soil cores at various sites in Armadale (samples were taken from pairs of adjacent sites, to assess the validity of the pattern). F = site of foundry.

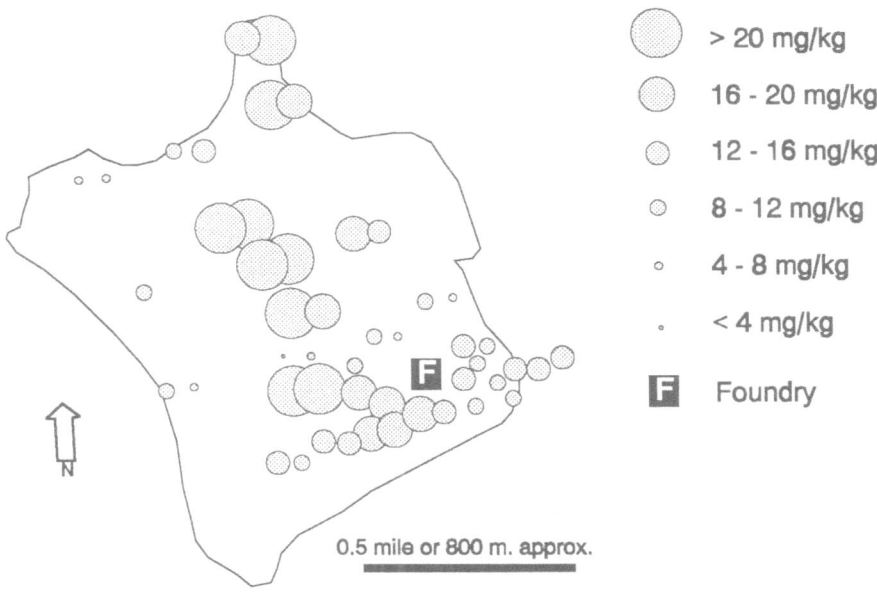

In Bathgate, there were no houses to the west of the foundry. The results from the use of the LTSs and from the wind-tunnel study showed that a hill north-east of the foundry deflected the prevailing south-westerly winds, so that air pollutants from the foundry were deposited on the housing zone directly east of the foundry, where a cluster of lung cancer was present only in 1966-1976, i.e., following the change in the metallurgical processes of the local steel foundry (Maps 11.3 and 11.4) (Lloyd et al., 1985b).

The elongated pollution pattern for Kirkintilloch indicated the channelling effect imposed by the canal valley on the prevailing winds and thereby on the foundry's pollution plume (Smith et al., 1987). The highest values of most metals were found east of the iron foundry and demonstrated the umbrella effect by being at some distance from the foundry (Map 11.5).

The zone containing the maximal pollution also contained the only census enumeration district with a statistically significantly elevated SMR. The zonal SMRs declined from the most polluted zones in the north to the least polluted zones in the south and this gradient persisted after standardizing for social class. High values for nickel were found in the soil cores from a separate zone where a long-vanished nickel smelter had operated.

Map 11.3: Map of Bathgate showing sites of the active foundry (F1) and of residences of lung cancer cases in Bathgate in 1960-1965; zone A = maximal pollution, zones B + C = other zones. (A hill is situated on the northern side of the town.)

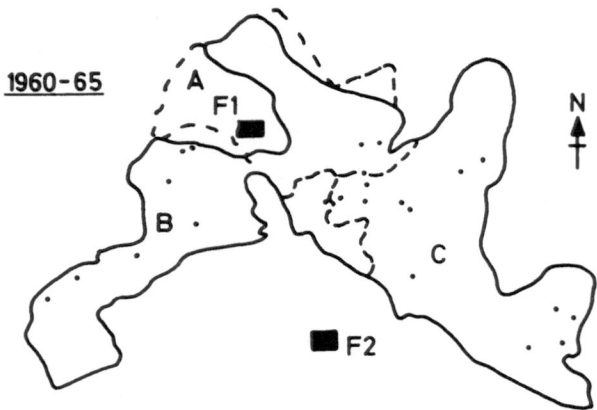

Map 11.4: Map of Bathgate showing sites of the active foundry (F1) and of residences of lung cancer cases in Bathgate in 1966-1976; zone A = maximal pollution, zones B + C = other zones. (A hill is situated on the northern side of the town.)

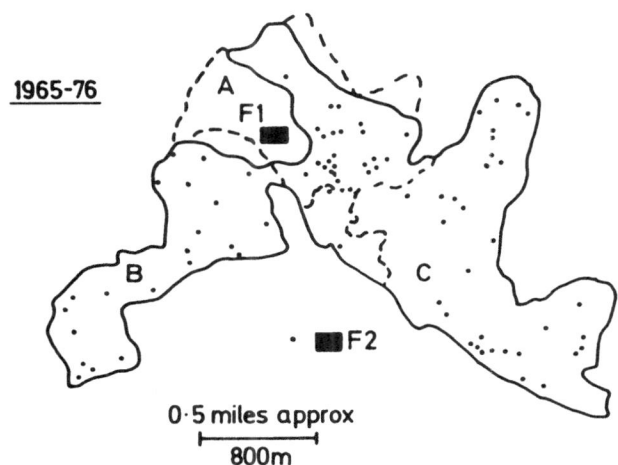

In Armadale, the sex ratios of births were high during the late 1960s; the exceptional and statistically significantly high value of 2.21 in 1967 was attained by no other comparable community in Scotland within five years on either side of that date. Within the town, the highest ratio (of 340, with 22 births) was in 1967 in the zone with the large cluster of lung cancer during 1968-1976, i.e., close to and downwind from the foundry. The year 1967 marked the start of the increased registration of lung cancer, preceding by one year the onset of high mortality (Lloyd et al., 1985a). The annual fertility rates from 1950 to 1973 showed no abnormal trends.

Within Bathgate, the highest sex ratio of births (1.45, with 299 births) in 1964-1970 was found in the zone with the highest pollution from a foundry and with the cluster of lung cancer (Lloyd et al., 1985b).

Map 11.5: Map of Kirkintilloch: black circles = top 10 values of iron and manganese (Fe and Mn) in soil cores at sampling sites; triangles = values ranked 11 to 20; open circles = sites with lower values; X = centroid of the only census district with a statistically high SMR for lung cancer in 1966-1976; active iron foundry (black rectangle) between zones A and B (aggregates of census districts) in the north; inactive iron foundry (cross-hatched rectangle) in zone B; zones C to E = comparison zones. The northern boundaries of zones A and B form the canal valley; scale of map = 3 km approximately from east to west.

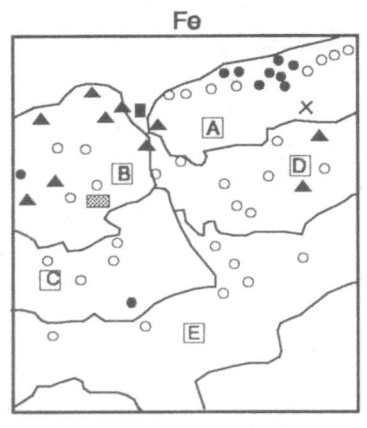

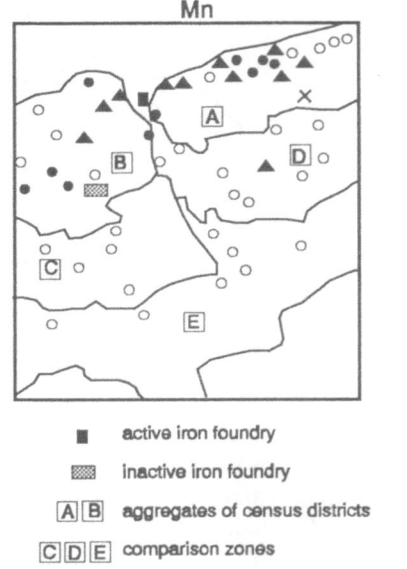

- top 10 values of Iron and magnese (Fe and Mg) in soil cores at sampling sites

▲ values ranked 11 to 20

○ sites with lower values

× Centroid of the only census district with a statistically high SMR for lung cancer in 1966-76

■ active iron foundry

▓ inactive iron foundry

Ⓐ Ⓑ aggregates of census districts

Ⓒ Ⓓ Ⓔ comparison zones

1.5 km

11.4 Conclusions and discussion

The geographical analysis of the spatial distributions of lung cancer in several Scottish towns and of air pollution data from low-technology sampling suggested a possible link between metal pollution of the ambient air and non-occupationally related clusters of lung cancer in housing areas near certain types of foundry. The verification of such a relationship would extended previous findings of a causal relationship between metal fumes and lung cancer in occupational settings to the general environment. In Armadale and Bathgate, however, the unusual patterns in time and space of the epidemic requires invoking a mechanism of short-latency carcinogenesis, for example, from a tumour promoter having been released into the atmosphere during changes in the metallurgical process of the local foundries in the early 1960s (Lloyd et al., 1985a). In all of these towns, the pattern of mortality agreed with that of pollution only after the influence of geography on air flow patterns had been assessed.

The LTSs provided valuable data on the spatial distribution of air pollution in these foundry towns: they demonstrated that the zones of maximal pollution coincided with the clusters of lung cancer, thereby assisting the interpretation of the spatial patterns of the mortality. The large number of the LTS sites in the town allowed the discovery of geographical patterns for pollution - the effects of peculiarities of topography - that had not previously been suspected. Examples included: the flow of pollution to the northern area of Armadale; the deflection of the prevailing southwesterly winds by a hill in Bathgate so that pollutants from the foundry affected the residential areas directly east of the foundry; and the distorted umbrella effect at Kirkintilloch, due to the influence of the canal valley. The value of sampling soil cores in investigations where historical sources of metallic air pollution are of epidemiological relevance was underlined by the high concentrations of nickel detected in an area of Kirkintilloch, which indicated the site of a long-demolished nickel refinery.

The reliability of the LTS method was indicated by the comparability of the pollution patterns in Armadale shown clearly on most types of LTS; the validity of these patterns was supported by results of wind-tunnel studies. The low cost of LTSs was a further advantage: under £1000 for the materials for the many types of sampler in the complex, 20-month Armadale study; and far less in most studies where two types of LTS would suffice. If more detailed data from conventional samplers had been required also, the spatial information gained from the use of LTSs would have shown the optimal sites for conventional samplers to be located.

LTSs have also been used for determining the geography of air pollution with other types of industrial pollution - for example, in a Scottish coal-mining town exposed to fumes from a smouldering coal tip (Lloyd et al., 1988b). Researchers elsewhere have also used similarly simple techniques to monitor the spatial distributions of other types of air and soil pollutants, including organic chemicals (Martin & Coughtrey, 1982; Villeneuve et al., 1988).

Can obstetric epidemiological parameters provide early warning of environmental toxicity? The positive results for the sex ratios of births in the studies in Armadale and Bathgate were certainly related in space and time to changes in specific metallurgical processes in those foundries. Further, a statistically significantly high sex ratio of 1.11 during 1979-1983 was noted for the large town of Motherwell, which contains much of British Steel's operations in Scotland (Lloyd et al., 1987). However, unpublished research showed that high sex ratios were not found near the iron foundries in Kirkintilloch; so the association appeared to be limited to areas with pollution from only some types of metallurgical process. Nevertheless, in the neighbourhood of waste incinerators in central Scotland, where toxins including chlorinated hydrocarbons and metals have been found, several abnormal obstetric parameters have been found:

(i)   some unusual sex ratios of human births;
(ii)   a statistically significant excess of human twinning, also found in the cattle herds of farms; and
(iii)   relatively high incidences of microphthalmos in human populations and of congenital abnormalities of calves reported by local dairy farmers (Lloyd et al., 1988c; Williams et al., 1988).

In some cases, therefore, obstetric epidemiology may have a sentinel role to play in uncovering the effects of environmental toxins.

Nevertheless, abnormal sex ratios of births have been linked also with environmental factors other than industrial pollution (Hytten, 1982). On a comparatively large geographical scale, moderate imbalances of sex ratios in communities have been found in Scotland during the quinquennia around 1961 and 1971, with the frequency of raised ratios highest on the east coast, intermediate on the west coast and lowest inland (Lloyd et al., 1987). In Newfoundland, where communities on the east coast had shown higher rates of disease than those on the west, high sex ratios for young children (used as proxy for births) in the communities at the 1976 and 1981 censuses were statistically significantly more common on the east coast than on the west coast. The mechanism for these geographical patterns is unknown but may be related to diet. Further systematic studies are clearly required to clarify the potential value of obstetric data sources for indicating environmental toxicity.

The experience from these studies demonstrates that information on the geographical distribution of disease and geographical information on (potential) exposures or their source can be of use to epidemiologists interested in investigating links between environment and health. The main issues are the need for a multidisciplinary approach and careful planning, especially for choosing priorities.

Acknowledgements

These studies were undertaken by a team that included Dr F. Gailey, Dr M.M. Lloyd, Mr G.H. Smith and Dr F.L.R. Williams.

References

Gailey, F.A.Y. & O.L. Lloyd (1986). Methodological investigations into low technology monitoring of atmospheric metal pollution. I. The effects of sampler size on metal concentrations. 2. The effects of length of exposure on metal concentrations. 3. The degree of consistency of the metal concentrations. *Environ Pollution B*, 12: 41-59; 61-74; 85-109.

Hytten, F.E. (1982). Commentary. Boys and girls, *Br J Obstet Gynaecol*, **89**: 97-99.

Lawson A. (in press). GLIM and normalizing constant models in spatial and directional data analysis. *Comput, Stat and Data Anal*. Special Issue.

Lloyd, O.L. (1978). Respiratory cancer clustering associated with localized industrial air pollution. *Lancet*, **8059**: 318-320.

Lloyd, O.L. (1981). Mortality in a small industrial town: problems of analysis and interpretation. *In*: A. W. Gardner, ed. *Current approaches to occupational health.* Bristol, John Wright, pp. 283-310.

Lloyd, O.L. & J. MacDonald (1984). Continuous epidemiological mapping - a needed public health watchdog. *Public Health (Lond)*, 98: 321-326.

Lloyd, O.L., G. Smith, M.M. Lloyd, Y. Holland & F.A. Gailey (1985a). Raised mortality from lung cancer and high sex ratios of births associated with industrial pollution, *Br J Ind Med*, **42**: 475-480.

Lloyd, O.L., R. Barclay, M.M. Lloyd & Armadale Group Practice (1985b). Lung cancer and other health problems in a Scottish industrial town: a review. *Ambio*, **14**: 322-8.

Lloyd, O.L. & F.A.Y. Gailey (1987). Techniques of low-technology sampling of air pollution by metals: a comparison of concentrations and map patterns. *Br J Ind Med*, **44**: 494-504.

Lloyd, O.L., F.L.R. Williams, W.G. Berry & C. du V. Florey, ed. (1987). *An atlas of mortality in Scotland*. London, Croom Helm.

Lloyd, M.M., O.L. Lloyd & W.R. Lyster (1988a). Slugs and snails against sugar and spice [sex ratios at birth]. *Br Med J*, **297**: 1627-1628.

Lloyd, O.L., F.L.R. Williams & G.H. Smith (1988). Patterns of pollution in two industrial communities determined by the use of soil cores and moss bags, *Environ Health Scotland*, **2**(12): 2-6.

Lloyd, O.L., M.M. Lloyd, F.L.R. Williams & A. Lawson (1988c). Twinning in cattle and in humans exposed to air pollution from incinerators. *Br J Ind Med*, **45**: 556-561.

Martin, M.H. & P.J. Coughtrey (1982). Biological monitoring of heavy metal pollution. London, *Applied Science*.

Smith, G.H., O.L. Lloyd & F. Hubbard (1986). Soil arsenic in Armadale, Scotland. *Arch Environ Health*, **41**: 120-122.

Smith, G.H., F.L.R. Williams & O.L. Lloyd (1987). Lung cancer and air pollution from iron foundries in a Scottish town: an epidemiological and environmental study. *Br J Ind Med*, **44**: 795-802.

Villeneuve, J.P., E. Fogelqvist & C. Cattini (1988). Lichens as bioindicators for atmospheric pollution by chlorinated hydrocarbons. *Chemosphere*, **17**: 399-403.

Williams, F.L.R., O.L. Lloyd & W.G. Berry (1987). Mortality from non-malignant respiratory diseases in Scotland between 1959 and 1983. *Ambio*, 16(4): 206-210.

Williams, F.L.R. & O.L. Lloyd (1988). The epidemic of respiratory cancer in the town of Armadale: the use of long-term epidemiological surveillance to test a causal hypothesis. *Public Health (Lond)*, **102**: 531-538.

Williams, F.L.R., O.L. Lloyd & M.M. Lloyd (1988). Animal and human disease near waste incinerators: a review. *In*: K. Sumino, ed. *Environmental and occupational chemical hazards*. COFM National University of Singapore & International Centre for Medical Research, Kobe University School of Medicine, Kishimoto, Japan no. 8.

Yule, F.A. & O.L. Lloyd (1984). An index of atmospheric pollution survey in Armadale, central Scotland. *Water, Air Soil Pollution*, **22**: 27-45.

Owen L. Lloyd
Department of Community and Family Medicine
Chinese University of Hong Kong
Lek Yuen Health Centre, Shatin
New Territories, Hong Kong

12   ROAD TRAFFIC ACCIDENTS INVOLVING CHILDREN IN NORTH-EAST ENGLAND

Simon Raybould and Sean Walsh

Abstract
Road traffic accidents are a very serious cause of morbidity (and mortality) among children. The data are therefore worthy of examination, and in addition the nature of the data lends itself to geographical information system (GIS) analysis. The data for north-east England are analysed in this chapter, using very simple GIS techniques. Some interesting results are presented, which are used to illustrate the utility of GIS analysis of data in the field of health and environment, even when using the most basic of techniques. The caveats of the work and the data are also discussed.

12.1 Introduction

Accidents involving road vehicles are one of the most important causes of morbidity among children in the developed world. In Great Britain, for example, approximately 9000 serious or fatal accidents occur every year. Clearly, any method that is likely to result in a reduction of these numbers is worthy of consideration.

In this chapter, some very basic geographical information system (GIS) techniques are used to undertake a relatively simple analysis of childhood road traffic accident (RTA) data. At this stage the work is of a fairly simple and exploratory nature, but nevertheless, it has already exposed previously unsuspected facets of the data that are pertinent to RTA prevention. Furthermore, the uncomplicated nature (in principle at least) of this example of the use of GIS in the field of health and environment provides an excellent entrée into the potential of such works.

RTAs for children are recorded in two locations, each site holding different data. These datasets are briefly described, and the way in which they are matched is also outlined. This results in a single and comprehensive dataset of childhood RTAs both in terms of the data for each RTA held and the coverage of RTAs in the dataset. Locational data for each RTA are then processed to provide information about the ward of both the accident itself and the home address of the child. Descriptive data about the child and the accident are used to create subsets of data. Once obtained, the ward allocation data are very briefly analysed and the importance of the GIS processing clearly shown.

M. J. C. de Lepper et al. (eds.), The Added Value of Geographical Information Systems in Public and Environmental Health, 181–188.
© 1995 Kluwer Academic Publishers.

end of the chapter. The majority of the chapter however, is a description of the initial work undertaken into studying RTA patterns.

12.2 The data

RTA data are provided from two sources. Primarily, information came from the police, with additional information being provided by local authorities. Northumbria Police Authority covers the counties of Northumberland and Tyne & Wear in the north-eastern part of England. In general terms, Tyne & Wear is an urban metropolitan county that is relatively densely populated. Northumberland, conversely, is predominantly rural, with a few built-up towns. The total population of the two counties is approximately 1,431,000, a significant proportion of the national total. The area covered by Northumbria Police Authority was therefore taken as the study area.

Obviously, as a combination of only two counties, this area is not completely representative, so its use as a study area implies that there will be errors and uncertainties in subsequent analysis due to boundary effects. The most obvious source of such problems is the scenario where an RTA is lost from the analysis because either the home address of the child (or children) involved or the actual accident location lies outside of the study area boundaries. The importance of such cases will become more apparent in later sections. Given the absolute size of the area and the scale relative to the country as a whole, the likelihood of such problem cases was deemed to be acceptably small.

A consistent finding of previous research is that the police under-ascertain cases known to hospitals as road traffic casualties. The overall underreporting rate is as high as 33% but decreases for serious injuries to 19% and appears to be complete for deaths. The police believe that the category of accident that is most problematic is that in which children on pedal cycles are injured in an accident that does not involve any other vehicle. In any event, there is reason to suppose that the distribution of any omitted cases is random and any error introduced is self-cancelling. The data may be therefore regarded as representative and reliable in terms of coverage.

Information on accidents that involve injury to any party is coded at police headquarters from a paper format onto a computer format such as the STATS 19 system. These data are stored on the police computer in the traffic engineers department of the relevant local authority, each case indexed by a unique accident and casualty reference number. Normally, the only geographically referenced variable on the STATS 19 system is the grid reference of the accident site in the form of a ten- digit grid reference. This has a spatial resolution of 10 m.

As part of a special collaborative study between the Department of Child Health, University of Newcastle upon Tyne and Northumbria Police Authority, the police headquarters provided the postcode of the home address along with the accident and casualty reference number for every child under 16 years who was being transferred onto the STATS 19 system.

By means of linking these cases with those on the police computer, the postcode of the home address of injured children on the STATS 19 system could be ascertained.

Postcodes contain, on average, only 15 household addresses and consist of all, or part of, a single side of a single street. The addresses in any given postcode are spatially contiguous. Clearly therefore, the postcode has an inherently geographical element. The Postcode Address File published by the Post Office encapsulates this by providing a lookup table between postcodes and the grid reference of that postcode, with a spatial resolution of 100 m. This lookup is not perfect however; the level of error and the nature of the geographical bias and error endemic to the Postcode Address File is dealt with elsewhere. By the judicious use of relational functions, it is therefore possible to obtain, for each child, accurate and spatially sensitive information about the location of the accident and of the home address, in the form of grid references - clearly information appropriate for input into GIS - as well as such non-locational data as age and sex of the casualty, the severity of the injuries received, journey purpose and a verbal description of the nature of the accident. Obviously, additional information such as the level of neighbourhood deprivation near to the home address can be added at will through GIS. Similarly, more complex information such as the total distance of high-grade roads within a certain radius of the accident can be added, provided that the necessary data are available.

Such a case-by-case dataset was constructed for the two-year period July 1988 to June 1990, giving a total of 2,423 cases. In the process, a number of cases were lost e.g., where no match could be found with the local authority data from STATS 19 or vice versa. Further errors in the data were due to mistyping other than in the reference codes. By definition, there is no way to calculate the rate of such errors, but a map of grid references for the accidents shows several in the sea.

Generally, however, the distribution of accident locations reflects the population density of the study area, with the majority of RTAs occurring in the urban parts of Tyne & Wear and a very much lower density across most of the rest of the study area.

12.3 The GIS input

All the preceding work described in this chapter, while easy enough to handle in a GIS by relational matching, could also have been carried out using a variety of other software ranging from statistical analysis packages such as SPSSx, spreadsheets and similar software such as DB4 or Ingress, to specially written software in FORTRAN or C++. The subsequent work described here however, could only conveniently be carried out in a GIS: in this case, commercially available package ARC-INFO.

The grid references of the accidents were input to a point-in-polygon routine (or PiP). This is a simple algorithm in principle that, for every point, returns the polygon in which that point lies. In this case the algorithm informed, for every RTA, the ward in which it lies. Wards are the second smallest unit for that census data are available in Great Britain. There are 10,444 in England, Scotland and Wales and 238 in the study area. Their significance in this chapter lies not in the fact that other data are available at the ward scale (wards are a poorly defined areal unit for many research purposes) but that historically, wards are the unit for that rates of childhood RTAs have been reported. Crucially, the numerators for such rate calculations have been taken as the RTAs themselves and the denominators for such calculations have been taken from the population census, which is taken in the United Kingdom every 10 years. The 1991 census data is not yet available, and so calculations have been made based upon 1981 data. Obviously, these data suffer from significant ageing problems whereby, for example, children involved in an RTA who are less than 10 years old will not be noted in the census at all. This is not as serious a problem as it may sound, however, as the gross pattern of the population distribution is relatively stable. Work is currently underway to look at the use of general practitioner registers as a more contemporary surrogate. This chapter investigates the effect of a far more invidious problem, however: that such denominator data is based upon the place of home residence. In other words, while the numerator data are based upon the location of the RTA itself, RTA literature has previously used denominator data based upon the location of the place of residence of children - who are quite possibly not the population of that RTAs are a subset.

In short, the numerator and denominator operands in the literature are on a different base. It is therefore desirable to perform one of two operations. Either the denominator data can be modified in some way to transfer children back to the wards from which accident victims come or the victims themselves can be transferred to the wards in which they are resident, rather than in which they have the accident.

While the former could only be achieved by some sort of proration based upon the latter in any case, a transferring of the casualty cases back to their residential wards is probably more desirable in terms of an investigation of the non-mechanical causes of childhood

RTAs (i.e., the causes such as attitudes and experience of the child, home environment, etc. rather than road type and design, speed of car and visibility, etc.). Such a transfer is uniquely possible in this work due to the previously described link of home postcode to grid reference and enumeration district, as enumeration districts fit perfectly into wards.

Map 12.1 shows a simple choropleth map based upon quintiles of wards, where the mapped variable is the rate of childhood RTAs with the numerator based upon accident location and a denominator from the census.

Map 12.2 shows a similar map, this time with the numerator values based upon the ward of residence of the accident victims.

The darkest cross-hatched wards are those that have the highest rate of RTAs. Clearly there is a strong similarity between the maps; a closer examination shows a not insignificant number of differences.

These simple visualization techniques are another example of the utility of GIS. While, admittedly, quintile-based choropleth maps are not a particularly sophisticated method of visual communication, their power should not be underestimated, particularly in the field of epidemiology, where their application is relatively new, once placed in the context of interactive map composition (and if desired, the ability to visually superimpose additional locational data such as a road network).

Map 12.1:   Rate of childhood road traffic accidents: based upon accident location

Map 12.2:   Rate of childhood road traffic accidents: based upon residence location

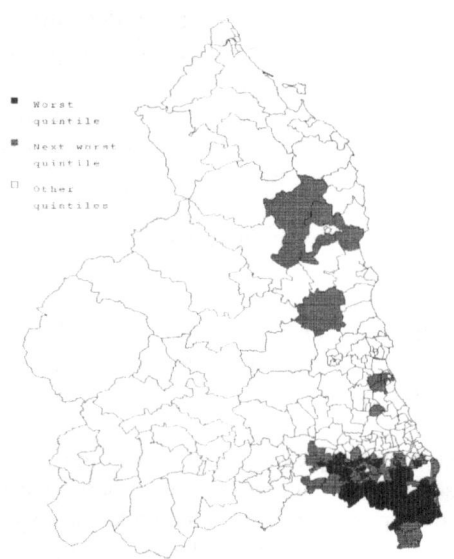

To aid the interpretation of Maps 12.1 and 12.2, which are, after all, designed for comparison to each other, the GIS can be further called upon. Most GIS have a graphics module that is connected in some way to a database handling facility. By use of the latter it is easy to give the former a variety of modified versions of the data to manipulate: one example of this is sieve mapping.

Using the relational database element of the GIS package ARC-INFO (INFO), wards were examined to determine that quintile they lay in with respect to Maps 12.1 and 12.2. In so far as previous work on childhood RTA prevention has tended to be based upon data similar to that presented in Map 12.1, but with subsequent intervention based on the assumptions of Map 12.2 (in that education is given in schools near areas of high risk, but to the children who live there and not those who are injured there), it is pertinent to ask whether the pattern shown in Map 12.2 is similar to that shown in Map 12.1. The most simple approach is to label wards appropriately if they are in the top quintile for both Maps 12.1 and 12.2. Map 12.3 shows such wards shaded.

The reader may judge whether the number of wards shaded in Map 12.3 is higher or lower, but clearly, accident prevention initiatives targeted at top quintile wards in Map

12.1 that are not shaded in Map 12.3 represent an inefficient use of resources. Only about 60% of childhood RTAs occur in the ward in which the child in question is resident.

Map 12.3:   Rate of childhood road traffic accidents

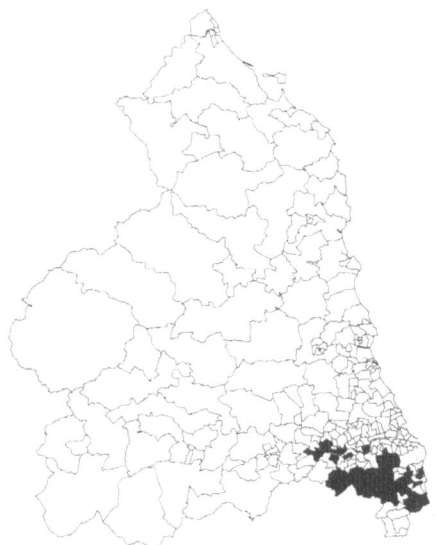

12.4 Discussion

Obviously this is a somewhat simple example of the use of a GIS in a health and environment field but one that clearly shows the utility of the approach. Striking visualization is possible (that has a surprisingly great effect when presented to medical practitioners) as well as a few of the more basic techniques. Important and yet previously unasked questions have been answered about the robustness of previous work in the field. Sufficient interest has been generated to warrant further work, almost because of, rather than despite, the GIS simplicity, in that techniques are easily explained in principle to practitioners in the field.

It is equally obvious that the work presented in this chapter is only a very simple beginning and much more work remains to be done. Some analysis, for example, of the time of day of the RTA needs little if any GIS input. Other work, such as an examination of distance-to-accident figures or cluster analysis of RTAs, need limited GIS work. The more complex and, ultimately, more rewarding work, however, lies clearly in the field of

GIS. Examples might include network analysis of road patterns or public transport facilities, optimal siting calculations for road crossings or assessment of the risks involved in journeys to school.

Acknowledgements

The authors are grateful to Dr Steve Jarviss and Mr Mike Coombes for their help in setting up the data transfer protocols and to Ms Jacqueline Nicol for work in the GIS software.

Simon Raybould
Centre for Urban and Regional Development Studies
University of Newcastle upon Tyne
Newcastle upon Tyne NE1 7RU
United Kingdom

Sean Walsh
Department of Child Health
University of Newcastle upon Tyne
Newcastle upon Tyne NE1 7RU
United Kingdom

13 GEOGRAPHICAL SOFTWARE APPLICATIONS FOR HEALTH SECTOR PLANNING: EXPERIENCES FROM A STUDY FOR FAMINE MANAGEMENT

Debarati Guha-Sapir

Abstract

The use of geographical techniques in the health sector has been an issue of increasing interest. Along with facilities in graphic presentations rendered easy by new software techniques in computers, mapping and related interpretations of numerical data has captured the attention of health care specialists as an useful tool. The stumbling block, however, has been the potential for analysis and the concrete uses that such a marriage would bring. The resolution and the accuracy of the data used in the health sector is generally far below the standards required for the proper application of geographical tools. Terminologies and underlying concepts also differ sufficiently to make interdisciplinary initiatives difficult. Despite these hurdles, in 1987, within a study financed by the World Health Organization (WHO) on information systems for famine management in Africa, the WHO Collaborating Centre for Research on Disaster Epidemiology (Brussels, Belgium) experimented with a specific application of geographical techniques for health sector planning. Although applications of geographical techniques were not the principle objective of the study, the potential for geographical information systems for health priority-setting and optimal planning, especially in situations where a rapid response is required, was considered worth a certain investment in research and development for appropriate applications.

13.1 Background and objectives of the study

Information systems, in the context of famines are, principally, of two kinds: early-warning systems and management systems. The former have had great success in the last two years. The latter have been, on the contrary, visibly neglected. The purpose of this case study, called consolidated information system for famine management in Africa (CISFAM) was to design a centralized information service for programme managers to make educated and rapid decisions for famine interventions and policy. The system would respond to the information needs of government, nongovernment and intergovernment agencies for planning, targeting and policy-making in famine response programmes. It would serve as a centralized source where pre-selected information on sectors such as agriculture, meteorology or economy, in addition to health, would be quickly and easily available, thereby eliminating the need for planners to go to different specialized agencies for information.

The rationale for this study was based on four conditions dominating rapid and effective famine response action.

M. J. C. de Lepper et al. (eds.), The Added Value of Geographical Information Systems in Public and Environmental Health, 189–199.
© 1995 Kluwer Academic Publishers.

First, resource constraints were getting increasingly serious with grave implications for continued international assistance for control of famine conditions in Africa. Second, while the health sector is focal in famine relief, the crisis is essentially a multisectoral problem and requires multidisciplinary data for rapid programme planning and resource allocation. Third, large quantities of data exist in specialized agencies of the larger United Nations family and in national archives. Most of these remained unused in emergencies, since the international databases are frequently in sophisticated forms that are inaccessible to the uninitiated and the national ones are non-standardized and non-computerized, making retrieval of selected information difficult. Finally, nongovernmental organizations were observed to frequently have regular data reporting systems that were not adequately processed or used either by themselves or by the governments with whom they work. These agencies formed a potentially important repository of local data.

It was also increasingly recognized that, in planning health care services for famine conditions, where rapidity of decision-making is critical, information from several sectors besides health is required and, furthermore, available for use in digested forms that allow interpretation and analysis by health specialists. The study therefore addressed essentially the issue of information for rapid response planning in situations of mass displacements or acute local food shortages. Based on field experiences, five additional sectors besides health were selected as being relevant. These were demography; agriculture; logistics and infrastructure; socioeconomic; and environment and meteorology.

The specific objectives of the study were:

(i)   to conceptualize and design a ready-to-use information system for transfer to the national health authorities or the relief and recovery management unit;
(ii) to identify and develop image-, graphic- and map-linked databases for quick interpretation and operational decision-making.

This chapter focuses on the second objective, its experience in the application of geographical information system (GIS) techniques.

13.2 Data aspects

13.2.1 Database design

The structure of the system was defined by country blocks. Information by category, as available, was collected for each country and any additional information generated by special survey and studies generated by other bodies was appended in the informational annexes to the country data block. These annexes were restricted to only the surveys dealing with food, nutrition and health. The system was not designed to create any new sources of information but rather to enhance the utilization of information held in existing databases. The project therefore examined databanks and information systems of the United Nations agencies and allied bodies and found that many useful sources of information and databases were frequently unknown and even less utilized by most of the implementing agencies. A very small proportion of the potential users, including national governments showed any knowledge of the internationally maintained systems and databases and fewer acknowledged ever requesting or receiving any information from them. Several databases were examined, including both electronically maintained and data on cards, reports and other forms of hard copy. Certain countries had better data reporting than others and the variability was significant. This was problematic for standardization of data format across countries. In addition, specific sectors had better data collection systems than others.

Meteorological, ecological and climate databases provided the best quality information in terms of reliability, coverage, accuracy and time series. The Climate System Monitoring (CSM) databases of the World Meteorological Organization (WMO), initiated as a response to the occurrence of significant climate anomalies over the last decades, provided useful information when associated with adverse socioeconomic effects. The WMO databases carried synthesized information on climate anomalies, rainfall variability and vegetation data by small geographical areas. It had available time-series data for the past 110 years on African rainfall. The data, however, were technical in nature and were not divided by the political and administrative boundaries of the countries. Similarly, the soils and temperature databases of Africa held by the United Nations Environment Programme were also reliable sources. Both sets required some interpretation and adjustment to be classified most meaningfully by international and national boundaries.

The health sector information, on the other hand, was disappointing in both coverage, quality and continuity. Physical resources, such as hospitals, dispensaries, health centres and skilled personnel enumerated in the data source agencies, were the only items with relatively regular reporting. However, a major caveat in these items was that hospitals and health centres were frequently inoperative in reality and therefore, the value of this data was questionable for rapid response planning. For crisis management and long-term planning, data on the existence and functioning capacity of institutions in affected areas can be key to efficient response. Several sample surveys on nutritional status and incidence of nutritional deficiency diseases were available, but no official figures were reported on an ongoing basis. Continuous nutritional data collection in this field was generally undertaken only by large nongovernmental organizations, which did this as an incidental by-product of their principal activities.

Archival data were also uneven in coverage over time and quality varied. The cholera and yellow fever data transferred from cards filed on the weekly epidemiological reports were perhaps the only regular incidence data to be reported to WHO by province. The data were considered to be underreported and the magnitude of this underreporting was unknown. The problem became aggravated when the data were compiled by different agencies. Estimates differed for the same disease according to the source (government, nongovernmental or international) and, on occasion, delayed reporting caused inexplicable increases in the incidence of diseases that were not due to the extension of the reporting period.

13.2.2 Data quality

Data limitations are a primary consideration in interdisciplinary approaches to analysis. While they should not block exploratory research, it is essential to examine well the caveats such that they can be accounted for in the development of applications. Limitations become additionally problematic when data from different sectors have to be worked into a GIS application. For example, the denominators of indices differ as well as time and space units. Some examples of these limitations are described below.

The comparability of the statistics in Africa, both temporal and spatial, was mainly limited by the use of non-comparable definitions. For example, in agriculture, under traditional African conditions, it is not always clear what would be the main occupation of the individual. Persons can be occupied in different occupations at different times of the year, creating definitional ambiguities especially for risk assessment. In certain countries of the region, farmers frequently work away from their holdings during a large part of the year, for example, on plantations, in mines or nearby towns. Furthermore, the fact that a man can have several wives who independently cultivate separate pieces of land although the land may customarily belong to the head of household creates problems for both demographic and agricultural information.

In general, the problem in this project was more the lack of usability of data rather than their inaccessibility. In other words, much of the data existed, but processes to reformulate them in ways that could be entered in models or images were complicated. In addition, conceptual (between sectors) and numeric compatibility were also problematic. For example, time intervals of data that are of critical importance were at different scales for epidemiological data compared with ecological or climate data. This meant aggregated scores had to be developed to standardise time frames between variables. Similarly, health data was often hospital-based, which imposed certain limitations in its interpretation. The population or geographical area it represented was not compatible with data from other sectors that were more precise. Economic or environmental data, on the other hand, tended to be population-based and were not subject to the same constraints.

The data also suffered from limitations due to their selectivity. Hospital statistics, although an accurate source of diagnostic information, present a serious bias arising from selectivity in relation to factors such as location, type of disease, provision of health facilities, age and sex and socioeconomic status of the patient. The population served was frequently unknown. It was therefore not possible to generalize the hospital data to a community or a geographical unit. These statistics could give an inaccurate appreciation of the prevalence of morbidity, if extrapolated. They could, nevertheless, measure the relative distribution of diseases in the areas covered and be treated as valuable adjuncts to mortality statistics, suggesting priorities for provision of more medical facilities and efficient medical care.

For purposes where rapidity is the primary objective, timeliness of statistical information is *sine qua non*. But this can be maintained almost only at the expense of accuracy, while improvements in quality require more time and increase the cost of the information.

13.3 A GIS for famine management: technical aspects

Like most computer-based information systems, CISFAM had a series of basic features: data capture, storage and retrieval, analysis, output and display. Data capture involved putting information into the computer and organizing it in memory. The key element was how GIS software handled each piece of information. Retrieval, the reversal of storage, recovered ordered data from the memory and searched information with certain characteristics (e.g., all hospitals less than 20 km from a paved road or with maternity wards) for use in the development of maps or in estimating risk. A special feature of CISFAM was its capacity to analyse as well as display the results in images. This involved the retrieval of data files or parts of them in any combination and analysing them to generate composite maps. These display possibilities would help planners manipulate and prioritize needs for resource allocation by changing risk factors and therefore shift priority emphasis on the maps.

The GIS aspect of the project depended on two basic techniques of data use: overlays and statistics. In consultation with the staff of UNEP-GRID, two main directions were identified for exploration. One was overlaying data planes for composite images. For example, for any defined area, datasets could be overlaid in an electronic version of stacking maps with different information for the same area. The greater the number of datasets, the greater the number of possible comparisons. For example, in a study of malaria control, the software would compare water distribution, health centres, domestic animal distribution and endemic malaria incidence by overlaying these datasets one by one. In emergency health planning, this had special implications in pinpointing vector-breeding sites, location of displaced population camps or temporary food or health care distribution points as well as arable land area and firewood availability.

The second was the potential of spatial modelling techniques using composite databases of numerical and image data. For example, certain diseases with biological, environmental and human spatial parameters were modelled to reproduce a multidimensional distribution of the disease over the area of interest.

13.4 Experimental application of GIS technology to health planning: the case of Senegal

One of the most important developments in affordable computer technology has been the emergence of GIS, and today, many types of GIS are currently in use. For CISFAM, the GIS system used integrated tabular (i.e., vital statistics) and thematic (i.e., ecological) maps with digital satellite data (i.e., vegetation) in a geographicalal context that allowed analysis of physical, biological, economic and social information on visual planes. The feasibility of linking these different types of data was tested by developing a model for detailed estimation of population density in Senegal.

In the absence of data, the development of a population density dataplane for Senegal was undertaken using a model with specified parameters that were hypothesized to determine the population distribution. Appropriate coefficients were estimated by regression analysis using a combination of extrapolated data and data from countries with similar ecological conditions. Thus, relative distance from roads according to the type of road, relative distance to a city or town according to its size and administrative importance, presence of water sources, climate and topography were all considered using different techniques in the model. The latitude and longitude of major transportation routes and 118 towns in Senegal were digitized from 1:500,000 scale maps obtained from the Institute Géographique National, Paris. Each of the towns was assigned a rank from one (village) to five (metropolitan area), based upon the relative size of each. The Earth Resources Laboratory Application Software (ELAS) was used at UNEP/GRID to develop six raster or gridded cell (pixels) dataplanes generated at 30 seconds of latitude and longitude resolution (approximately one square kilometre) one for the transportation network and one for each of the five ranked town sizes. Dataplanes for distances away from each town and road were then developed using the ELAS overlay DIST. Topography, climate, desertification and surface water distribution were introduced as satellite images. The final estimated population density dataplane was generated by the summation of each of these distances indexed through tables with different weights for the roads and relative size of towns as contained in the following chart using the ELAS overlay DBAS. Hypothesized coefficients of influence of other variables were included in the model. The iterations were made successively, as a function of constraints to produce topographic distributions of the variables of interest. The results are shown in Maps 13.1, 13.2 and 13.3.

13.5 Conclusions

CISFMA was founded on the concept of management information systems for health. Although this concept is no longer controversial, a certain fuzziness exists about the elements and purposes of such systems. Data of many kinds are fed into a management information system, but unless selectively compiled and processed, they convey little other than standard archival data. The value of data and statistics in a management information system is judged by their usefulness in programme evaluation and policy formulation.

Today, technology can almost certainly be counted on to cut down the costs of producing and collecting figures (e.g., processing). However, it should be recalled that data compilation is only the start of a whole chain of processes, which include interpretation of the data collected and statistical analysis and research, all of which are entirely dependent on the basic conceptualization of the initiative and on human ingenuity for interpretation. The reconciliation of the needs for international comparability and hence conformity with the standards laid down by the international agencies and the specific domestic needs of an individual country can pose difficulties. This requires a certain amount of ground work in standardization and common definitions and criteria between the collaborating agencies.

The importance of multisectoral data needs for the health sector underscores the interest in using image-processing techniques. Currently the problem is not so much that data from the different health-related sectors are unavailable. On the contrary, there is a very large number of existing databases. What is more, the issue is the underexploitation of this data. The digestion of the different data into formats that are interpretable by health policy-makers would constitute an extremely efficient tool towards the better use of resources. Today, the technically difficult formats in which specialized geographical, meteorological and ecological data are maintained and the discipline-specific terminology pose major stumbling blocks to interdisciplinary use.

Map 13.1: Estimation of population density; the UNEP/GRID model applied to
Senegal, integrating major cities and transportation routes

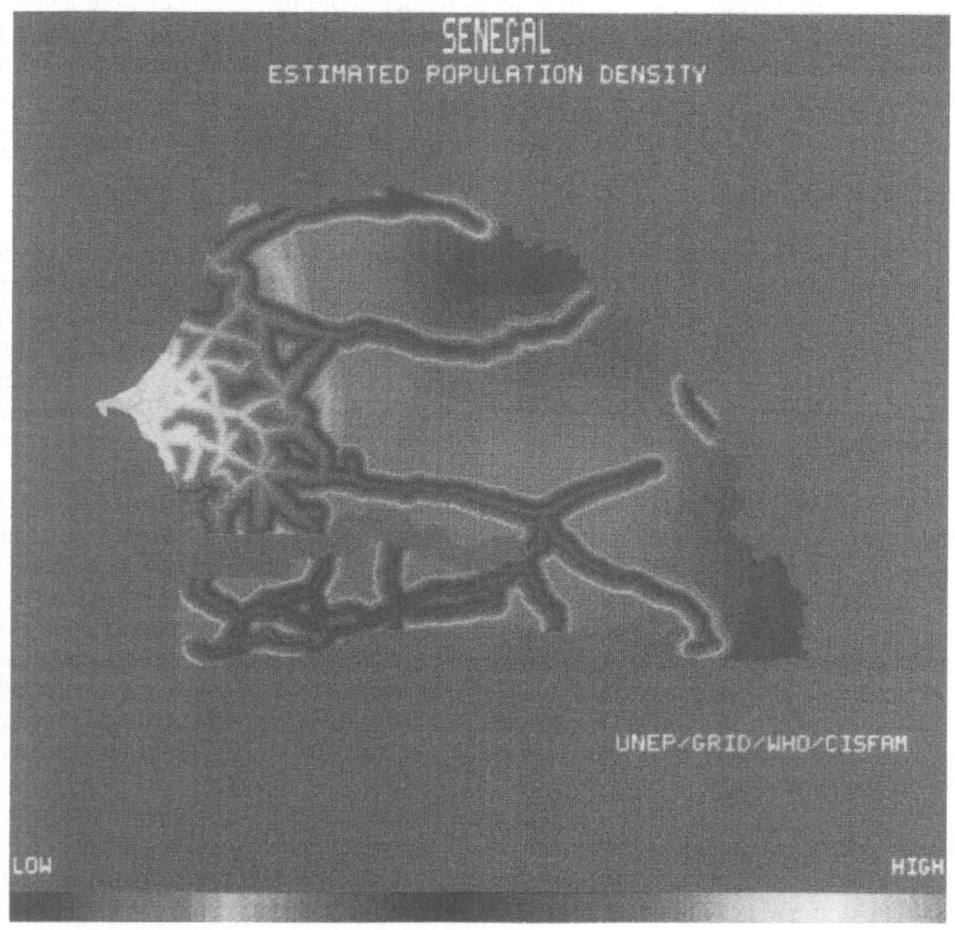

Map 13.2: Modelled distribution of health care accessibility in Senegal

Map 13.3: Desertification hazard in Senegal

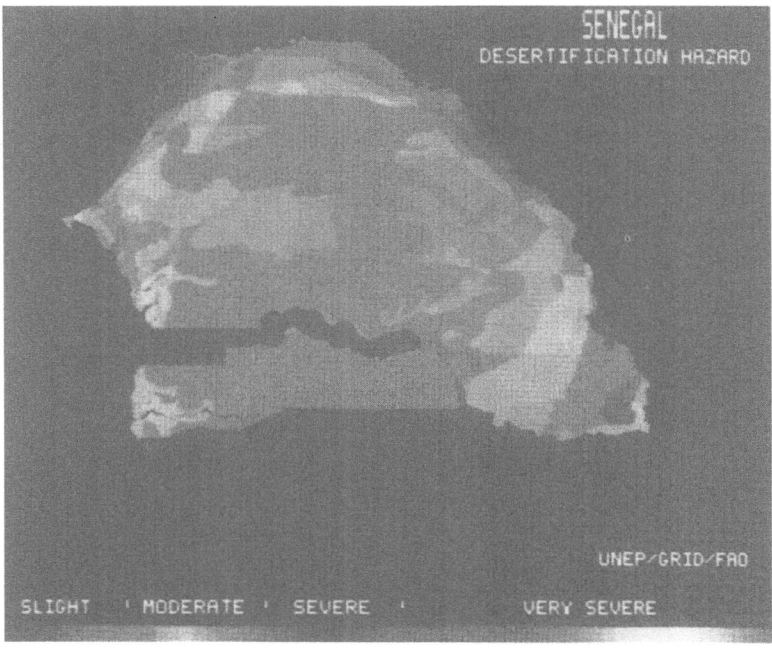

In addition, the constraint on monitoring health is not the inability to provide reliable and relevant indicators. This can be done with a degree of confidence. The key element is the appreciation on the part of international and national policy-makers of the importance of collecting and using relevant data to monitor conditions and plan programmes and the will to use tools, techniques and materials that are already available to them.

It is clear that spatial modelling is a feasible alternative to the use of sector-specific information in discrete or purely statistical format. A propitious marriage between statistical modelling and geographical techniques for image-processing can produce extremely powerful and attractive results. These products can not only facilitate the planning and optimization of health care services but provide the potential for risk analysis of communities for health conditions related to environmental as well as non-environmental characteristics. The composite visual presentation of what would otherwise be disparate collection of large datasets, has significant advantages for policy making. They provide convincing material for policy-makers to take decisions. The problems of data compatibility between those drawn from the health sector and those collected from remote-sensed mechanisms can be difficult but not insurmountable. Estimated data can be used successfully, as proved by this experiment. Their use requires preliminary analysis (such as a system of equations) to estimate the coefficients of parameters that are expected to influence the variable of interest. The software and computer requirements today are also no longer prohibitive in cost. However, a constructive framework wherein scientists from different disciplines can work together is essential. Also essential is a large measure of creativity and imagination, since the terrain is so far practically untouched, as disciplines tend to work within their boundaries.

In the final analysis, it still remains to be demonstrated, on a larger scale than was possible in the CISFAM study, that these tools may be used fruitfully for health planning. A series of specific applications in selected health problem areas, undertaken within the same conceptual base, would advance significantly what is today a nascent idea, still in the domain of theoretical discussion.

References

Carter, D., G.W. Heath, G. Hovmork & H. Sax (1989). Space applications for disaster mitigation and management. *Acta Astronomatica*, **19**: 229-249.
Friedman, M. (1990). Putting the data to work. *Dev Forum*, **18**: 12-13.
Graeme, T. (1986). Satellite imagery: a broader view of the earth and its resources. *UNDRO News*, 11-16.

Hugh-Jones, M. (1989). Applications of remote sensing is essential to the identification of the habitats of parasites and disease vectors. *Parasitol Today*, **5**: 244-251.

Linthicum, K.J., C.L. Bailey, G.F. Davies & C.J. Tucker (1987). Detection of rift valley fever viral activity in Kenya by satellite remote sensing imagery. *Science*, **235**: 1656-1659.

Paul, C.K. & A.C. Mascarenhas (1981). Remote sensing indevelopment. *Science*, **214**: 139-145.

UN Environment Programme. Putting the data to work.

Wortman, Sterling (1980). World food and nutrition: the scientific and technological base. *Science*, **209**: 157-164.

Debarati Guha-Sapir
WHO Collaborating Centre for Disaster Epidemiology
Department of Epidemiology
University of Louvain
School of Public Health
EPID. - U.C.L. 30.34
Clos Chapelle-aux-Champs 30
B-1200 Brussels
Belgium

14 GEOGRAPHICAL INFORMATION SYSTEMS: A NEW TOOL IN THE FIGHT AGAINST SCHISTOSOMIASIS

Steven S. Yoon

Abstract

Schistosomiasis is a group of parasitic diseases caused by *Schistosoma* parasites associated with bodies of fresh water affecting more than 200 million people in 76 countries. The disease is present on many continents, including Africa, Asia and Latin America. Like many other parasitic diseases, schistosomiasis has adapted itself to the particular ecology of the habitat: the presence of fresh water is necessary for transmission. The control measures for schistosomiasis include chemotherapy, health education and molluscacide use (snail control). In order to actively identify and control the disease, it is vital that up-to-date information on the epidemiology of the disease be available to the control programme manager. The Schistosomiasis Unit of the Division of Control of Tropical Diseases of the World Health Organization has been exploring the use of geographic information system (GIS) for monitoring the epidemiology of schistosomiasis. Initial studies suggest that GIS can be useful in monitoring schistosomiasis epidemiology and assisting programme managers in their control efforts. However, in order to fully realize the usefulness of GIS, the quality of epidemiological data has to improve and additional information has to be incorporated into GIS to examine the relationship between various ecological and disease-related variables.

14.1 History of GIS in public health

Geographic information systems (GIS) are a relatively new tool that are becoming more widely used in the field of public health. Their previous limitations, including the lack of user-friendly software and hardware necessary to run GIS software, have all but disappeared. With the ever-increasing power of personal computers, desktop GIS will likely be used in more innovative and useful ways.

Interestingly, public health workers were using maps for epidemiological work long before the invention of GIS. One of the first known occurrences of mapping disease was during the yellow fever outbreak in Europe during the late eighteenth and early nineteenth century (Meade et al., 1988). And, well-known and much referenced epidemiological work done by John Snow in the United Kingdom in the middle of nineteenth century also used geographical representation of disease incidence to investigate the source of a deadly cholera epidemic.

Today, with the availability of powerful personal computers and sophisticated user-friendly software, what used to be a tedious and approximate process of manually

M. J. C. de Lepper et al. (eds.), The Added Value of Geographical Information Systems in Public and Environmental Health, 201–213.

drawing maps and displaying data is less time consuming and more exact. In addition, several other factors are attracting health specialists to map-based data presentation and analysis. First, map-based presentation of human disease data can be both informative and intuitive, since human diseases tend to follow certain spatial and temporal patterns. Second, the information presented using maps may be easier for non-health specialists to comprehend. This is an important issue in the instances when health resource allocation decisions are made by politicians and non health-specialists. Third, the tools for analysis of spatial information (GIS) have become easy enough to use by non-GIS specialists for their work. Finally, GIS allows the display, analysis and understanding of data from diverse sources in what could be called a holistic approach to data management, analysis and presentation.

The Schistosomiasis Unit of the Division of Control of Tropical Diseases (CTD/SCH) of the World Health Organization (WHO) has been interested in exploring geographical techniques for the monitoring of the global epidemiology of schistosomiasis, presentation of epidemiological data and the analysis of disease data. This chapter briefly discusses the SCH Units recent activities and experiences.

14.2 Schistosomiasis

Schistosomiasis is currently endemic in 76 countries around the world. It is estimated that more than 200 million people residing in urban, rural and agricultural areas are infected among 600 million people who are exposed to infection (WHO, 1985). These 600 million people are exposed to infection because of poor socio-economic situations and substandard hygienic practices.

There are actually many species of *Schistosomiasis*, four of which are particularly important because of their effect on humans: *S. haematobium, S. mansoni, S. japonicum* and *S. intercalatum*. Although schistosomiasis does not often lead to death, it does cause disability and affects children in particular. In general, 60-70% of all infected people are 10-14 years of age. In addition, the peak prevalence and intensity of infection generally occurs in children 10-14 years of age (for *S. haematobium* and *S. mansoni*). Primarily because of its effects on growing children, the disease is an important public health problem.

14.2.1   Life-cycle of Schistosoma

The life-cycle of *Schistosoma* has two phases: sexual and asexual. The sexual phase involves the pairing of adult male and female worms in the venous blood vessels of the intestine (*S. mansoni* and *S. japonicum*), the bladder (*S. haematobium*) or the rectum (*S. intercalatum*). The eggs are passed from the human body through urine or faeces.

When excreted eggs come in contact with water, miracidia hatch out of the egg. Miracidia are ciliated (covered with fine hair) organisms. The movement of the cilia propels the miracidium, which, in order to develop further, must come into contact with the proper fresh water snail. *S. mansoni*, *S. haematobium* and *S. intercalatum* all develop in aquatic snails, which live entirely in the water. *S. japonicum* develops in amphibious snails, which can live on land as well as in the water. Inside the snail, the asexual phase of the life cycle takes place. Each miracidium develops into thousands of cercariae and are released into the water. Cercariae must come into contact with a suitable mammalian host within 48 hours or they will die. If a suitable mammalian host is found, cercariae will penetrate the skin of the host and begin the sexual phase of the life cycle. Each cercariae, after entering the hosts body, develops into a single adult worm, either male or female. The time required for the worms to mature, to mate and to start to lay eggs is between 30 and 45 days.

The cercaria enters the hosts body through the skin while the host is immersed in water. Any activity in cercaria-infected water can result in penetration. The risk of infection is proportional to the type and duration of the activity: the larger the area of the body in contact with water and the longer the water contact, the higher the risk of infection. This is one of the reasons why children are at high risk: they tend to play in water frequently for long periods of time.

14.2.2 Geographical distribution

Different species of *Schistosoma* are endemic in different parts of the world. In some areas of the world, only one type of *Schistosoma* is present. In other places, two or more types of schistosomiasis may be present. The distribution of schistosomiasis is neither uniform nor contiguous. However, one of the limiting factors in its distribution is the presence of fresh water (Yoon & Mott, 1991b).

The geographical distribution is wide-ranging from China, the Philippines and Indonesia, (*S. japonicum*), through the Arabian Peninsula and nearby states, the Sudan and the Nile valley and delta, numerous countries of the north African littoral and the whole of sub-Saharan Africa (*S. mansoni* and *S. haematobium*); to the Americas - Brazil; Suriname; Venezuela and certain Caribbean islands (*S. mansoni*). As agricultural, hydroelectric and other water resource development projects expand in the endemic countries, transmission of schistosomiasis spreads and the degree of transmission is intensified.

S. haematobium, the cause of urinary schistosomiasis, is endemic in 55 countries of the Eastern Mediterranean and African Regions. *S. intercalatum*, the cause of rectal schistosomiasis, is increasingly recognized in central Africa. It is now endemic in Cameroon, Central African Republic, Chad, Congo, Equatorial Guinea, Gabon, São Tomé and Principe and Zaire. *S. mansoni*, the cause of intestinal schistosomiasis, is endemic in 52 countries of the Region of the Americas, the African Region and the Eastern Mediterranean Region. *S. japonicum*, which also causes intestinal schistosomiasis, is endemic in countries of the Western Pacific Region. In six countries, all three species of *Schistosoma* are endemic. Both *S. mansoni* and *S. haematobium* are endemic in 41 countries of the Africa Region or the Eastern Mediterranean Region.

14.3 Uses of GIS in the control of schistosomiasis

There are three obvious uses of GIS:

(i)   as a surveillance and monitoring tool;
(ii)   as an analytical tool; and
(iii)   as a presentation tool.

By allowing the user to rapidly retrieve and display data associated with a geographical area, GIS can be considered an ideal tool for the surveillance and monitoring of tropical diseases. GIS can easily display, for example, prevalence levels of schistosomiasis for each town around a body of water using different colours. Using longitudinal data, GIS can be used to identify the towns with increasing prevalence over a period of time.

Surveillance and monitoring of geographical transmission data, such as villages with high prevalence rates and possible water transmission sites, help programme managers to combat the disease. Since an individual becomes infected with schistosomiasis during contact with infected water (there can be no other means of transmission), the presence of infected snails would determine whether a particular water site is a source of transmission. Although the distribution of snails around a body of water can be homogeneous, studies have shown that the distribution of infected snails can vary greatly even among sites located closely together (Barreto, 1991). This fact makes the knowledge of a precise pattern of infected snail distribution important for control efforts.

GIS is also an excellent tool for analysis of existing data and modelling of schistosomiasis epidemiology. Schistosomiasis, like many other parasitic diseases, is a disease with a complex human-parasite interaction. Many factors affect the relationship, including socioeconomic and ecological factors (Farooq et al., 1966; Dalton & Pole, 1978). Environmental variables such as rainfall can drastically affect the snail population, which in turn can affect the disease prevalence. Besides natural environmental factors, human-made water resource development such as dams and irrigation, urbanization of the human population and movement of nomadic and refugee populations within and across borders can affect the epidemiological diversity of schistosomiasis. Finally, the presence of other water-related diseases such as malaria can further complicate schistosomiasis control efforts.

So far, the efforts at examining all the factors that make up the life cycle of *Schistosoma* and using the information for control purposes have been limited. Most of the effort has gone into the creation of statistical tables relating the significance of one variable to another. Information presented in this format, however, is difficult to comprehend and does not serve its purpose of assisting in control efforts. GIS could, on the other hand, provide the necessary tools to make the information more understandable and therefore more useful. The possible influences of different factors with different spatial influence can be integrated by using a technique known as overlay. By analysing the data using overlays, what used to take several figure tables can be explained using a single map. With simplified data analysis and presentation, programme managers can better respond to changing situations. In addition, GIS provides the tools that can assist in integrating the control efforts of different diseases.

A personal computer (PC)-based GIS combines an easy-to-understand graphic map-based presentation with an easy-to-use query mechanism and the dynamic user-interaction of PC databases. By combining these important capabilities, GIS can provide the schistosomiasis programme manager with timely and pertinent information. Additionally, data stored in an electronic format can be modified, updated and disseminated easily.

The measurement of the presence of a disease in a given population is usually expressed as one of two numbers: 1) prevalence or 2) incidence. The measurement of the presence of schistosomiasis is usually expressed as prevalence and, in some instances, also as the intensity of infection. The source of such figures is usually the ministry of health for each country. However, due to resource limitations, the figures available from the ministries are often of poor quality. There are several reasons for this:

(i)   insufficient data collection activity;
(ii)   improper aggregation of data; and
(iii)   delays between the collection and publication of data.

Quite often, when resources are limited, not enough data is collected for management purposes or, more commonly, the wrong data is collected. The data used to provide information about control programmes is only as good as the process used to collect the data.

The distribution and epidemiology of schistosomiasis vary greatly with geographical, demographic and socioeconomic factors. Within a geographical or political boundary, the prevalence of schistosomiasis can vary among people of different age groups, sex or occupations. Prevalence figures are usually an aggregation of data from a number of smaller areas. These figures can be misleading, since some areas may be free of disease while others are heavily infected. Although summary data can be useful for presentations to those not involved in day-to-day control activities, for programme managers, more detailed data is necessary for operational decisions. It is important for the people in charge of the control programme to know which towns have the highest disease prevalence. Furthermore, if data are not processed in a timely manner, they cannot be used to make operational decisions.

14.4 Activities in the Schistosomiasis Unit of the Division of Control of Tropical Diseases of WHO

14.4.1 The atlas of schistosomiasis

The Schistosomiasis Unit of the Division of Control of Tropical Diseases of WHO (WHOs SCH Unit) has been working on a geographical approach to the analysis of the epidemiology and control of schistosomiasis since 1981. There were several reasons for the initial interest in a geographical approach. First, the relationship between various political and physical boundaries and the distribution of schistosomiasis were observed, but the underlying causes were unknown. Second, the epidemiological diversity of schistosomiasis indicated that the aggregation of distribution data was important down to the family level, but there was no easy way of working with this kind of data. Third, because schistosomiasis was a disease related to the development process in many countries, the monitoring of changes in physical environment was important. These reasons, along with others, prompted the SCH Unit to look for different approaches for monitoring and controlling schistosomiasis.

One of the first activities the SCH Unit carried out that was geographically oriented was the publication of the *Atlas of the global distribution of schistosomiasis* (Doumenge et al., 1987). The development of the atlas was an attempt at providing global disease information in an easy-to-understand geographical format. The atlas includes maps of 76 schistosomiasis-endemic countries, indicating the location of villages and towns with schistosomiasis data. Also provided along with the maps are tables of detailed data on the actual prevalence level of schistosomiasis in each town or village, the case detection method used and the sources of the data, including the date of publication and authors of the study. The data included in the atlas come mainly from published studies in journals or in some instances, from the ministries of health of each country.

Unfortunately, there were several serious shortcomings to the atlas and its potential use as an analytical tool. First, since the maps were drawn by hand, they cannot be updated easily to show changes in the prevalence of the disease. The cost of redrawing the maps and reprinting the atlas would be prohibitively expensive. Second, although prevalence data were available to the reader, it was not easy to compare data from different countries, from different areas in the same country and from different time periods. Third, the data could not be used directly by the user for analysis since they first had to be entered into the computer.

14.4.2 Personal computer-based global database

In order to simplify the access to global epidemiological information on schistosomiasis, the SCH Unit next produced a PC-based schistosomiasis database (Yoon & Mott, 1991a). In addition to the disease information for each endemic country, the database contains information on government control activities, the names and location of the people responsible for the national control programme and information on how the control programme is financed. Information on water development projects in each country is also available from the database. The data included in the global schistosomiasis database are from the *Atlas of the global distribution of schistosomiasis* as well as the information provided by the ministries of health of endemic countries.

Although the database was designed to be used by programme managers for comparative purposes, in some instances the way the data are presented limits their usefulness. For example, each country has only a single figure listed for the prevalence. As a result, the user cannot see the distribution of the disease within a country. Unfortunately, with this database, there is no simple way of presenting more detailed information on each country without overwhelming the user with large tables of numbers. The best way to overcome this limitation is to use some method of geographical presentation and analysis.

14.5 GIS activities in the Schistosomiasis Unit of the Division of Control of Tropical Diseases of WHO

GIS improves on the shortcomings of the paper-based atlas and takes the computer-based database concept several steps further. As a preliminary step to incorporating the schistosomiasis data into a GIS, simple PC-based mapping software was purchased to see whether it would meet the criteria of an easy-to-use yet powerful analysis tool. After the data were entered into the computer and a number of maps were digitized, the data from a number of different countries were examined. Using the software, maps with point data overlays indicating the prevalence of schistosomiasis were generated. In addition, a filtering function was used to display the points satisfying a given criteria, such as prevalence levels above a certain number.

Although the maps were quite useful, it became evident that the software used, MapInfo (MapInfo Corporation, Troy, New York, USA), was not powerful enough to perform more sophisticated analysis. The software could display point and boundary data, but it could not take the data and perform statistical analysis, such as distribution and averaging. In addition, the software did not have a built-in modelling language nor have the capacity to perform multiple-database queries or complex data and overlay analysis. After a number of different types of GIS software were reviewed, a more powerful PC-based GIS was selected to carry out the next stage of the investigation.

Although several important criteria were established for the next phase of the project, the most important was the ease of use of the software, both for entering and accessing the data. The WHO SCH Unit did not have the resources to commit a full-time person to run the software, nor was it able to purchase a new hardware platform for GIS. In addition, the software had to be able to import the data already entered and digitized for MapInfo. With these criteria in mind, the software selected for the next phase was SPANS (Tydac Technologies, Canada). SPANS is a PC- and work station-based product with an easy-to-use menu and has a suite of sophisticated functions for the integration, analysis and modelling of geographical data. It requires OS/2 on the PC.

The maps used in the atlas were unique because they were drawn specifically for use in the atlas. This meant that the maps were not standardized to any particular scale. For example, an atlas page may contain a map of China, covering an area approximately 5,000 km by 4,000 km while another page may contain a map of St Lucia, covering a 30 km by 60 km area (Maps 14.1 and 14.2). Even then, some of the larger countries showed noticeable distortion when displayed on the screen. In addition to the national boundaries, districts within each country, major roads, major rivers, major dams, agricultural areas and villages and towns with available schistosomiasis data were digitized.

After the maps were digitized, schistosomiasis data for each country were entered into a dBase-compatible file. A file was created for each country, with fields that listed the name of the districts, towns, schistosomiasis prevalence, the source of the data and the authors and the title of the publication. The files were then prepared for geocoding using MapInfo, which involved the creation of additional fields for x and y coordinates.

Once the files were prepared for geocoding, the software was used to locate the villages and towns on the maps and the corresponding data were entered in the database.

For some of the countries, different symbols were used to display the change in prevalence over a period of time. By categorizing the amount of change and using different colours and sizes for the symbol, it was possible to easily pinpoint the areas with an increase in prevalence. This information will be valuable for programme managers who will closely examine the areas for possible causes of the increase.

In another exercise, data from a country were used to demonstrate the clustering effect of high prevalence areas. First, the major agricultural development areas, as defined in the atlas, were digitized, along with the villages and towns. The prevalence data from each village or town was grouped into categories, and differently sized icons were assigned to each group. By overlaying the prevalence data icons on top of the agricultural development area map, it was easy to demonstrate that the locations with high prevalence tended to cluster in or around the agricultural development areas.

These exercises and initial experiments with GIS have shown promise in dealing with the dynamics of schistosomiasis epidemiology. Some additional exercises may be useful in further defining the exact role GIS can play in the control of schistosomiasis and other tropical diseases.

14.6 Conclusion and future goals

In order for data to be shared among various groups involved in schistosomiasis control, appropriate hardware and software have to be found for each level of the hierarchy. The levels of the hierarchy include, but are not necessarily limited to: 1) WHO headquarters in Geneva, Switzerland, 2) the WHO regional offices, 3) the ministry of health in each country, 4) district or regional offices in charge of schistosomiasis control and 5) field teams. Once SPANS was chosen as the software to use at the headquarters level, the SCH Unit sought to identify a system that could be used at the country level by the schistosomiasis control teams in Africa, Asia, the Middle East and South America. The Unit looked for software with a simple user interface and minimal hardware and software requirements. Although the Macintosh environment is more user-friendly than the IBM MS and PC-DOS environment, the IBM-PC platform was chosen because of its availability in developing countries.

Map 14.1: Distribution of *Schistosoma haematobium* and *Schistosoma mansoni* in Egypt

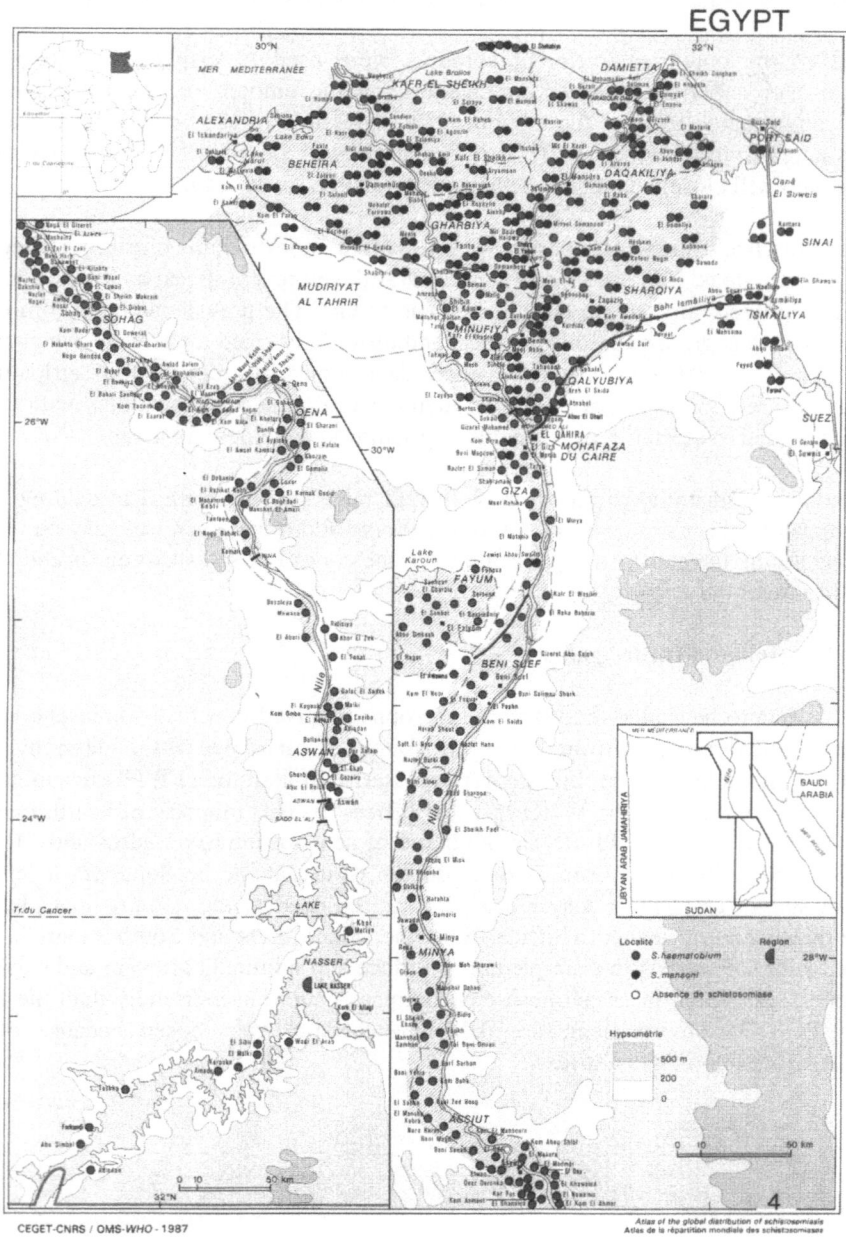

Source:Doumenge et al. (1987) <u>Atlas of the global distribution of schistosomiasis</u>

Map 14.2   Distribution of *Schistosoma japonicum* in China and Japan

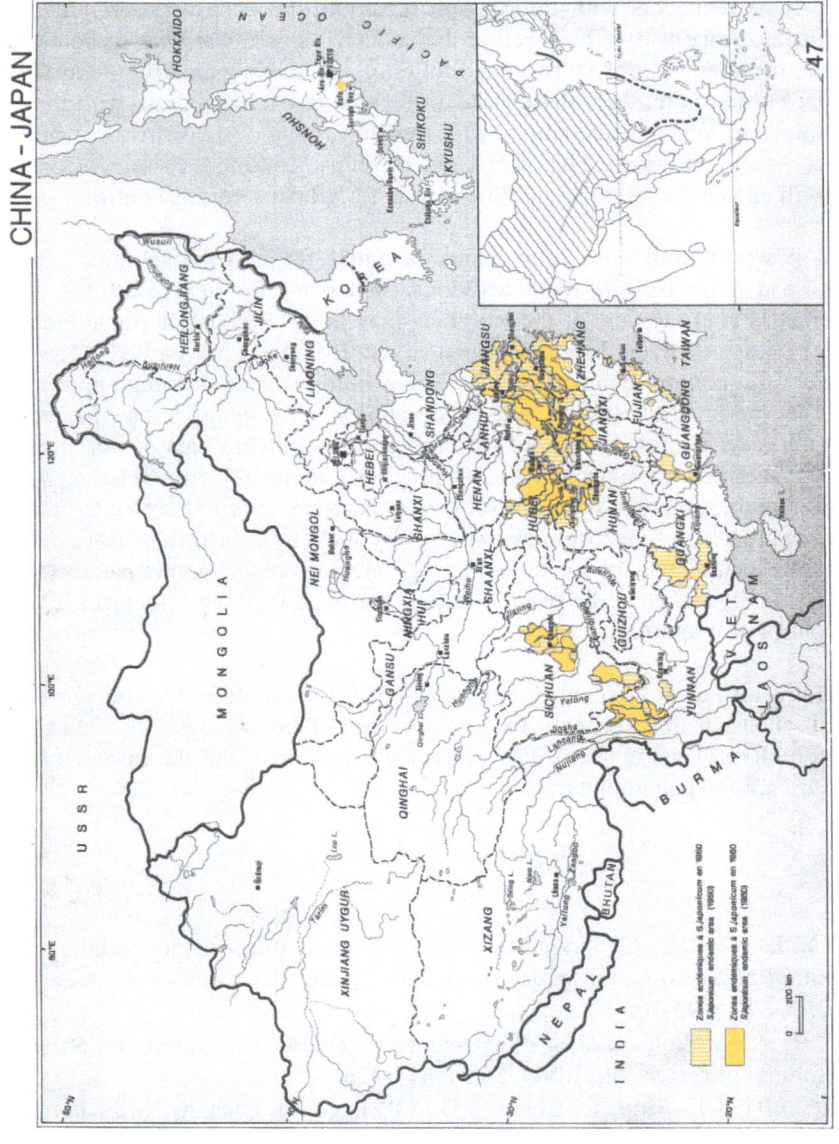

Source:Doumenge et al. (1987) Atlas of the global distribution of schistosomiasis

The main criteria for the software to be used at the country level is the ability to provide the programme managers with information regarding the control programmes in their own countries. Since it is to be based in the country, the software has to be easy-to-use and robust, meaning that the software is foolproof and stable enough to be used by non-specialists. Furthermore, the software has to be able to exchange data with the SPANS software in use at WHO headquarters. Unfortunately, none of the software available met all the necessary conditions. The SCH Unit, therefore, continues to search for a proper tool that will enable the hierarchical flow of data for schistosomiasis control.

GIS is a powerful tool with tremendous potential for assisting control programme managers and scientists in their activities. What is the next step for GIS and schistosomiasis control? One of the activities may be to establish a viable hierarchy of data flow between different levels of management, from the level of WHO headquarters to the level of in-country control projects. This would mean identifying appropriate GIS software that can be used at the various levels and establishing guidelines for the kind of data and the level of detail required for each level. Allowing easy integration of data from different levels of management, from global to local, will assist in delivering control programmes that are most cost-effective. Another activity may be to take a more integrated approach to schistosomiasis control by looking at factors that are outside the typical public health areas by including ecological and socioeconomic parameters in the model. Eventually, GIS can be used as a tool that will facilitate the success of multi-disease control programmes.

However, it is also clear from our initial experiences that, in order for GIS projects to be successful, attention must be paid to the collection of both geographical and attribute data, the problem of aggregating data of varying resolution and the necessary training and education of the potential users.

References

Barreto, M.L. (1991). Geographical and socioeconomic factors relating to the distribution of *Schistosoma mansoni* infection in an urban area of north-east Brazil. *Bul WHO*, **69**(1): 93-102.

Dalton, P.R. & D. Pole (1978). Water-contact patterns in relation to Schistosoma haematobium infection. *Bul WHO*, **56**(3): 417-426.

Doumenge, J.P., K.E. Mott, C. Cheung, D. Villenave, O. Chapuis, M.F. Perrin & G. Reaud-Thomas (1987). *Atlas of the global distribution of schistosomiasis*. Bordeaux, Presses Universitaires de Bordeaux.

Farooq, M., J. Nielsen, S.A. Samaam, M.B. Mallah & A.A. Allam (1966). The epidemiology of *Schistosoma haematobium* and *S. mansoni* infections in the Egypt-49 project area. 3. Prevalence of bilharziasis in relation to certain environmental factors. *Bul WHO*, **35**: 319-330.

Meade, M., J. Florin & W. Gesler (1988). *Medical geography*. New York, Guilford Press.

WHO (1985). *The control of schistosomiasis*. Report of a WHO Expert Committee. Geneva, World Health Organization, Technical Report Series No. 728.

Yoon, S.S. & K.E. Mott (1991a). Dr Schisto, an information stack for Zoom.

Yoon, S.S. & K.E. Mott (1991b). Global schistosomiasis database: practical considerations in the design of a user-friendly database. *Methods Informatics Med*, **30**: 127-131.

Steven S. Yoon
Miriam Hospital
Department of Geographical Medicine
164 Summit Ave.
Providence, RI 02906
United States of America

15 GEOGRAPHICAL INFORMATION SYSTEMS AND SPATIAL EPIDEMIOLOGY: MODELLING THE POSSIBLE ASSOCIATION BETWEEN CANCER OF THE LARYNX AND INCINERATION IN NORTH-WEST ENGLAND

Anthony C. Gatrell and Christine E. Dunn

Abstract

We consider some of the spatial statistical and epidemiological literature on tests for the clustering of disease events. We then turn to a consideration of the possible health effects resulting from exposure to the products of waste incineration, since proximity to such incinerators may give rise to a locally elevatied incidence of disease. One possibility raised is incidence of cancer of the larynx. After reviewing the epidemiology of this disease, especially its possible links to environmental contamination, we look at the relation between laryngeal carcinoma and proximity to both hospital and industrial waste incinerators. Particular interest centres on analysing these data using methods from the theory of spatial point processes. We suggest how to embed such methods into a conventional geographical information system environment.

15.1 Introduction

There is a rich tradition of geographical research aiming to understand the links between environment and disease (Learmonth, 1988). With the advent of concern about possible ill effects from exposure to ionizin gradiation, from air pollution and from poor water quality, research by medical geographers has focused recent attention on industrial and chemical environments as sources of exposure to environmental pollution. Attempts are being made to assess the possible health consequences of such pollutants. With the emergence of geographical information systems (GIS) as an enabling technology, the growing maturity of quantitative geography (especially statistical modelling) and the increasing availability of spatially referenced epidemiological data, epidemiological research conducted by geographers has expanded substantially since the late 1980s. Inevitably, and quite properly, much of this work involves collaborative efforts between geographers with analytical expertise and medical scientists with their clinical and public health backgrounds (see Gatrell et al. (1991a) and Openshaw et al. (1987) for examples of such collaboration).

The present chapter seeks to do the following. First, it reviews briefly some of the spatial statistical and epidemiological literature on clustering. It then discusses one possible source of environmental pollution, incineration of waste. Dispersion of potentially toxic contaminants away from such point sources may lead to localized exposure and locally

M. J. C. de Lepper et al. (eds.), The Added Value of Geographical Information Systems in Public and Environmental Health, 215–235.
© *1995 Kluwer Academic Publishers.*

raised incidence of chronic disease. Examining this hypothesis forms the basis of subsequent empirical work. Since this empirical work focuses on one, fairly rare, respiratory cancer, that of the larynx, we look next briefly at the epidemiology of the disease, in particular at any evidence suggesting that environmental pollution might be implicated as a risk factor. Using data for parts of north-west England, we employ GIS methods to define our own areas of possible exposure around incinerators and model the incidence of laryngeal cancer around these and a set of control sites. Because of the inherent arbitrariness of areal units, however, we offer an alternative approach based on the theory of spatial point processes, indicating how this can be embedded in a GIS framework.

15.2 Spatial epidemiology and the detection of clusters

Spatial epidemiology can be defined as a set of "studies of disease causation and prevention which adopt a distinctly analytical spatial perspective" (Thomas, 1990, p. 1). Much of this work has its origins in studies of disease diffusion (which, surprisingly, does not seem to have attracted much attention from GIS specialists), such as that by Cliff et al. (1986). Recently, however, much of what might be called spatial epidemiology has focused on attempts to detect or model spatial non-randomness in epidemiological data.

Disease data available to the spatial epidemiologist can take two forms. First, and more commonly, we have data aggregated to a set of areal units; maybe counties, districts, or, in Great Britain, electoral wards. We might then wish to know whether the disease incidence or prevalence in any area was statistically significantly higher than in other areas. Second, we might have a set of individual records, including sufficient information that allows us to represent these individuals as a set of point events. We may then wish to ask whether these data give evidence of departure from non-randomness; if so, we might then want to know where such clusters lie in space or we might want to fit some spatial model to the data. We summarize some of the literature on both area-based and point-based approaches. Regardless of whether we use areas or points, however, we should distinguish between tests for clustering as a phenomenon (i.e., that seek to establish whether there is any evidence of clustering in the data) and tests that seek to detect clusters (i.e., where hot-spots of disease might lie); see Besag & Newell (1991) for further comment on this distinction.

Traditionally, area-based approaches have sought to show whether the observed count of cases in an area is significantly different from the expected number; thus, Poisson probabilities are frequently computed (Muir et al., 1990). Alternatively, measures of relative risk, such as observed/expected ratios or standardized incidence ratios, are computed. A battery of tests is available to detect whether there is any clustering of areas with similar risk. Such autocorrelation tests (Cliff & Haggett, 1988) have been used by geographers for many years and are now finding their way into GIS via links made to proprietary systems (Kehris, 1990; Ding & Fotheringham, 1992).

Recently, researchers have expressed concern that simply mapping P values or standardized ratios is inappropriate, since low expected counts can give inflated measures of relative risk (Besag & Newell, 1991). The observed number of cases will often vary dramatically from area to area and the precision of the estimate in each area varies; we therefore need some way of taking this into account. Clayton & Kaldor (1987) have proposed methods for solving these problems. One method is to modify the estimate of risk in an area according to the mean relative risk in the study region as a whole. An alternative, spatial approach is to smooth the estimate of risk in an area so that its value is partly a function of values in surrounding areas; this is akin to recognizing some autocorrelation structure in the data. We look forward to seeing these so-called empirical Bayes methods adopted as part of the standard armoury of spatial analysis in a GIS framework (see Cartwright et al. (1990) for an application to leukaemia data).

Other work using areal units has adopted a more focused approach (to use the expression of Besag & Newell 1991), such that a hypothesized point source of pollution is defined and attempts are made to see whether risk in a set of areal units declines with distance. Small areas, such as wards or enumeration districts in Britain, are used in order that expected counts (using age-sex disaggregated population data) may be obtained. Examples include work by Bithell & Stone (1989) and Alexander et al. (1991a), much of it generated in response to the debate over links between reprocessing of nuclear wastes at Sellafield in Cumbria and the incidence of childhood leukaemia. Statistical approaches are summarized and discussed in Hills & Alexander (1989). One problem with all this work, however, is that the results are sensitive to the given configuration of boundaries and to the distance zones chosen for analysis; redefining these may lead to very different maps and different inferences. For these reasons, much contemporary analysis is now stressing the virtues of approaches based on analysing the configuration of point events.

As with area-based approaches, we can distinguish between methods used simply to detect the existence or otherwise, of clustering and those that seek to identify the locations of such clusters. Among the second set of approaches, we can further distinguish methods that search, on an inductive basis, for the locations of significant clusters from methods that posit a putative point source of pollution and then test to see whether disease cases are clustered around it.

Cuzick & Edwards (1990) and Diggle & Chetwynd (1991) have devised point-based approaches for testing for the existence of clustering. Cuzick and Edwards take a set of controls drawn from the population at risk and then derive statistics according to whether the nearest neighbours of cases are other cases or controls; clustering may be said to exist if the observed number of case-case pairs is significantly higher than that expected under the null hypothesis of random intermixing of cases and controls. Diggle & Chetwynds approach requires the estimation of a K-function for both cases and controls; this is a measure that describes the relative positions of pairs of points. Subtraction of the two functions yields a third that will reveal evidence of case clustering relative to controls. Gatrell et al. (1991b) have applied the method to data on motor neuron disease in Lancashire and South Cumbria.

Among methods that seek to locate significant clusters on a map, one of the best known is the geographical analysis machine (Openshaw et al., 1988), though this is not strictly point-based as it requires areal data in order to compute expected counts. In essence, it works by defining a circle of fixed radius over each point of a square grid, counting the number of cases falling within each circle and computing an expected count for each circle; if the observed count is significantly in excess of the expected the circle is plotted on a map. The procedure is then repeated for an increasing range of circle sizes, leading to some striking visual displays. Besag & Newell (1991) have sought to set the geographical analysis machine on a firmer statistical footing.

Diggle (1990) has devised a new model for assessing whether a disease of low incidence has an elevated incidence in the vicinity of a suspected point source of pollution. Since it is used in Section 15.6 below we discuss it in some detail here. It assumes that the data are available for a set of point locations, represented by a pair of coordinates (x_1, x_2) or, more compactly, as a vector (x).

The model postulates that the local intensity (density of disease cases), denoted as $\lambda (x)$ is a function of the following: first, the overall incidence of the disease in the region of interest, which we may denote as ρ; second, the background population, $\lambda_0 (x)$; and third, some function that describes the way in which incidence is thought to vary with distance, $f(x, \alpha, \beta)$. Formally:

$$\lambda (x) = \rho\lambda_0(x) \ f(x, \alpha, \beta)$$

$$(1)$$

We may parameterize f(.) as:

$$f(x, a, b) = 1 + \alpha \exp (- \beta \ d^2)$$

$$(2)$$

where d is distance from the hypothesized point source and α and β are parameters to be estimated. Clearly, interest centres on the parameters α and β (an intercept and distance decay term, respectively), for if they are not significantly different from zero then $\lambda(x)$ simply equals the intensity of background population, scaled by ρ. This scenario represents the null hypothesis of no distance decay effect. To the extent that α and β are significantly different from zero, the alternative hypothesis of an association with the putative point source is accepted.

The model is estimated using maximum likelihood methods (Diggle 1990; Diggle et al., 1990). Clearly, in order to fit the model we need data that allow us to represent both the intensity of the disease cases and that of the background population. In the empirical work described later, we consider what information is required on the latter; it may be appropriate to use data on other diseases as controls. But regardless of what it comprises (and the model assumes it is point-based information) how do we estimate intensity? The method Diggle proposes uses what is known as kernel estimation, which in simple terms means the estimation of a continuous function from discrete data (see Silverman (1986), for a most readable introduction). Here, the estimate of the intensity of the controls or background population at any point on a grid is a weighted sum of the distances between that location and the set of controls. Diggle (1990) discusses the choice of weights and how this estimate of intensity may be smoothed optimally.

The model generates a log-likelihood function, $L(\alpha, \beta)$ of the parameters, and fitting the model gives estimates of these parameters that maximize this log-likelihood. In other words, given the empirical data, what are the values of α and β that are most likely to have given rise to those data? We may then compare $L(\alpha, \beta)$, the value of the log-likelihood given α and β, with $L(0, 0)$, the value of the log-likelihood when α and β are zero (the null model). Twice the difference between these two values of L is evaluated according to the chi-square distribution with two degrees of freedom. If this exceeds the critical value, then the null hypothesis is rejected.

15.3 Incineration and health

It would seem that methods such as those discussed above could be of considerable use to researchers interested in assessing whether exposure to hazardous wastes is associated with ill health. This postulated association has been the focus of a considerable body of work (reviewed critically in a recent report by the British Medical Association (1991). The majority of this empirical work has looked at exposure to contaminated groundwater. Of interest here, however, is the inhalation of toxic substances as a result of waste incineration. Such research has been given new urgency inasmuch as landfill is being recognized as unsuitable for much hazardous waste disposal; new capacity is being sought for incineration, resulting in many new planning applications to construct plants designed to burn chemical wastes at high temperatures. Such applications have, inevitably, generated public concern over possible health implications. To what extent are such concerns justified?

Since incineration is the controlled combustion of waste products ranging from garden refuse to human remains, a full examination of the process requires consideration of: chemical waste incinerators; municipal incinerators; sewage sludge incinerators; hospital incinerators; and even crematoria. No comprehensive study of the possible health effects of any of these has been attempted; certainly, there is a notable absence of any good empirical studies. What evidence there is on health effects is therefore largely indirect, in that the emissions of such facilities have been quite well characterized; the uncertainty arises over the possible health responses to such doses.

Emissions from incinerators that cause concern arise as a result of incomplete combustion; if the temperature is not sufficiently high or if the material is not given adequate time to burn, then toxic chemical wastes will not be sufficiently destroyed and new chemical compounds may form. In municipal incinerators, for instance, organics such as polychlorinated dioxins and furans have been measured. Some of these are known to be extremely toxic to animals, though their danger to humans is still the subject of disagreement (Bailar, 1991). Acid gases, such as hydrochloric acid, and toxic metals (e.g., arsenic, cadmium, mercury, nickel and lead) are also known to be emitted by some incinerators.

Much of the debate concerning incineration and health was prompted by concern about possible raised incidence of congenital eye malformations around two chemical waste incinerators, one in Scotland (now closed) the other in South Wales (Gatrell & Lovett, 1991). Two official studies found no evidence to support the anecdotal evidence, but it is clear from other epidemiological work (summarized in British Medical Association 1991, pp. 132-133) that there are serious concerns about under- and mis-reporting of cases. In addition, around the Scottish plant there was evidence of high levels of twinning, both in cattle and human populations, and the suggestion was made that this is linked to the release of chlorinated organic compounds (Lloyd et al., 1988).

Crematoria have also been the subject of recent concern. The combustion of wood and polyvinyl chloride (used in coffin linings) is known to produce the most toxic of the dioxins (2,3,7,8 - TCDD), while mercury used in amalgam fillings has generated emissions of potentially hazardous mercury vapour from the flues (Hay, 1986; Mills 1990; Kunzler & Andree, 1991). Of course, whether these are serious is a moot point, given the low volumes of material disposed of, at least compared with chemical and municipal waste incinerators.

Of particular interest in view of some of the empirical work described below are hospital incinerators. Since hospital wastes comprise a high proportion of chlorinated plastics, and since it is recognized (Her Majestys Inspectorate of Pollution, 1990) that hospital incinerators do not consistently burn at sufficiently high temperatures to destroy the wastes, we can expect stack emissions and ash to contain dioxins, furans and other organic compounds, as well as some inorganics such as mercury. A study conducted by the National Society for Clean Air (NSCA, 1988) confirmed that many environmental health officers in Britain were worried about health impacts.

15.4 Epidemiology of cancer of the larynx

The carcinogenic properties of toxic metals emitted by incinerators are well known, though the long-term effects of the chlorinated organics are less clear. Given the work described below on seeking to model the relationship between proximity to incinerators and a rare respiratory cancer, it is important to review briefly what is known about the epidemiology of this disease.

There seems little doubt, on the basis of several epidemiological studies (Rothman et al., 1980; Burch et al., 1981; Guenel et al., 1988; Kleinsasser, 1988), that the main risk factors implicated in cancer of the larynx are smoking and alcohol consumption. Some work (Guenel et al., 1988) suggests that the effect of both factors is synergistic; in other words, the risks of developing the cancer multiply when both alcohol and tobacco are consumed in excess.

What evidence is there of occupational or other external environmental exposure to carcinogens? In particular, what evidence is there of exposure to the products of incomplete combustion? Lynch et al. (1979) found a raised incidence of the cancer among those manufacturing isopropyl alcohol; the high concentrations of diethyl sulfate were implicated as the carcinogen. In a study in Buffalo of patients admitted to hospital in a ten-year period, Viadana et al. (1976) examined the relationship between all cancers and those exposed to inhalation of chemicals (such as barbers, leather workers, painters and chemical operatives) or combustion products (the latter group including bakers, bus drivers, and firefighters, among others). There were significantly elevated risks for those in the former group but not the latter; the risk for barbers was particularly high and thought to be linked to inhalation of fumes from cosmetics. No workers employed in the waste disposal industry were investigated. Flanders et al. (1984) detected elevated risks in some textile workers but could not ascertain with any precision the substances to which they had been exposed; in addition to fibres, these might have included machine oils and even compounds of nickel, chromium or cadmium (Flanders et al., 1984, p. 31).

Asbestos fibres are also thought to be a causative factor (Stell & McGill, 1975) and it seems plausible that poor disposal of asbestos waste by incineration might release asbestos fibres into the atmosphere. Although there are likely to be confounding variables, such as smoking, there is evidence from some studies (Freifeld, 1977) that individuals with no history of smoking but exposed to asbestos develop the cancer. Burch et al., (1981), in a carefully designed case-control study in Ontario, found evidence to support the role of asbestos exposure, though exposure to nickel did not appear to be significant. Their work also suggested elevated risks for those who had worked in foundries and metal-processing and had been exposed to fumes.

In the most thorough synthesis available of the epidemiological literature, Kleinsasser (1988) refers to the role of polycyclic aromatic hydrocarbons (PAH). These are known carcinogens whose presence in tar from cigarettes is the likely explanation for the etiological role of cigarette consumption. He notes that PAH are produced in several industrial processes, being by-products of all sorts of combustion. Interestingly, a plausible biochemical explanation of the possible association between PAH and related hydrocarbons, and laryngeal cancer has emerged. These hydrocarbons are activated in cells by an enzyme called aryl hydrocarbon hydroxylase, thereby causing changes in DNA. It transpires that the inducibility of this enzyme is substantially raised in patients with laryngeal carcinoma (Trell et al., 1976).

It is quite clear from this brief review that a study of the epidemiology of the disease, especially from a geographical perspective, is fraught with difficulties. First, the disease is one of low incidence and any attempt to study it on a small scale will result in low numbers. Second, any attempt to examine a relationship to putative sources of pollution needs to take account of the range of confounding variables (most significantly, alcohol and tobacco consumption). Third, defining exposure is extremely difficult; it may be easier in the workplace, but simply using place of residence as a surrogate is far from perfect. Last, the disease reaches its peak incidence among those aged 55-65 years and most of the individuals will have moved, perhaps many times, before settling in the residential locations used as the addresses at time of diagnosis (and used in later epidemiological analyses in the absence of detailed life histories), locations that may have little or nothing to do with exposure.

Of course, all of these problems are familiar in epidemiological work. We can do our best to minimize them, but many will remain to cloud investigations and subsequent interpretation. With these caveats in mind, we proceed to some empirical analysis of our own. Although no study has been reported that implicates working in or living near to, an incinerator as a risk factor for laryngeal cancer, it seems quite reasonable, on the basis of some of the carcinogens identified earlier, to infer that such exposure might be implicated. At the very least, we see no reason to believe, on an a priori basis, why the hypothesis is untenable.

15.5 Modelling cancer of the larynx in areal units

We obtained from the North West Regional Cancer Registry in Manchester data for north-west England on 37,900 cases of cancer of the larynx and lung, registered over a period of 11 years (1974-1984). Data included the cancer site, sex, age, year of notification, and unit postcode. The latter is used for geo-referencing; using the Central Postcode Directory, a machine-readable file of about 1.6 million records, each postcode can be matched to an Ordnance Survey grid reference that in England and Wales, has a resolution of 100 m (Gatrell, 1989; Gatrell et al., 1991a). We also obtained 1981 census data for all electoral wards in our study area; these were used to compute the expected incidence of each cancer by area. We explain shortly why we obtained data on cancer of the lung as well as larynx.

After the above data were imported into ARC/INFO, a point-in-polygon overlay operation on the coverage of cancers and the ward polygons allowed us to attach a ward code to each case. Programs were written in INFO that counted observed cases by ward and calculated expected numbers from age-specific rates for north-west England, together with standardized incidence ratios (SIRs) and approximate confidence intervals. We have reported elsewhere (Gatrell & Dunn, 1990) on the results, which showed evidence of spatial autocorrelation (Goodchild, 1985; Kehris, 1990) among SIRs for men but the pattern for female laryngeal cancer was essentially random. Fitting a statistical model to the data, whereby incidence was regressed on the proportion of the population in social classes 4 and 5 (a measure of social deprivation, used here as a crude proxy for smoking prevalence), suggested that this explanatory variable did account for a significant proportion of the variance.

However, none of this requires a GIS. We now seek to exploit the capabilities of a GIS in investigating the hypothesized link to incineration. To do this, we constructed a database of 44 hospital incinerators in the region. We also selected, at random, a set of 44 control sites. Although arbitrary, we defined areas possibly at risk from the incinerators to be circles of radius 2 km (dispersion calculations in other studies suggest that the highest levels of dioxin concentration occur about 1 km from the source of discharge). We recognize that this is crude; a circle overlay operation is simple in ARC/INFO, but implementing a full dispersion plume model for each site and then intersecting this with the cancer data has not yet been attempted (see Kingham (1992) for progress in this direction).

For both sets of circles we performed the following operations. First, each set of circles was overlaid separately with both the full point coverage of observed cases and the polygon coverage of ward boundaries. The point-in-polygon operation gave a count of cases by area (circle). The second overlay clipped out the wards or parts of wards, that each circle intersects. ARC/INFO calculated the area of each ward falling within the circle boundary. An estimate of the expected number of cases for each circle was found by multiplying the proportion of each ward area lying within the circle by the expected number for each ward; summing these values gave the expected number of cases for each circle. Estimates of social class for the circles were obtained in a similar manner. As an example, we show SRRs for the 44 hospital sites, for men only (Map 15.1).

Map 15.1: Standardized registration ratios (SRR) with confidence limits for 2-km radii around hospitals, for larynx cancer in men in north-western England

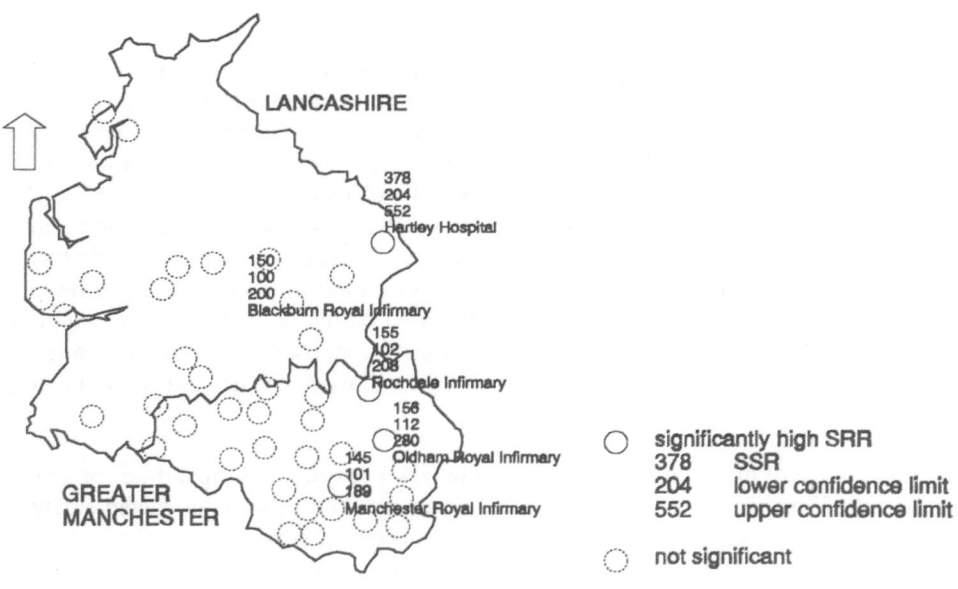

We now fit a model to the data, where the observations comprise the 44 hospitals and 44 control locations. For each observation we have the following information: an observed count of the number of cases (men and women) within the circle; an expected number; the estimated population in social classes 4 and 5 (manual occupations); and a dummy variable or factor, according to whether the circle represents a hospital or non-hospital site. Using ARCPLOT, it is simple to map a subset of the sites (e.g., hospitals only) on a shaded choropleth map of social class (Map 15.2).

Map 15.2: Percentage of population in social classes 4 and 5 in north-western England

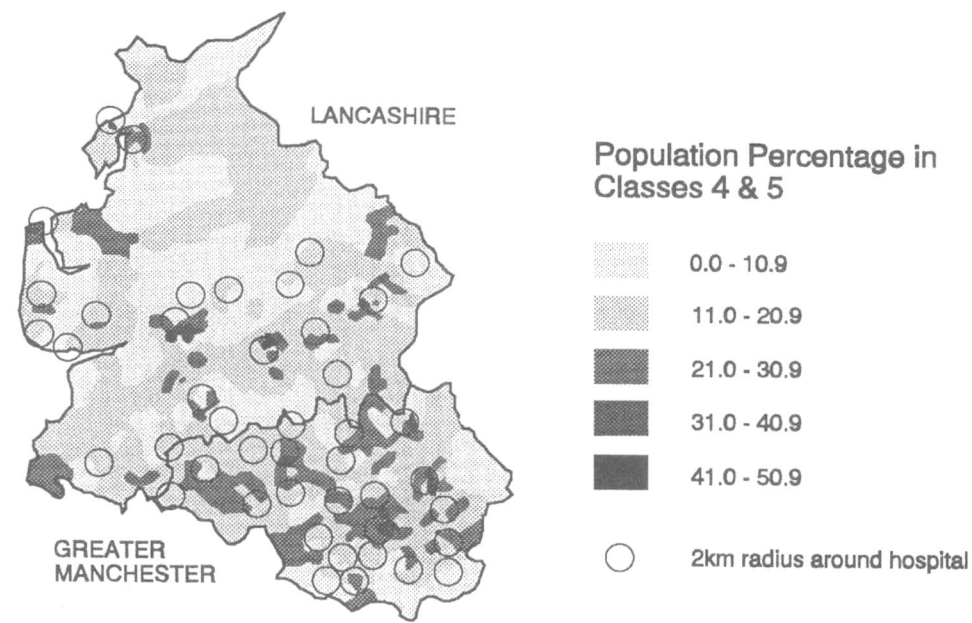

The model we fit belongs to the class of generalized linear models estimated using the package GLIM (Aitkin et al., 1989); specifically, we fit a Poisson regression model since we are dealing with small observed counts as our variable to be explained (Lovett et al., 1986). We begin by fitting a null model, associated with which is a measure of badness-of-fit known as scaled deviance. Variables are added in turn in order to effect a reduction in deviance. A chi-square test is used to assess whether the reduction in deviance is statistically significant. Estimates of the model parameters are evaluated using a *t*-test.

The results (Table 15.1) show that, having used the expected number of cases as an offset (in GLIM terminology), the social class variable is the next most important, giving significant reductions in deviance for both sexes. The addition of the factor (incinerator versus non-incinerator site) adds little or nothing to the explanation. As the parameter estimates reveal, the incinerator variable is not significant.

Table 15.1 Poisson regression results for hospitals and control locations

Men

Model	Deviance	Degrees of freedom	Reduction
Null	860.9	87	-
Offset	118.7	87	742.2
Social class	93.5	86	25.2
Incinerator	93.2	85	0.3

Model with parameter estimates (standard errors in parenthesis):
$$\ln Y = \ln E + 0.038 \text{ (social class)} - 0.043 \text{ (incinerator)}$$
$$\qquad\qquad (0.008) \qquad\qquad (0.083)$$

Women

Model	Deviance	Degrees of freedom	Reduction
Null	308.1	87	-
Offset	84.2	87	223.9
Social class	17.5	86	66.7
Incinerator	17.5	85	0.0

Model with parameter estimates (standard errors in parenthesis):
$$\ln Y = \ln E + 0.010 \text{ (social class)} - 0.600 \text{ (incinerator)}$$
$$\qquad\quad (0.002) \qquad\qquad (0.509)$$

Note: Y is the observed count of cases and E is the expected number of cases.

On the basis of these results, then, we cannot claim that proximity to a hospital incinerator is associated with an elevated incidence of laryngeal cancer. However, we have looked only at a single random sample of other sites. Moreover, other, non-specific illnesses may well be associated with such proximity, and further research should perhaps explore this.

15.6 Modelling cancer of the larynx as a spatial point process

Analysis by areal units, be these administrative or data collection zones such as electoral wards or buffer zones defined within a GIS, gives rise to some concern. The main worry is that the zones are inherently arbitrary; if the ward boundaries followed a different course or the exposure zones were given a different radius, a different set of results would emerge. We now offer an alternative approach, based on the theory of spatial point processes (Diggle 1983) and outlined above. Such an approach is possible here by virtue of the fact that we have a grid reference of 100-m resolution for each of our cases.

Rather than focus attention on one or more of the hospital incinerators, we use here as a putative point source of pollution an industrial waste incinerator. This facility operated between 1972 and 1980 at a site about 2 km south-west of the small town of Coppull in Lancashire. It was used for the disposal of a wide range of wastes, both liquids (including oils and solvents) and solids. The opening of the plant preceded the introduction of reasonably rigorous pollution regulations (the Control of Pollution Act from 1974) and concern was expressed at the time, and subsequent to the plants closure, about its environmental and health impact.

In earlier work involving one of the authors (Diggle et al., 1990) we fitted the model outlined in Section 2 to the distribution of laryngeal cancer in the Lancashire Health Authority, within which the plant was located. We summarize these findings briefly here before considering extensions.

As noted earlier, the model (equation 1) requires information on background population or (subject to a scale factor) the intensity of cases we would expect to observe if the null hypothesis of no distance decay effect is true. In Diggle et al. (1990) we used data on the distribution of a much commoner cancer, that of the lung. The distribution of this is shown in Figure 15.1 (there are 978 observations) and may be compared with that of laryngeal cancer (Figure 15.2). Note the visual aggregation of cases of laryngeal cancer near the incinerator; four cases lie within 1 km and five within 2 km of the site. This is out of a total of 58 cases in the 11-year period of study.

Figure 15.1:   Lung cancer cases in the Chorley-Ribble area

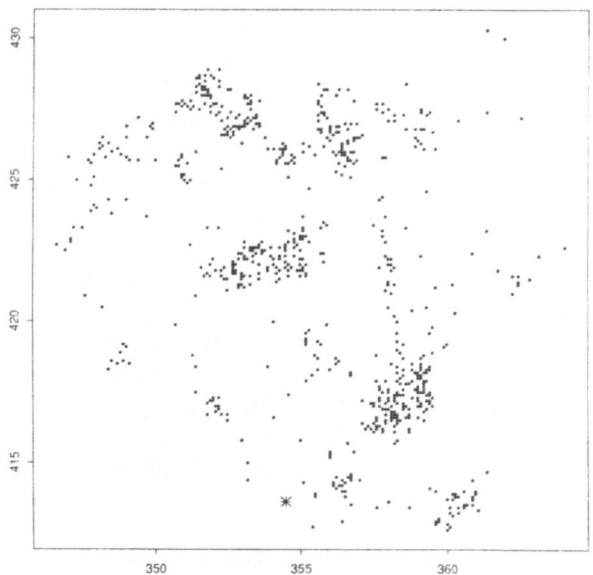

Figure 15.2:   Larynx cancer cases in the Chorley-Ribble area

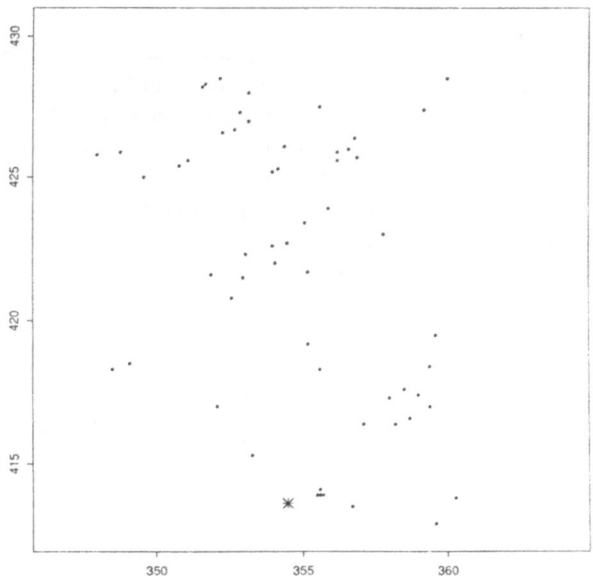

What justification is there for using another cancer case as a control? First, the age and sex distributions of the diseases are broadly similar. Second, smoking is a major risk factor for both. Smith et al. (1988) have reviewed at length the possibilities for using other cancer cases as controls in cancer epidemiology. Alternative controls are used below.

The plot of lung cancers seems adequately to mirror population density, with clumps corresponding to centres of population. This is perhaps made clearer in the map of smoothed lung cancers (Figure 15.3), produced using the kernel methods mentioned earlier.

Figure 15.3:   Kernel-smoothed density estimate for lung cancers

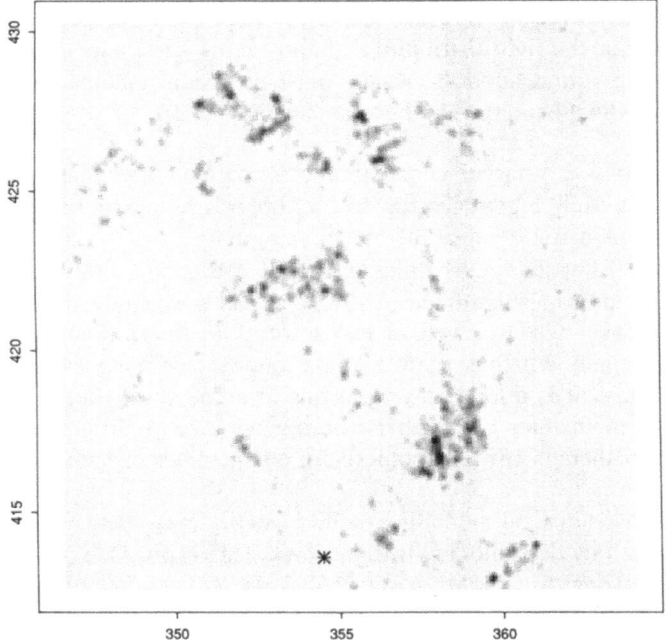

Fitting the model gave values of _ = 25.3 and β = 0.95, with L(_, β) = -394.59. The null log-likelihood is -399.36 and as twice this difference exceeds the critical value of χ^2 (*P*=0.05, 2 df), we can say that the null hypothesis is rejected and there is a significant association with proximity to the incinerator. However, as Diggle et al. (1990) noted, removal of two cases removes the significance of the fit; addition of another nearby increases its significance dramatically. The influence of small numbers is clear.

The authors did not go on to consider other measures of background population. We have fitted the model using another common cancer, stomach, as controls; there are 398 of these in the Lancashire Health Authority. As with the lung cancers, the distribution of these corresponds, in broad terms, to the natural heterogeneity of population distribution in this area. Fitting the model again gave a null log-likelihood of -347.21 and a maximum log-likelihood of -342.68; again, twice this difference is well in excess of the critical value.

Another possibility is to use the distribution of all unit postcodes as a surrogate for population distribution. In general, a unit postcode relates to about 15-17 households, though there will be substantial variation about this mean. Although this is rather crude (and we need ways of biasing the selection of a sample of these postcodes so that they match more closely the age distribution of the set of cases) we took all 4,389 unit postcodes in the region as controls. Again, our conclusion remains unaltered. The null log-likelihood is -486.44 and the maximized value is -482.16.

What of the distribution of larynx cancer in a wider region? Note that the incinerator lies in the south of our study region (Figure 15.2). Suppose we centre our region of interest on the site of the incinerator and fit the model using lung cancer as a control. Defining a region 30 km × 30 km gives 261 cases of larynx cancer and 5029 lung cancers. The model generates a null log-likelihood of -2224.50 and a maximized value of -2219.59. Twice this difference (9.82) is well in excess of 5.99, the critical value, leaving our conclusion unchanged. We thus seem to have quite convincing evidence of elevated incidence. Of course, it is a long way from this to argue for a causal role of pollution from this specific incinerator and further still to generalize to all incineration processes. But if nothing else, there is now a hypothesis for others to test elsewhere.

None of this spatial statistical modelling requires a GIS. However, it would seem useful to try to embed such models into a GIS framework. This is precisely what recent work at the north-west Regional Research Laboratory has done. The model we have outlined is written in standard FORTRAN. We have available to us the object code of ARC/INFO, written in the same language. One of our colleagues, Barry Rowlingson, has linked the code for the model with the ARC/INFO object code, such that the points for the cases and controls (already defined as ARC/INFO coverages) are accessed from the GIS. As a result, to run the model one simply issues a new ARC command together with parameters required by the model. The results appear in a separate window on the SUN SPARC work station. The procedure is fully interactive, so that users can specify with a cursor their choice of location of interest as a possible point source (e.g., a hospital incinerator, crematorium or nuclear reprocessing facility, depending on the research problem) and have it evaluated immediately. Clearly, some care needs to be exercized here; only locations of a priori interest ought perhaps to be evaluated. However, the model can identify locations of significantly elevated incidence and these may indeed be worthy of closer investigation by public health specialists.

Rowlingson has also generalized the model-fitting procedure to search automatically for other locations in the study region that may be significant. This involves fitting the model sequentially to a series of locations on a lattice. The results for the single health authority studied originally have shown that, while isolated pockets of laryngeal cancer occur elsewhere, the model yields a significant fit only in the immediate vicinity of the incinerator.

15.7 Conclusions and prospects for further work

Several extensions are being sought to this work. One is to develop the model along one or more lines. For instance, we might want to fit a model that posited the existence of multiple point sources of pollution; this is clearly of importance in major areas of industry, where it is potentially difficult to separate out effects due to one or more sources. Another, more pressing, extension is to add further covariates to the model. In the case of laryngeal cancer, for instance, we might wish to add some information available only at an aggregate areal level, such as census data on social class or some suitable surrogate for smoking behaviour. This we hope to report on in future work.

More generally, other work involving the north-west Regional Research Laboratory (Diggle & Rowlingson, 1991) has succeeded in embedding much of the spatial point process modelling into an interactive statistical environment, using the package (and, in effect, programming language) S-plus. In a windowing environment this allows a map (or maps) of the point events to be displayed, *K*-functions and other descriptions of the process(es) to be depicted in another window and, if relevant, smoothed maps of intensity in yet another window. Although it perhaps fails to meet the definition of a GIS in the strict sense, as a tool for spatial analysis in epidemiology it is truly impressive.

Lastly, we might ask questions about clustering in other kinds of spaces (Gatrell, 1991). For instance, suppose we were dealing with road traffic accidents. We would wish to know if there was clustering in the space defined by the road network (Bailey, personal communication, 1991) since, by definition, road traffic accidents must occur on the network. Similarly, appropriate control locations might need to be defined. Current research efforts are addressing this problem. If we can exploit these network structures in the GIS, then we will begin to see genuine links being made between spatial epidemiology and GIS.

Acknowledgements

We thank the Economic and Social Research Council and the Nuffield Foundation for financial support. The cancer data were made available through the kindness of the north-west Regional Cancer Registry. Bob Daly, Vangelis Kehris, Isobel Naumann and Barry Rowlingson gave invaluable support. Peter Diggle is thanked for inspiration and guidance on point process modelling.

References

Aitkin, M., D. Anderson, B.J. Francis & J. Hinde (1989). *Statistical modelling in GLIM.* Oxford, Oxford University Press.

Alexander, F.E., P.A. McKinney & R.A. Cartwright (1991a). The pattern of childhood and related adult malignancies near Kingston-upon-Hull. *J Public Health Med*, **13**: 96-100.

Alexander, F.E., T.J. Ricketts, J. Williams & R.A. Cartwright (1991b). Methods of mapping and identifying small clusters of rare diseases with applications to geographical epidemiology. *Geogr Anal*, **23**: 158-173.

Bailar, J. (1991). How dangerous are dioxins? *N Engl J Med*, **324**: 260-262.

Besag, J. & J. Newell (1991). The detection of clusters in rare diseases, *J R Stat Soc A* **154**: 143-55.

Bithell, J.F. & R.A. Stone (1989). On statistical methods for analysing the geographical distribution of cancer cases near nuclear installations. *J Epidemiol Community Health*, **43**: 79-85.

British Medical Association (1991). *Hazardous waste and human health*, Oxford, Oxford University Press.

Burch, J.D., G.R. Howe, A.B. Miler & R. Semenciw (1981). Tobacco, alcohol, asbestos and nickel in the etiology of cancer of the larynx: a case-control study. *J Nat Cancer Inst*, **67**: 1219-1224.

Cartwright, R.A., F.E. Alexander, P.A. McKinney & T.J. Ricketts (1990). *Leukaemia and lymphoma: an atlas of distribution within areas of England and Wales 1984-1988.* Leeds, Leukaemia Research Fund Centre, University of Leeds.

Clayton, D. & J. Kaldor (1987). Empirical Bayes estimates of age-standardized relative risks for use in disease mapping. *Biometrics*, **43**: 671-81.

Cliff, A.D. & P. Haggett (1988). *Atlas of disease distributions: analytical approaches to epidemiological data.* Oxford, Blackwell.

Cliff, A.D., P. Haggett & J.K. Ord (1986). *Spatial aspects of influenza epidemics.* London, Pion.

Cuzick. J. & R. Edwards (1990). Spatial clustering for inhomogeneous populations. *J R Stat Soc*, B, **52**: 73-104.

Diggle, P.J. (1983). *Statistical analysis of spatial point processes.* London, Academic Press.

Diggle, P.J. (1990). A point process modelling approach to raised incidence of a rare phenomenon in the vicinity of a pre-specified point. *J R Stat Soc*, A, **153**: 349-362.

Diggle, P.J., A.C. Gatrell & A.A. Lovett (1990). Modelling the prevalence of cancer of the larynx in part of Lancashire: a new methodology for spatial epidemiology. *In*: R.W. Thomas, ed. *Spatial epidemiol.* London, Pion pp. 35-47.

Diggle, P.J. & A. Chetwynd (1991). Second-order analysis of spatial clustering for inhomogeneous populations. *Biometrics*, **47**: 1155-1164.

Diggle, P.J. & B.S. Rowlingson (1991). *Spatial point process modelling in an interactive S-plus environment.* Technical report, Lancaster, Department of Mathematics, Lancaster University.

Ding, Y. & A.S. Fotheringham (1992). The integration of spatial analysis and GIS Comput. *Environ Urban Systems*, **16**: 3-19.

Flanders, W.D., C.I. Cann, K.J. Rothman & M.P. Fried (1984). Work-related risk factors for laryngeal cancer. *Am J Epidemiol*, **119**: 23-32.

Freifeld, S. (1977). Asbestos exposure and laryngeal carcinoma. *JAMA*, **238**: 1280.

Gatrell, A.C. (1989). On the spatial representation and accuracy of address-based data in the United Kingdom. *Int J Geogr Inf Systems*, **3**: 335-348.

Gatrell, A.C. (1991). Concepts of space and geographical data. *In*: D. Maguire, M. Goodchild & D. Rhind, ed. *Geographical information systems: overview, principles and applications.* London, Longman pp. 119-134.

Gatrell, A.C. & C.E. Dunn (1990). GIS in epidemiological research: analysing cancer of the larynx in north-west England. *In*: J. Harts, H.F.L. Ottens & H.J. Scholten, ed. *EGIS 90 proceedings.* Utrecht, EGIS Foundation pp. 346-355.

Gatrell, A.C. & A.A. Lovett (1991). Congenital eye malformations and incineration in Britain: a geographical analysis. *In*: R. Akhtar, ed. *Environ health.* New Delhi, Asish Publishing, pp. 419-430.

Gatrell, A.C., C.E. Dunn & P.J. Boyle (1991). The relative utility of the Central Postcode Directory and Pinpoint Address Code in applications of geographical information systems. *Environ Planning* A, **23**: 1447-1458.

Gatrell, A.C.. J.D. Mitchell, H.N. Gibson & P.J. Diggle (1991). Tests for spatial clustering in epidemiology: with special reference to motor neurone disease. *In*: C.F. Rose, ed. *New evidence in ALS.* London, Smith-Gordon.

Goodchild, M. (1985). *Spatial Autocorrelation.* Norwich, CATMOG, Geo Books.

Guenel, P., J.F. Chastang, D. Luce, A. Leclerc & J. Brugere (1988). A study of the interaction of alcohol drinking and tobacco smoking among French cases of laryngeal cancer. *J Epidemiol Community Health*, 42: 350-54.

Hay, A. (1986). Cremation and the environment. *Pharos: J Cremation Soc of Great Britain*, **52**: 16-25.

Her Majesty's Inspectorate of Pollution (1990). *Second annual report 1988-89*. London, HMSO.

Hills, M. & F.E. Alexander (1989). Statistical methods used in assessing the risk of disease near a source of possible environmental pollution: a review. *J R Stat Soc A*, **152**: 353-363.

Kehris, E. (1990). *Autocorrelation statistics within an ARC/INFO environment*. Lancaster, Lancaster University, North West Regional Research Laboratory, Research Report No. 16.

Kingham, S. (1992). *Respiratory health in Preston: a GIS approach*. PhD research in progress, Lancaster, Department of Geography, Lancaster University.

Kleinsasser, O. (1988). *Tumours of the larynx and hypopharynx*. New York, Thienne Medical Publications.

Kunzler, P. & M. Andree (1991). More mercury from crematorium chimney, *Nature*, **349**: 746-7.

Learmonth, A. (1988). *Disease ecology*. Oxford, Basil Blackwell.

Lloyd, O., M.M. Lloyd, F.L.R. Williams & A. Lawson (1988). Twinning in human populations and in cattle exposed to air pollution from incinerators. *Br J Ind Med*, **45**: 556-60.

Lovett, A.A., C.G. Bentham & R. Flowerdew (1986). Analysing geographic variations in mortality using Poisson regression: the example of ischaemic heart disease in England and Wales 1969-1973, *Soc Sci Med*, **23**: 935-943.

Lynch. J., N.M. Hanis, M.G. Bird, K.J. Murray & J.P. Walsh (1979). An association of upper respiratory cancer with exposure to diethyl sulphate. *J Occupational Med*, **21**: 333-341.

Mills, A. (1990). Mercury and crematorium chimney. *Nature*, **346**: 615.

Muir, K.R., S.E. Parkes. J.R. Mann, M. Stevens, A.H. Cameron, F. Raafat, P.J. Darbyshire, D.R. Ingram, A. Davis & D. Gascoigne (1990). "Clustering" - real or apparent? Probability maps of childhood cancer in the West Midlands Health Authority region. *Int J Epidemiol*, **19**: 853-859.

NSCA (1988). *Air pollution from crown property*. Brighton, National Society for Clean Air.

Openshaw, S., M. Charlton, C. Wymer & A.W. Craft (1987). A Mark 1 geographical analysis machine for the automated analysis of point data sets. *Int J Geogr Inf Systems*, 1: **335**-358.

Rothman, K.J., C.I. Cann, D. Flanders & M.P. Fried (1980). Epidemiology of laryngeal cancer, *Epidemiol Rev*, **2**: 195-209.

Silverman, B.W. (1986). *Density estimation for statistics and data analysis*. London, Chapman and Hall.

Smith, A.H., N.E. Pearce & P.W. Callas (1988). Cancer case-control studies with other cancers as controls. *Int J Epidemiol*, **17**: 298-306.

Stell, P.M. & T. McGill (1975). Exposure to asbestos and laryngeal carcinoma. *J Laryngol*, **89**: 513-517.

Thomas, R.W., ed. (1990). *Spatial epidemiology*, London, Pion.

Trell, E., R. Korsgaard, B. Hood, P. Kitzing, G. Norr & B.G. Simonss (1976). Aryl hydrocarbon hydroxylase inducibility and laryngeal carcinoma. *Lancet*. July 17: 140.

Viadana, E., I.D.J. Bross & L. Hoten (1976). Cancer experience of men exposed to inhalation of chemicals or to combustion products. *J Occup Med*, **18**: 787-792.

Anthony C. Gatrell
North-West Regional Research Laboratory
Lancaster University
Lancaster LA1 4YB
United Kingdom

Christine E. Dunn
Department of Geography
University of Durham
Durham DH1 3LE
United Kingdom

16 THE POTENTIAL ROLE OF GEOGRAPHICAL INFORMATION SYSTEMS TECHNOLOGY IN AIR TOXICS RISK ASSESSMENT, COMMUNICATION AND MANAGEMENT

T.J. Moore

Abstract
The California Air Toxics "Hot Spots" Information and Assessment Act of 1987 (Assembly Bill 2588) requires statewide implementation of a programme to identify carcinogenic and noncarcinogenic health risks associated with air toxics emitted by industry. Some of the programme's immediate objectives are to assess the health risks associated with the total emissions from each regulated facility, to identify the degree of public exposure and to notify any public potentially exposed to significant risk levels. The extent of the database will also allow regulatory agencies to eventually develop and evaluate risk management strategies. This chapter discusses the application of a geographic information system (GIS) during the preparation of an Assembly Bill 2588 multipathway air toxics health risk assessment. A case study provides insight on the development of a health risk assessment database and system integration with dispersion and risk assessment modelling elements. The value and potential of a GIS in the data analysis and management throughout the risk assessment process are emphasized. Finally, the potential use of a GIS to communicate risk and assess cumulative (regional) impacts for risk management decision-making is addressed.

16.1 Introduction

The California Air Toxics "Hot Spots" Information and Assessment Act of 1987, or Assembly Bill (AB) 2588, requires statewide implementation of a programme to identify potential carcinogenic and noncarcinogenic health risks associated with facility-specific air toxics emissions. The California AB 2588 multipathway air toxics health risk assessment programme is a unique state endeavour to inventory and assess the health risks associated with air toxics emissions and, when necessary, to notify the public of potential health risks.

California has some of the greatest pollution problems within the United States which, coupled with active environmental lobbies and regulatory agencies, has made the state a leader in the development of environmental technologies, standards and programmes. Thus, the AB 2588 risk assessment process is important to a wider audience, particularly due to:

(i)   the fact that no state or federal agency has required health risk assessments for such a broad range of individual sources;

M. J. C. de Lepper et al. (eds.), The Added Value of Geographical Information Systems in Public and Environmental Health, 237–262.
© 1995 Kluwer Academic Publishers.

(ii)   the increased environmental consciousness, throughout the United States, of the public and industry; and,

(iii)   the significant role of air toxics in the new Clean Air Act Amendments of 1990.

This aggressive action on the part of legislatures and regulators is increasing the size of the regulated community, the number of compliance requirements and the environmental database on air toxics emissions and air quality. Thus, improved information management techniques will be valuable to industry, environmental consulting firms, the public and regulatory agencies alike.

The functionality of a geographic information system (GIS) can greatly improve the quality and efficiency of spatial analysis and data management necessary for complex air toxics risk assessment. Also, GIS can:

(i)   enhance methods to perform alternative technical approaches;

(ii)   improve risk communication efforts; and,

(iii) provide an invaluable means to compile and analyse cumulative (multiple-facility) risk assessment data for risk management decision-making.

This chapter presents an introduction to the requirements of AB 2588 and the manner in which a GIS was used to complete an AB 2588 health risk assessment. Some implications of this technology to the possible expansion of the risk assessment methods are also presented. To limit the scope of discussion, this chapter focuses on carcinogenic risk assessment. The chapter continues with a discussion of the possible benefits of a GIS in risk communication and risk management.

16.2 The AB 2588 process

16.2.1 The regulatory programme

To place the role of GIS in a proper context, it is useful to review the AB 2588 process. The "Hot Spots" Information and Assessment Act is an aggressive and ambitious effort by the State of California to assess the potential health risks associated with air toxics emissions. The programme is broadly guided by policies and procedures disseminated by the California Air Pollution Control Officers Association (CAPCOA, a steering committee including scientists and policy-makers from 11 air pollution control districts (APCDs)), the California Air Resources Board (ARB) and the California Office of Environmental Health Hazard Assessment (OEHHA). While overall technical support is provided by ARB and OEHHA, the programme is administered by each APCD.

The regulatory compliance schedule takes a phased approach, which is dependent on the potential criteria type (total organic gases, nitrogen oxides, sulfur oxides or particulate matter) emissions from each facility. The three general phases of implementation are:

Phase I :   greater than 25 tonnes/year of criteria pollutants;
Phase II :   10-25 tonnes/year of criteria pollutants;
Phase III: less than 10 tonnes/year of criteria pollutants.

In general, facilities must initiate programme activities in either 1989, 1990 or 1991, depending on the facility classification (Phase I, II or III, respectively). Also, the legislation requires the operator of each facility to biennially re-evaluate and update, if necessary, the air toxics emissions inventory and health risk assessment.

The scope of programme activities includes the following:

(i)   preparation of toxics emissions inventory plans (TEIPs);
(ii)   preparation of toxics emissions inventory reports (TEIRs);
(iii)   district prioritization (high, medium and low), for risk assessment, of all facilities based on each facility emissions inventory;
(iv)   air toxics risk assessment (currently required only for high-priority facilities) within 150-180 days of prioritization;
(v)   ranking of facilities by predicted risk;
(vi)   public notification of significant risk levels.

16.2.2 The guideline risk assessment

The core of the programme is the health risk assessment. The risk assessment guidelines are disseminated by CAPCOA and revised as warranted. At this writing, the January 1992 edition of the guidelines are used in practice (CAPCOA 1992). The local APCDs can provide additional guidelines for facilities within their jurisdiction. The guideline risk assessment is designed to allow regulators to compare and rank facilities relative to each other based on the potential health risks associated with emissions.

The four basic components of the multipathway health risk assessment are hazard identification, dose-response assessment, exposure assessment and risk characterization. The following is a brief description of each element:

(i)   Hazard identification: the identification of potentially hazardous substances and the estimation of potential emissions.

(ii)   Dose-response: the evaluation of toxicological studies based on animal or human data, epidemiological research or other data, to develop carcinogenic unit risk factors or noncarcinogenic acceptable or threshold exposure levels.

(iii) Exposure: the identification of potential pollutant-specific pathways and the modelling of pollution transport through the environment. A multipathway analysis includes exposure through air, soil, water, plants, etc. The US Environmental Protection Agency (EPA) (1989) has described exposure assessment as "the determination or estimation (qualitative or quantitative) of the magnitude, frequency, duration and route of exposure".

(iv)   Risk: the assessment of potential long-term and short-term health risks associated with all pollutants by merging the information from each element above.

Within the guidance, three levels of effort for a risk assessment exist: screening, refined and alternative. A screening or refined level of effort includes many worst-case or conservative assumptions. A screening risk assessment treats emissions and meteorological conditions in a very simplistic manner. A refined risk assessment will use more realistic emissions or meteorological assumptions and data. Finally, an alternative risk assessment will generally use more plausible input parameter assumptions and a stochastic method to present the breadth of statistical uncertainty.

The GIS risk assessment case study presented herein will be based on procedures and systems developed to perform refined health risk assessments. The systems developed can be adapted to perform alternative assessments. While some discussion will be given to alternative assessments and the utility of GIS in this capacity, the case study discusses the integration of GIS into a detailed refined carcinogenic risk assessment. Therefore, the following discussion outlines the programme requirements of a refined carcinogenic health risk assessment.

In most districts, the EPAs industrial source complex (ISC) model is used to estimate the nature and extent of the atmospheric dispersion. This Gaussian plume model is widely used in air quality regulatory applications to estimate air quality impacts from point-, area- and volume-type emission sources. Critical tasks associated with dispersion modelling are the location of sources, facility boundaries, nearby buildings, elevated terrain and geographical features.

Since cancer and other health effects can afflict certain individuals more readily, sensitive receptors must be identified. These receptors include schools, hospitals, nursing homes, day care centres and retirement centres.

Multipathway exposure must be considered for certain compounds, such as polycyclic aromatic hydrocarbons or total chlorinated dibenzo-*p*-dioxins/dibenzofurans. The interrelationship of the pathways is depicted in Figure 16.1. The assumptions and equations used to model these pathways are based on the regulatory guidance. Multipathway assessments require that the deposition and fate of pollutants be estimated for a variety of media — water, soil, plant and animal ingestion, mothers milk, etc. Numerous assumptions are made to limit the extent of the site-specific investigations. For example, the potential surface water pathways must be identified.

Figure 16.1:   AB 2588 exposure routes

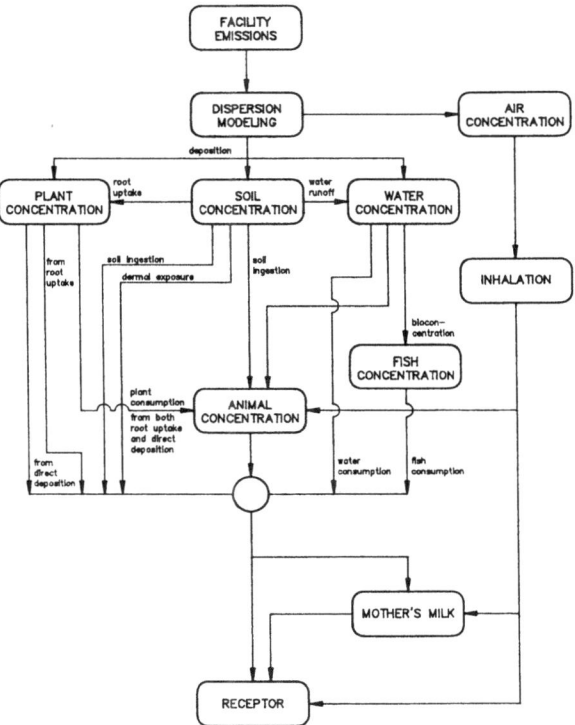

Source: CAPCOA (1992)

However, the multipathway assessment for exposure through water or fish ingestion is generally limited to a qualitative review, because a quantitative assessment requires a significantly greater effort than allowed in the regulatory schedule. For the plant and animal pathways, lacking any other information and as specified in Sears (1992) and Tran & Murphy (1992), the AB 2588 exposure assessment assumes that 15% of vegetables and 50% of meat are homegrown or locally produced.

The cancer risks are estimated for individual and population-wide exposures. The maximum excess lifetime cancer risk is estimated for each model receptor (hypothetical individual). The excess cancer risk is a measure of the probability of contracting cancer in a lifetime (the number of expected excess cancer cases per million people exposed; lifetime defined as 70 years). The excess cancer burden is the expected number of excess cancer cases for the total population exposed.

Due to considerable uncertainty in health risk assessment and the demanding regulatory schedule, the programmes refined health risk assessment procedures were derived to calculate worst-plausible risks. For example, a lifetime exposure assumes that an individual remains fixed at a location for 70 years, with no mobility whatsoever. This assumption and others, tend to compound the overprediction of potential risks. The combination of a lack of information and an accelerated schedule to analyse the data results in the regulatory preference for an overly health-conservative approach.

16.3 The integration of GIS into risk assessment

A GIS is a tool that allows the user to manage spatial data on a computer. A GIS provides the user with capabilities to attach attributes to points, lines and areas on a digital map. These attributes can then be used in data calculations, database queries and graphic presentation of the data. GIS technology can be a critical part of solutions to many environmental assessment problems including, as exemplified here, air toxics risk assessment.

16.3.1 Systems and data

The PC ARC/INFO GIS software package (version 3.4D) was used [developed by Environmental Systems Research Institute (ESRI), Redlands, CA, USA. The software was installed and executed on an 80386-based, 20 MHz PC with a tape backup system and 350-MB hard drive. Since this version of the PC ARC/INFO software is compatible with dBASE, dBASE (version 4.1) was used to manipulate all data and attribute files.

The ACE2588 (Assessment of Chemical Exposure for AB 2588) computer programme, as described in Tran & Murphy (1992) and Sears (1992), was used to perform the health risk assessment. This computer programme implements the recommended regulatory guidance for performing the exposure and risk assessment. The programme was developed to be modular. Figure 16.2 presents the links between the data files, the ISC dispersion model and the ACE2588 programme. The ISC dispersion analysis is conducted with a unit emission rate for all sources; each source is a separate model source group. The ACE2588 programme post-processes the dispersion models binary concentration output file by using the estimated emission rates for each source and then generates an output file with a wide variety of risk diagnostics.

Other software and data used in the risk assessments and linked to the GIS include:

(i)   AutoCAD: some facility plot plan revisions;
(ii)   SURFER Graphics: risk isopleth generation;
(iii) Etak MapBase: detailed digital map data with street address ranges;
(iv)   state government offices: street address data for schools, hospitals, daycare centers, etc.;
(v)   urban decision systems: census tract centroid and population data.

16.3.2 A case study in Kern County, California

Although the recently completed regulatory risk assessments are in the domain of public information, a hypothetical case has been prepared and presented in this section, to show the many elements of the role of GIS in air toxics risk assessment. This hypothetical facility has been arbitrarily placed in a location north-west of Bakersfield, in Kern County, California. To limit the scope of the discussion, this case study will focus on the carcinogenic risk analysis of a health risk assessment.

Modelling input

Much of the information on each individual device or stack is contained in the facilitys TEIR or can be acquired by staff at the facility. This includes stack height, diameter, exit velocity, flow rate and emissions of each AB 2588 pollutant that must be quantified. However, the location of each emission point was sometimes not well defined in the TEIP or the TEIR. In some cases, only a hard-copy facility drawing could be obtained. The essential information was extracted from the hard copy to generate an AutoCAD facility drawing that contains the location of dispersion model emission sources and the facility fenceline. This scaled drawing is exported to the GIS for UTM (Universal Transverse Mercator) coordinate data generation.

The facility plot plan coverage is positioned on the Etak base map by reference to several manually interpolated UTM coordinate pairs on the facility fenceline (fenceline corners, for example). Then, to check consistency, UTM coordinate pairs for points along the fenceline are digitally interpolated and the coordinates are compared with manually interpolated estimates from hard-copy maps [such as United States Geological Survey (USGS) 7.5 minute maps or facility plot plans].

Once the facility coverage is confirmed with respect to the Etak and USGS base maps, the source and fenceline UTM coordinate pair generation can begin. The source location UTM coordinates are generated by the GIS in a database file by reference to the source number for each point in the facility coverage. The source UTM coordinates are exported in a format ready for the ISC dispersion model. The source identification number provides a unique reference for each source in the modelling analysis. Since the assessments are facility-wide and the size of some facilities is potentially quite large, the number of sources can be considerable. In modelling analyses with a large number of sources, a GIS greatly improves the efficiency, accuracy and consistency in which source locations can be identified.

The receptor grid used to calculate health risks at specific geographical locations is generated by input of two UTM coordinate pairs into a computer programme. The two coordinate pairs identify the opposite corners of a rectangular or square, outer boundary of the receptor grid. The programme then generates an equally spaced grid of model receptors with UTM coordinates. Since the determination of occupational exposures and risks is not the goal of the regulatory programme, those receptors within the facility property can be selected automatically with the GIS, by use of the fenceline in the facility coverage and deleted from the receptor grid. Another programme is used to add receptors along the fenceline, at an equal interval.

The combination of the receptor grid generation computer programme and the GIS provided an effective means to flexibly generate receptor grids. This proved quite useful for difficult facility configurations. In some cases, the fenceline was represented by several nearby closed polygons in a plot plan (i.e., noncontiguous operations of the same facility). Further, many facility fencelines were not represented as simple polygons. The GIS provides a reliable and easy method to generate nearby model receptors and to ensure that on site receptors are not used inadvertently.

Figure 16.2:   Modelling flow diagram

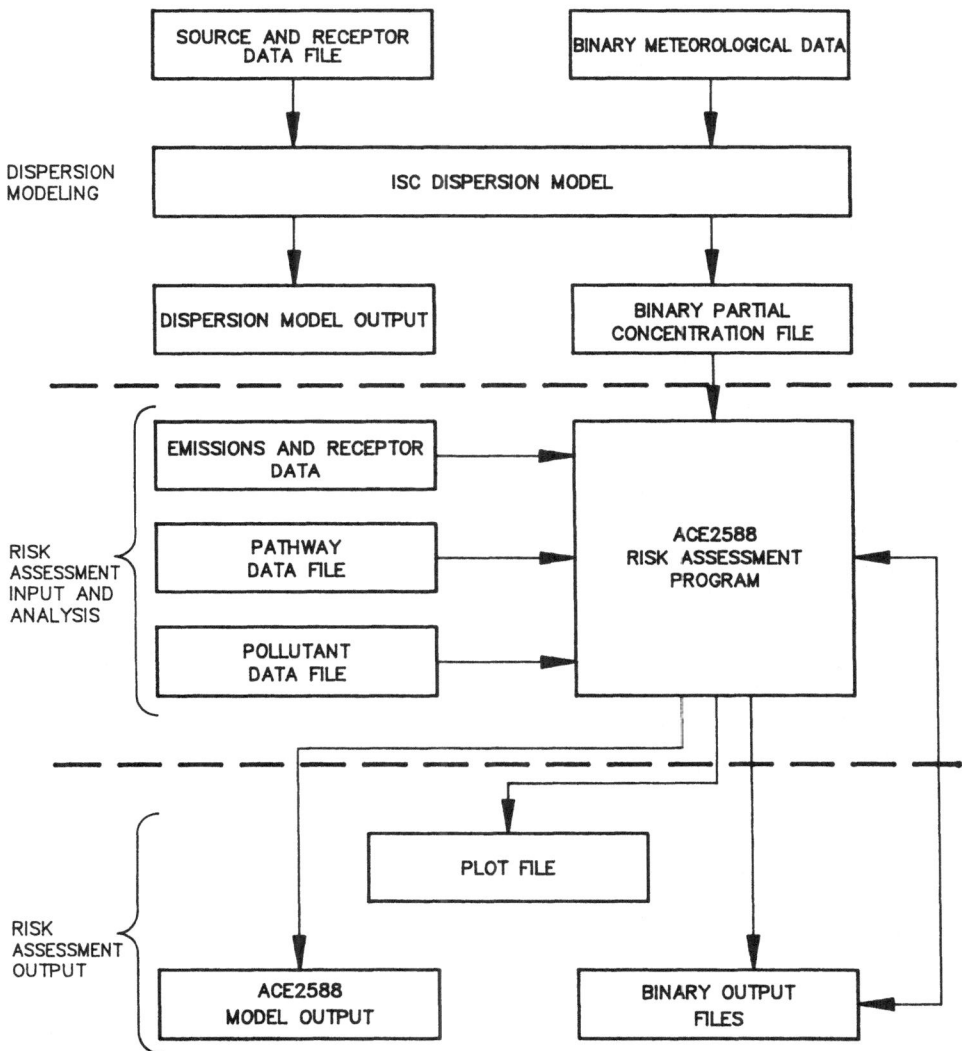

Another benefit of this combination was the ability to generate refined receptor grids centred around the coarse grid modelling MEI (maximally exposed individual) receptor. The initial modelling with a coarse receptor grid (e.g., 200-m receptor spacing) will identify a peak cancer risk. For this receptor, in preparation for further investigation, a refined receptor grid (say, 50-m spacing) can quickly be centred on this point. The GIS provides environmental modellers with an important tool to vary and check receptor grid placements.

The siting of model receptors at potentially sensitive locations can also be performed with the GIS. These receptors provide the computer models with the ability to assess the health risks to potentially sensitive individuals. The sensitive receptor database must first be parsed to ensure consistency with street-naming conventions and the best possible address-matching scenario. The records in the address data file of sensitive receptors are geocoded (automatically positioned and registered to a coordinate system) to points on the Etak map, based on the address range information associated with each road-type feature. The computer algorithm positions the receptors along the roadway by the use of linear interpolation within the address range. For one facility near Bakersfield, approximately 260 sensitive locations were identified out of approximately 300 for the surrounding area. Approximately 40 sensitive receptors were placed interactively within the GIS by manual inspection. Generally, the siting of these receptors could not be automated due to a lack of address range data for rural regions or for some highways, street name discrepancies or other minor problems.

To address potential population-wide health concerns, model receptors are located at census tract centroids. The census tract centroids can be positioned in the GIS by use of a utility to convert latitude and longitude data in the census file to UTM coordinates. By use of the GIS, only the centroids in the defined study area are selected from the county centroid database. The centroid data appeared to be in good agreement with the Etak base maps census tract boundaries.

A computer programme was created to analyse, merge and format the detailed emissions and source information into a model-ready data file for ISC and ACE2588. The model receptor coordinates and associated information, such as census tract population, are also exported from the GIS in a suitable format for the models.

Modelling output

After the dispersion modelling and risk analysis, isopleths are generated and exported to the GIS in a DXF format. The isopleth coverage is positioned in the GIS by means of a coordinate reference to the hard-copy isopleth output.

Map 16.1 displays an example of a detailed map of the predicted worst-plausible excess cancer risk isopleths associated with the operational emissions of a facility. The map identifies surrounding features, such as railroads, highways and surface streets, to better communicate the extent of potential risks. Additionally, the map identifies the location of sensitive receptors in the vicinity. The boundary analysis capabilities of a GIS allow a risk assessment team to automatically select and report identifying information on only the receptors located within the zone of impact (for example, the area within the one-in-one-million, worst-plausible risk isopleth). Table 16.1 displays information on the receptors located within the zone of impact (1.0×10^{-6} risk isopleth) in this example. The identification of sensitive receptors located within a bright line isopleth is becoming accepted regulatory policy by local APCDs.

Map 16.1 also displays the surface water locations (dashed lines) relative to the potential zone of impact. This information may be useful as regulatory agencies review the risk assessment reports and determine further action. If a facility has emissions of multipathway pollutants and is located near surface waters, predicted cancer risk values may be given even greater consideration during risk management. As the regulatory programme's risk assessment methods develop and mature, this type of scenario may require a detailed water pathway risk assessment.

Map 16.2 displays the census tract boundaries (dashed lines) with respect to the modelled cancer risk isopleths and the facility fenceline. The population-wide excess cancer burden value is calculated by multiplying the excess cancer risk value by the population value for each centroid location within the zone of impact and summing the product for all locations. Typically, the significant regulatory cancer burden threshold is 0.5 cancer cases. The computer model will perform this calculation, provided that the centroid locations and populations are input in the receptor data file. This figure provides a comparative visualization of the population and worst-plausible health risk distributions.

16.4 Implications of GIS technology for air toxics risk assessment

In these regulatory risk assessments, GIS technology proved quite valuable to the effective management of data. The above examples demonstrate some important GIS functionality that improved data input, analysis and output. However, a beneficial contribution of the system has been the improvement of overall quality assurance. When many projects are being completed concurrently under a compressed schedule, the importance of the data management task is clear. For example, one project required a risk assessment for four expansive, noncontiguous sites. The total area of the modelled operations encompassed in excess of approximately 90 km^2 and included more than 550 model sources. This effort would have been significantly more complicated without computerized spatial analysis capabilities.

The GIS also improves the ability of a risk assessment team to modify various alternative assumptions and present other scenarios, particularly based on alternative exposure data. For instance, the guideline risk assessment assumes a lifetime continuous exposure (70 years, absolute fixed mobility) in the calculation of cancer risk. However, many times a facility will be located in the middle of an industrial area, a commercial or business centre or a recreational activity centre. The lifetime continuous exposure assumption is extremely conservative for these locations. By a first-order linear assumption, the GIS will allow the risk assessment team to adjust the risk at a receptor based on the type of land use. For example, Map 16.2 presents areas near the modelled facility that are highlighted, hypothetically, as industrial or commercial property. The areas may be determined by aerial photogrammetry or other data. All predicted cancer risk values at receptor points that lie within the boundaries of the designated industrial or commercial zones could be multiplied by a factor of 0.14 to account for the exposure of an individual at an off-site worksite (eight hours per day, 240 days per year, 46 years versus a 70-year lifetime continuous exposure).

The population-wide health effects parameter (cancer burden) may be further evaluated if the refined assessment estimates a value in excess of 0.5. Digital census boundary data can be represented down to the census block group or even block, level. Thus, an excess cancer burden could be estimated by summing the product of risk and population at each block group or block centroid. A GIS is a useful tool to locate the centroids and present the analysis.

In the regulatory programmes risk assessment methods, the human exposure assumptions are simplified because there is either a lack of information or a lack of readily available quantification procedures. A GIS can improve the calculation methods for inhalation or multipathway exposures by integrating digital spatial information into exposure and risk calculations. For example, the programme's risk assessment procedures require facilities to identify potentially sensitive populations in the vicinity of a site. This is performed by locating nearby schools, nursing homes, retirement centres, etc. where young or elderly people will be located for prolonged periods. Since excess cancer risk can vary depending on age, the procedure attempts to address this in the decision-making process by identification of the potentially sensitive sites. However, for example, age-dependent population data may identify concentrations of elderly people in planned retirement

Map 16.1: Worst-plausible excess cancer risk (lifetime continuous) isopleths for a hypothetical facility

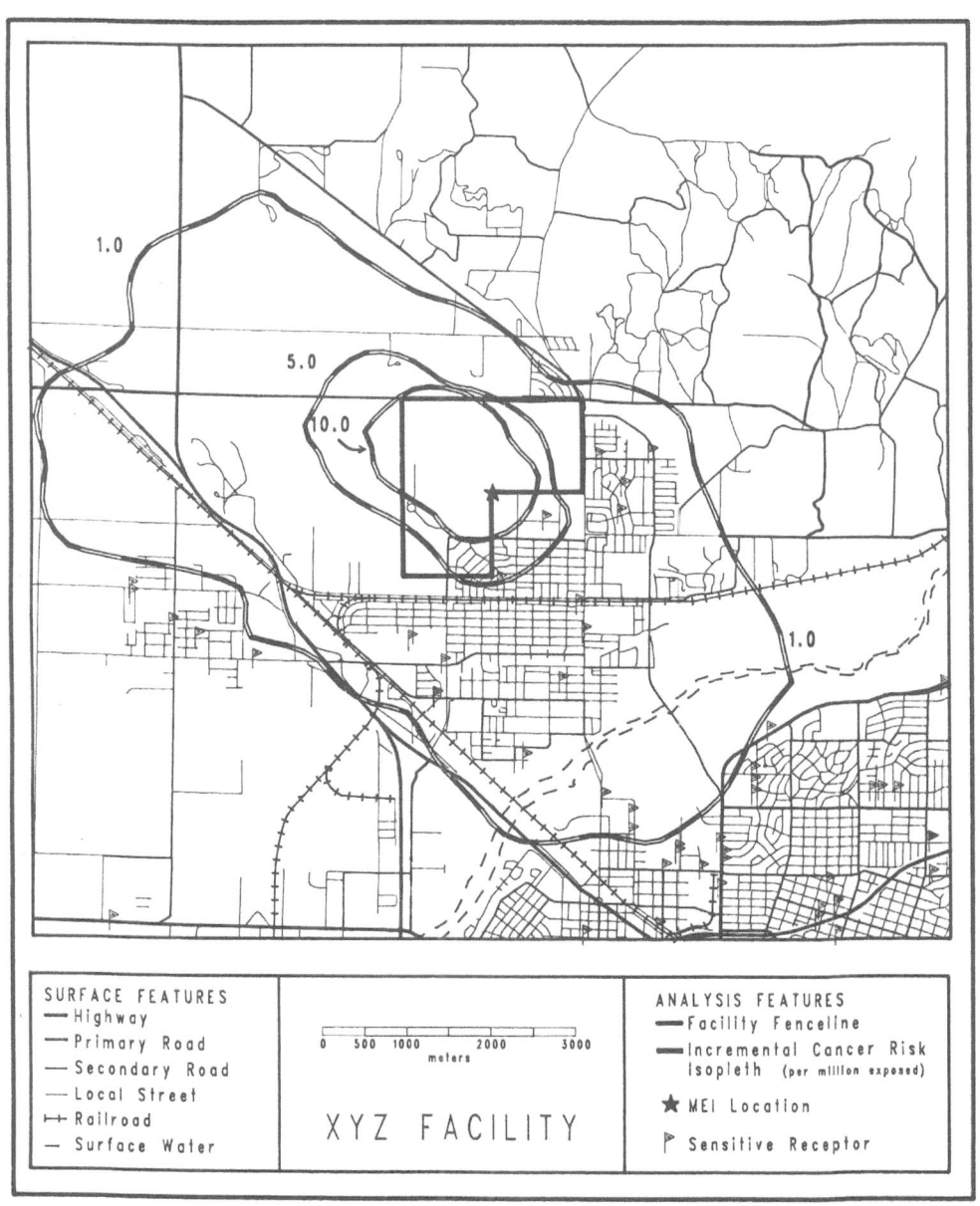

SURFACE FEATURES
— Highway
— Primary Road
— Secondary Road
— Local Street
⊢+ Railroad
— Surface Water

0   500 1000   2000   3000
meters

XYZ FACILITY

ANALYSIS FEATURES
━ Facility Fenceline
━ Incremental Cancer Risk
   Isopleth (per million exposed)
★ MEI Location
⊳ Sensitive Receptor

Map 16.2: Lifetime continuous excess cancer risk isopleths, census tract boundaries and
the hypothetical location of nearby commercial or industrial zones

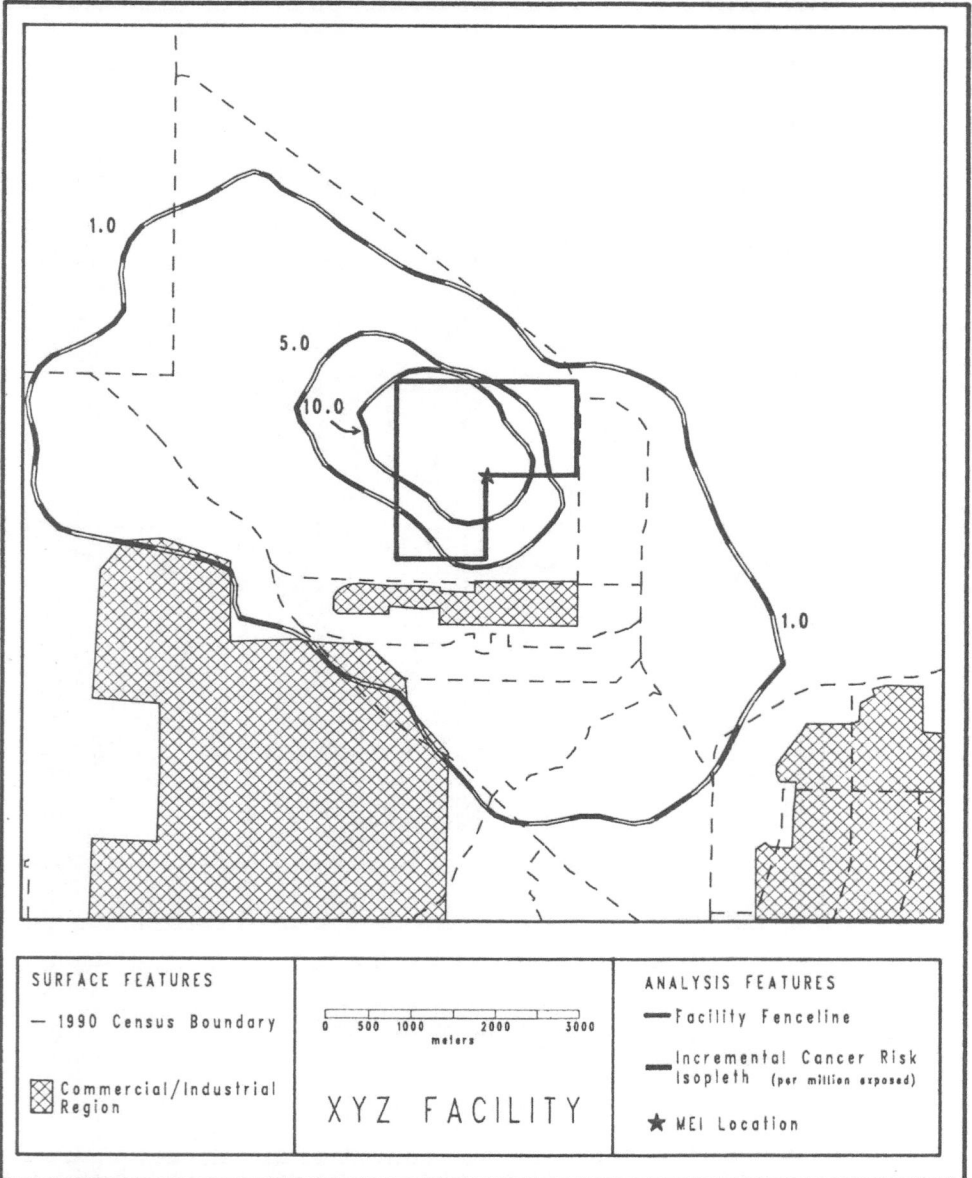

Table 16.1: Receptors located within the zone of impact of the operational emissions of a facility

Name	Address	City	Zip Code	UTM East	UTM North	Licensed Capacity
Kids World Preschool	3311 Manor Street	Oildale	93308	317413.7	3923033.0	75
Kids World Preschool & Day Care	3311 Manor Street	Bakersfield	93308	317413.7	3923033.0	25
Manor Heights Christian School	3311 Manor Street	Bakersfield	93308	317413.7	3923033.0	?
Living Waters Guest Home	300 Day Street	Bakersfield	93308	316803.5	3922986.0	5
Highland Elementary School	2900 Barnett Street	Bakersfield	93308	317093.9	3922620.0	?
Rascal Ranch Child Dev. Ctr. Preschool	222 Circle Drive	Bakersfield	93308	317048.8	3922286.0	48
North High School	300 Galaxy Avenue	Bakersfield	93308	316159.8	3922202.0	?
Allways Caring Board & Care	609 Linda Vista Drive	Bakersfield	93308	315621.1	3921468.0	6
Standard Elementary School	115 East Minner Avenue	Oildale	93308	316555.2	3921181.0	?
Skidgel's Private School	121 Ferguson Avenue	Bakersfield	93308	316623.7	3920843.0	?
North Beardsley Elementary School	900 Sanford Drive	Bakersfield	93308	314557.9	3920753.0	?
Bakersfield Community Hospital	901 Olive Drive	Bakersfield	93308	314954.0	3920467.0	64
Wilma's Board & Care	212 Belle West	Oildale	93308	316345.2	3920250.0	5
Beardsley Intermediate School	1005 Roberts Lane	Bakersfield	93308	314851.5	3920067.0	?
Beardsley Junior High School	1001 Roberts Lane	Bakersfield	93308	314862.6	3920067.0	17
Riverview Preschool	401 Willow Drive	Bakersfield	93308	315905.9	3919329.0	20
Willow Tree	401 Willow Drive	Bakersfield	93308	315905.9	3919329.0	15
Bethel Kiddie Korral	1418 West Columbus Avenue	Bakersfield	93301	316840.7	3918870.0	?
Bethel Apostolic Academy	1418 West Columbus Avenue	Bakersfield	93301	316840.7	3918870.0	60
NAPD-Socialization, Enrichment, Education, Dev. Program	4032 Jewett Avenue	Bakersfield	93301	317155.7	3918675.0	?
Hills (Stella I.) Elementary School	3800 Jewett Avenue	Bakersfield	93301	317155.8	3918445.0	?

? = Data not obtained from the California Department of Education

communities. Therefore, age-dependent population statistics by census tract or block group can assist in the identification of high-density elderly populations. These considerations are currently not addressed in the regulatory risk assessment methods.

For multipathway exposure and risk assessment, a GIS provides the capability to combine existing, multidisciplinary environmental data into an air toxics health evaluation. For example, to estimate the contribution of soil ingestion to the total predicted excess cancer risk, the pollutant concentration in the soil must be estimated. Currently, for a guideline risk assessment, the soil concentration is estimated by an equation following the form:

$$C_S = D \, X/(K_S \, M \, B \, T_E)$$

where:

C_S =   average soil concentration over the evaluation period ($\mu g/kg$)
D =   deposition on the effected soil area per day ($\mu g/m^2/day$)
X =   soil half-life integral function (days)
K_S =   soil elimination constant (day^{-1})
M =   soil mixing depth (m)
B =   soil bulk density (kg/m^3)
T_E =   total exposure period (days)

In a guideline risk assessment, the soil bulk density is assumed to be equal to 1,333 kg/m^3. The soil mixing depth is assumed to be 0.01 m, typical of a playground setting, for all receptor locations. The programme guidance acknowledges that soil mixing depths for agricultural areas may be assumed to be equal to 0.15 m. However, to be conservative and since the model does not consider the spatial variation of this parameter, the 0.01 m value is generally used as a default model input. Thus, spatially distributed digital soil data, such as soil density and mixing depth, can improve the soil concentration estimates and concomitant risk estimates so that they reflect site-specific conditions. This example typifies only one component of a detailed multipathway assessment. By using a GIS and digital data to flag receptor-specific characteristics, similar improvements to the calculation methods can be envisioned for other routes of exposure.

16.5 The potential for improved risk communication with GIS

In the scope of the regulatory programme, risk notification is a requirement for facilities with predicted off-site risks above a significant risk level. Many APCDs are currently developing risk notification guidelines and awaiting further guidance from the state agencies. Therefore, over the next few months and years, this process will be an experiment in large-scale risk communication. For example, as a result of the South Coast Air Quality Management Districts (SCAQMD) proposed risk notification guidelines, the District estimates that 0.6 to 1.2 million people could be notified by facilities during Phase I (the SCAQMD staff released these estimates at a February 1992 public workshop).

Some benefits of GIS to risk communication and risk notification are presented below. While very similar, risk notification is a subset of risk communication. Risk communication is the action of communicating scientific knowledge and technical data and the underlying assumptions, to the general public through two-way dialogue. Risk notification, being more mechanistic, is the process of sending explanatory letters to exposed individuals, publishing newspaper notices and holding town meetings.

16.5.1 Risk communication

Risk communication can be difficult, if not disastrous, when approached without planning or commitment. As with any communication, the communicator must know the audience to effectively transfer information and understanding. This principle is well known in product marketing and market research. And, maybe not surprisingly, GIS is often used in market research to analyse population characteristics and product retail potential. In a similar manner, a GIS can assist a facility operator in preparing for risk communication.

Trained risk communicators sometimes suggest developing and administering a survey to people in the affected community. The function of the survey is to acquire a population profile that will assist the risk communicator in addressing the concerns and attitudes within the population. A GIS incorporated into the survey data collection and analysis will be helpful in representing population profile statistics by city, zip code or neighbourhood. This information, combined with predicted health risk data, may indicate certain issues that must be addressed during preparation for risk communication. This survey approach can be used by a facility operator or even by a regulatory agency.

In the GIS marketplace, data on population characteristics are available from a number of sources, including government agencies and private vendors. These sources can provide detailed population data, with attributes that can be general (annual household income, age, education, race, primary language, sex, etc.) or very specific (the number of household members who have had cancer treatment in the past three years). While the quantity of digital population (and other) data continues to increase, the spatial unit to which these data are registered continues to decrease. Thus, data are available by city, zip code, census tract, census block group, census block and even household (in fact, according to Parker (1991) and Francica (1991), Lotus Development Corporation and Equifax, Inc. canceled plans to release Lotus Marketplace: Household, a CD-ROM database product of names, addresses and marketing information on approximately 100 million households in the United States, due to privacy concerns raised by consumers). A GIS not only allows the user to analyse and present these data, but the GIS user can create new data by combining information from the archived sources (e.g., with a GIS, race statistics by zip code can be easily generated from race statistics by census tract or block group). The functionality of a GIS and the wealth of available data, make archived data sources quite valuable in the characterization of audiences.

The risk communication audience sometimes has other concerns or perceptions based on past incidences, that should be anticipated. A risk communicator must understand the operational history of the facility and the context of public concern. According to S. Hill (personal communication, 1992) of the Bay Area Air Quality Management District (BAAQMD), the first APCD to initiate risk notification under the regulatory programme, the District noted that some audience concerns over health and welfare were raised based upon past odour nuisances. A facility may be able to use a GIS and the health risk assessment, to better display the lack of a correlation between the odour nuisances and predicted cancer risks. For example, the GIS could be used to geocode (by address) the location of the odour nuisance reports and to overlay the predicted cancer risk isopleths. In this way, a non-issue (relative to the technical aspects of the risk assessment) may be addressed by showing concern and background investigation and not readily dismissed. This may help to improve the sense of public partnership in the risk communication and risk management process.

Above all, risk communication professionals emphasize that the message to the audience must be focused and the key points should be limited. One of the best messages to communicate to the public is the facilitys ongoing air pollution control measures that have reduced emissions and the concomitant risks. While this may be communicated well with tables, bar charts, graphs and pie charts, a good way to show the overall improvement is through the generation and display of predicted risk-change isopleths and the shading of areas of reduced risk levels. All discussions of emissions and risk reductions cannot compare to the effectiveness of shaded zones on a map that show the location of streets, schools, etc.

16.5.2 Risk notification

While most APCDs and the state agencies are still developing risk notification guidelines, most proposed guidance focuses on communication of risk to significantly exposed individuals via letter, newspaper notice or public meeting.

GIS can improve the risk notification process as follows:

(i)   GIS-integrated health risk assessment is already map-based, so preparation of maps or plots for risk notification letters, notices or meetings can be done quickly and cost effectively.

(ii)   Residential and business mailing address criteria and lists can be developed specifically for the area within the significant risk isopleth(s).

(iii) Mailing address criteria and lists can be flagged by a spatial query on location with respect to risk isopleths to identify different risk notification requirements (most proposed or adopted risk notification guidance uses a tiered approach to risk notification requirements based on ranges of predicted risk; e.g., a long notice letter versus a short notice letter depending on the distribution of predicted risk).

(iv)   Based on residential and worker population distributions, roadway locations and meeting times, the most suitable geographicalally located public meeting places may be identified.

(v)   Based on population data, preparations may be made for spatially varying language requirements in written and verbal communications.

(vi)   The notification requirements can be modified by and responsive to community feedback based on returned public comments and other information on postcards originally mailed with the risk notification materials to households (e.g., a public meeting may be deemed appropriate by some criteria based on a review of response times and distributions).

16.6 The benefits of GIS in risk management

16.6.1 Background on risk management and the regulatory programme

Air toxics risk management is defined as any strategy developed and/or action taken to mitigate potential health risks from air toxics emissions. For a facility, risk management may involve: facility audits, installation of air pollution control equipment, material substitution or process modification. For regulatory agencies, risk management may involve the development of rules and regulations to reduce air toxics emissions and public risk notification requirements.

In either case, risk management must be based on a definition of significant risk. As stated by SCAQMD (1992), this definition may include consideration of the message to be conveyed to the public, background risks, previous regulatory decisions and other factors. For this programme, local APCDs must define the level of risk that is determined to be significant. At this writing, the general direction (proposed and adopted) is to use a cancer risk level of 10 to 100 in one million (1.0×10^{-5} to 1.0×10^{-4} excess cancer risk). According to the developing state and local regulatory guidance, these significant risk levels are to be compared with predicted facility-specific risks. The individuals within a significant risk isopleth, a bright line, are to be notified.

Thus, as is typical with other elements of the evolving regulatory programme, the significant risk levels and public notification guidelines are decentralized (i.e., the requirements are specific to each facility and its interaction with the surrounding community). Other decentralized risk assessment and management components include:

(i)   risk assessment modelling grid-spacing is facility-specific;
(ii)   no requirement to use computers for mapping and risk isopleth overlay;
(iii)   different receptor grid extent for each facility;
(iv)   defined significant risk levels may vary by APCD; and
(v)   sensitive receptors are identified by each facility.

This decentralized approach can create information management problems for regulatory agencies, industry and the public alike.

For regulatory agencies, the problems include:

(i)   merging massive quantities of data, from varying facility-specific receptor grids and from hundreds of risk assessments to develop risk management strategies;
(ii)   presenting overall regional risk data to the public; and,
(iii)   handling risk notification with a bright line isopleth.

For industry, the problems include:

(i)   managing and communicating risk for multiple facilities statewide, under varying significance thresholds and requirements; and,
(ii)   communicating bright line significant risk issues.

For the public, the problems include:

(i)   understanding the complicated issues and technical assumptions of the health risk assessment to interpret what the bright line isopleth approach to significant risk really means; and,

(ii)   avoiding undue fear or saturation apathy, if numerous notice letters from many nearby facilities arrive in the mail within a short time period.

16.6.2 An alternative approach to risk management with GIS

A centralized approach to risk management at an APCD or statewide level, using a GIS as a tool, may greatly improve the management of data and communication of information for all parties. The following paragraphs describe a proposed approach that could be adopted, for example, by a local air pollution control district.

Determination of significant risk

Based on a review of past regulatory decisions performed by Travis et al. (1987), significant individual regulatory risk levels have apparently been developed with consideration given to the population risk. In other words, the level of individual risk deemed significant by regulators in the past can be correlated with population density (e.g., high individual risks are more acceptable in areas with a small population).

A GIS can allow an APCD to consider the distribution of population and to spatially vary significant risk thresholds based on this parameter (in the SCAQMD, for example, the population density can differ greatly between the Los Angeles Basin, southern Orange County and Riverside County). Because AB 2588 developed as a community right-to-know law, this approach might help an APCD to follow regulatory precedence and to better meet the intent of the law.

Further, to better determine and communicate areas of significant risk to the public, one may argue effectively that the cumulative (region-wide) risks must be calculated. Although the facility-specific predicted risk levels are overly conservative (including any resulting cumulative risk values that used on a relative basis to determine not only facilities associated with higher risks but geographical areas as well. The problem with a facility-specific risk approach is that cumulative risks are not directly addressed and the prioritization of notification efforts may be unduly misdirected. As an example, why should a resident in one area be notified of a facility-specific predicted industrial cancer risk level of 15 in one million, which is also approximately the cumulative industrial cancer risk level for that area, when a resident in another area may be "exposed" to cumulative predicted industrial cancer risk levels one or two orders of magnitude higher?

Because of this scenario, cumulative risk levels should be addressed in determining which level of risk is significant. A GIS can not only improve the method in calculating and presenting cumulative risk levels, but it can also be integrated into a centralized regulatory programme for effective data and risk management.

Thiessen polygon analysis and the potential for cumulative risk management

Many data analysis and presentation techniques are possible with GIS technology, including Thiessen polygon analysis for thematic map generation from point data. A good description of a Thiessen polygon analysis is presented by Maggio & Long (1991). A summary of this technique is presented in Figure 16.3. This technique creates a polygon, from a seed point, which contains all points closer to the seed point than any other seed point in the grid. This approach must be used with care, because inadequate point sampling for some attributes may lead to very inaccurate representations. However, when applied correctly, this approach can display the spatial variation and uncertainty of an attribute over a region. This technique is also quite useful in displaying general spatial patterns of attribute change over time.

To develop a cumulative risk-based GIS using Thiessen polygon analysis, an APCD could:

(i)   locate sensitive receptors and census tract centroids district-wide, referenced to a suitable coordinate system, by GIS geocoding techniques;

(ii)   ensure that sensitive receptors are identified consistently throughout the basin;

(iii)   identify a district-wide irregular grid of receptors for air toxics (IGRAT), which uses census tract centroids and sensitive receptors as the sampling points and as the basis for risk assessment, risk communication and risk management;

(iv)   identify each facility in the basin using a GIS and, based on the total emissions and facility prioritization, determine the subset of IGRAT receptors to be used within any facility-specific modelling grids;

(v)   establish a requirement for each facility to report, in a computer-readable file, a company identification number and a breakdown of health risk by pollutant and emission source for each IGRAT receptor of concern; and

(vi)   link existing facility data on record, such as standard industrial classification (SIC) code, to the reported data.

The resulting database generated by these procedures would allow an APCD to better manage predicted risk data for risk communication and risk management. With this GIS database, the APCD could graphically display and analyse cumulative risk levels by thematic representations (i.e., shading of polygons). The cumulative risk data could be analysed and presented by pollutant, emission source type or industry type (SIC code). Further, facility-specific predicted risks can still be presented. This representation will also provide a good method to show the change in predicted risk levels over time, by various classifications.

To identify zones of significant risk, population density-based cumulative significant risk thresholds could be displayed as an overlay to the cumulative risk thematic map. Alternatively, GIS and spatial statistics could be used to identify areas that have predicted risks outside one or two standard deviations of the mean.

To communicate risk to the public, this GIS-based approach will allow cumulative risks to be displayed district-wide, by various specific classifications and graphically. These data can also be displayed with other map overlays, such as city boundaries or zip code areas.

By cross-referencing zip code to the risk data in a GIS environment, the APCD could develop a database for risk notification that could be the basis for a centralized risk notification programme. A list of facilities that contribute to significant risk levels, by zip code of the impact area, could be created by the APCD. This list, along with an APCD letter and a package of facility-specific notification letters, could be sent to individuals within the zip code. This method could reduce the total number of mailings to individuals. This approach also provides a way to ensure consistency and oversight in risk notification.

An added benefit to industry of this approach is that industrial facilities will not necessarily be singled out during the notification process. The APCD may decide to further the environment of industry cooperation and public partnership by developing community or regional risk reduction task forces based on cumulative risk levels, municipal boundaries and other local geography.

16.7 Conclusions

A geographical and spatial information system can greatly improve the calculation methods and risk estimates associated with a multipathway air toxics risk assessment. Through integration with environmental modelling, a GIS can be a vital instrument in the quality assurance and quality control of data and modelling input files. Additionally, a GIS provides many ways to represent and integrate multidisciplinary data from various resources to improve the risk assessment information available for regulatory policy-makers and the public. The strength of this technology in the detailed, objective treatment of spatial quantification and uncertainty, two very important elements that

Figure 16.3:   Steps in generating a Thiessen polygon analysis from a point coverage.
                5A) point coverage. 5B) points joined by lines forming triangles. 5C)
                perpendicular bisectors of the lines are added and they meet in each
                triangle. 5D) original lines are removed leaving Thiessen polygons. 5E)
                Thiessen polygons are shaded based on the value of a relevant attribute
                associated with the corresponding points (based on Maggio & Long,
                1991)

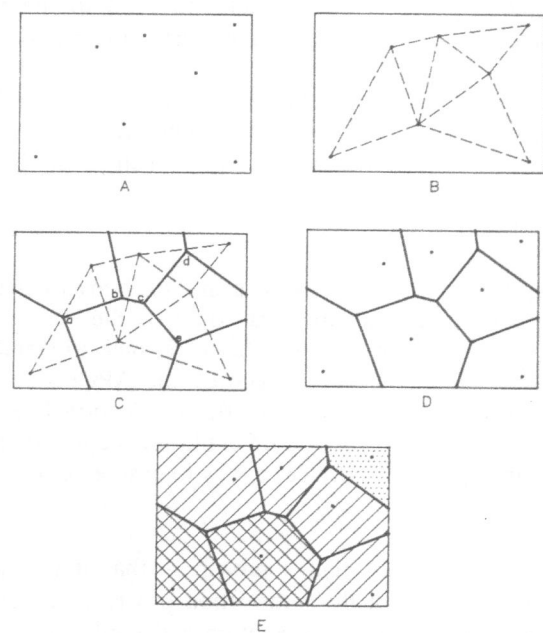

must be addressed in risk assessment, will assist in improving the methods. With the
increasing emphasis on air toxics in Title III of the new Clean Air Act Amendments of
1990 and developing state air toxics programmes throughout the United States, air toxics
health risk assessment continues to be an important and emerging issue awaiting new
GIS solutions.

As a result of growing regulatory interest in air toxics and risk-based analyses, risk
communication and risk management will also become increasingly more important to
regulatory agencies and industry. The EPA (1992) documents the current knowledge on
environmental equity and acknowledges that some population segments may share
disproportionately in environmental exposures and health risks. However, this report is

only the first step in a process to ensure that greater attention is given in the United States to risk communication and risk management. The EPA recognizes a need to gather more environmental and population data and to emphasize greater research. To this end, the report documents the initiation of some GIS-based EPA research by regional offices. However, as acknowledged by the agency, this research will be hindered by the current lack of health data. (For instance, as compared with governments of other western high-income countries, the United States is behind in the collection of mortality statistics by various socioeconomic variables.)

As GIS technology continues to expand into environmental applications, new data and data management and analysis techniques can improve the ways and the expediency, in which environmental scientists answer what-if questions. GIS technology can play a significant role in improving the state of the art, and dissolving the current boundaries between risk assessment, risk communication and risk management.

Acknowledgements

Jim Harvey and Mark Scop created all the figures for this chapter (with sources as noted in the captions, when appropriate). Also, Mr Harvey developed most of the computer programmes discussed and performed all of the GIS data manipulation necessary to complete the health risk assessments. This chapter is based on a paper presented at the Air and Waste Management Associations Annual Conference in Kansas City, Missouri, USA in June 1992. The views presented in this chapter are solely those of the author.

References

California Air Pollution Control Officers Association (1992). *Air Toxics "Hot Spots" programme risk assessment guidelines*. Cameron Park, CA, CAPCOA.

Environmental Protection Agency (1989). Risk assessment guidance for Superfund, Volume 1. *Human health evaluation manual (Part A)*. Interim final. Washington DC, USGPO. (EPA/540/1-89/002).

Environmental Protection Agency (1992). *Environmental equity: reducing risk for all communities*. Volume 1. Workgroup report to the Administrator. Washington DC, USGPO. (EPA230-R-92-008).

Francica, J.R. (1991). Taking a swim in the data pool. *GIS World,* 4(2): 94-96.

Maggio, R.C. & D.W. Long (1991). Developing thematic maps from point sampling using thiessen polygon analysis. In: *Proceedings of the GIS/LIS 91 Conference*. Volume 1. Bethesda, MD, American Society for Photogrammetry and Remote Sensing and American Congress on Surveying and Mapping, Washington DC,

Association of American Geographers and Urban and Regional Information Systems Association, Aurora, CO, AM/FM International, pp. 1-10.

Parker, H.D. (1991). GIS NEWSLINK: Lotus cancels "Marketplace". *GIS World,* **4**(2), 24.

Sears, R. (1992). *Supplement to the ACE2588 users guide, Version 92269.* Ojai, CA.

South Coast Air Quality Management District (1992). *Defining "significant health risk" and notification requirements for the Air Toxics "Hot Spots" Information and Assessment Act of 1987 (AB 2588): discussion of possible approaches.* Diamond Bar, CA, SCAQMD.

Tran, K.T. & T.M. Murphy (1992). *Users guide to the assessment of chemical exposure for AB 2588 (ACE2588) model, version 92092.* Woodland Hills, CA.

Travis, C.C., S.A. Richter, E.A.C. Crouch, R. Wilson & E.D. Klema (1987). Cancer risk management: a review of 132 federal regulatory decisions. *Environm Sci & Technol,* **21**(5): 415-420.

T.J. Moore
ENSR Consulting and Engineering
1220 Avenida Acaso
Camarillo, CA 93012
United States of America

PART V

VALUE ADDED BY GEOGRAPHICAL INFORMATION SYSTEMS

PART V

VALUE ADDED IN GEOGRAPHICAL
INFORMATION SYSTEMS

17 SPATIAL INFORMATION TO MAKE A DIFFERENCE: VALUE ADDED DECISION-MAKING IN THE HEALTH SECTOR WITH GEOGRAPHICAL INFORMATION SYSTEMS

William E. Bertrand and Nancy B. Mock

Abstract
This chapter addresses the problem of diffusing geographic information system (GIS) technology to policy-makers, planners and managers. Although GIS holds enormous potential for improving the public's health, its use still is limited. This is in part due to the sociopsychological differences between information producers (technicians) and the consumers of GIS products (public health policy-makers and managers). The design and implementation of GIS applications must be guided by the principle of user orientation. The elaboration of information products, the major outputs of these systems, must be carefully tailored to provide decision support to intended clients. Mechanisms also must be elaborated that assure close interaction between decision-makers and GIS technicians. Included among these are staffing of projects, information audits and client-technician forums. These approaches are elaborated and illustrated based on the authors' experience with such projects.

17.1 Introduction

The profound importance of organizing, analysing and presenting data spatially has been appreciated intuitively by public health professionals since the field's inception. All public health students are taught in introductory epidemiology to characterize disease distributions by three main criteria: person, place and time. One of the landmark epidemiological studies undertaken by John Snow in 1850 drew heavily on the spatial distribution of cholera as a means to generate hypotheses about its cause and to advocate his final conclusions. Unfortunately, however, until very recently, the technology available to utilize spatial analysis for public health research and practice had not progressed much beyond John Snow's original tools. The recent and rapid maturation of microcomputer technology and the parallel development of sophisticated data reduction techniques has changed this picture dramatically. The spatial methodological framework can now be made widely available to public health researchers, policy-makers and practitioners. Geographic information system (GIS) technology may be the decade's single most important analytic tool for the advancement of the health and wellbeing of the world's population and environment.

In its relatively short history, modern GIS already has proven itself as an excellent catalyst for interdisciplinary research and intersectoral collaboration directed towards achieving objectives of improved health status. This chapter documents the impressive strides that are being made in important interdisciplinary fields, including environmental

M. J. C. de Lepper et al. (eds.), The Added Value of Geographical Information Systems in Public and Environmental Health, 265–276.

epidemiology and nutrition policy. The ability to manipulate areal boundaries and to analyse relationships among attributes having differing spatial descriptions has greatly facilitated our capacity to examine apparent space-time disease clusters thought to be associated with environmental variables. Environmental, agronomic, agricultural, socioeconomic and health data are being synthesized and presented using GIS to more effectively pre-empt and address many complex public health problems. An example from the authors' experience has shown usefulness in providing information to alleviate food crises and to evaluate social and economic policies in the Sahel.

The purpose of this chapter is to identify the characteristics of GIS technology and applications that have specific relevance to decision-makers and to recommend strategies to more effectively incorporate this technology in practice and to improve its application. The underlying principle from which these strategies are elaborated is that the applications of GIS must primarily be user-driven.

17.2 Decision-makers and GIS

Information technology and the recent technology revolution intuitively appeal to policy and programme decision-makers. Modern managers are constantly being tantalized by new available databases that might assist them in making more informed decisions. To cite just a few, the United States 1990 raw census data is available on CD-ROM. Electronic libraries and bibliographical searches such as those supported by the National Library of Medicine in the United States (Medlars, a family of 37 different medical databases) are available to researchers and academics throughout the world. Potentially, physicians can access immediately large data banks to ascertain treatment and diagnostic protocols. Financial market information is available for database analysis on a real-time and historical basis. Decision makers in the developed world are faced with the dilemmas of how to effectively access, analyse and present this information.

GIS are particularly appealing to the programme manager because of their capability to transform data into information through the medium of spatial analysis and display. Spatially organized data are attractive to managers and administrators in part because so many of their decisions are specifically linked to geographical and administrative entities. One example from the authors experience documents major decisions made by central-level planning bodies allocating regional resources in a famine situation. Nutritional status and cereal price data presented together and organized by region have provided great insight into the food security problems of the Sahel. These insights have greatly increased the efficacy of decisions with respect to request for emergency relief, relative allocation formulas and, ultimately, improved management of the problem.

GIS provide the critical conceptual and methodological link that permits the storage, analysis and effective presentation of the growing number of databases becoming available to programme managers. Because the presentation method has particular appeal to politicians and managers, its use will grow even more rapidly as information technology makes it available to greater numbers of people through microcomputers.

17.3 The dilemma of information technology diffusion

The effective diffusion of innovation, however, depends on a number of factors that determine its adoption (Rogers, 1983). Innovation diffusion research illustrates that after the relatively small group (15%) of entrepreneurs or innovators in a population have adopted new technology, progress is delayed as the bulk of the population observes, tests and finally adopts the innovation. Successful adoption by this larger portion of the population largely depends upon a range of factors, including:

(i)   the characteristics, primarily education, of the innovating population;
(ii)   the characteristics of the innovation and its utility to the problem at hand;
(iii) the marketing and education regarding the benefits of the innovation.

The problem of accelerating the innovation process is not unique to GIS but is characteristic of technology in general. With GIS, however, there is a confounding factor in this process. Because of the technical complexity of GIS, GIS technicians have very little training and experience relating to managers of health care systems. For this reason, GIS methods do not systematically deliver what is needed or desired by consumers of information. The GIS technicians see their product as a working database made presentable in appropriate cartographic form. The decision-maker is more convinced by often less complex or technically sophisticated products that meet his or her needs. Often the elegance of a multi-colour photographic-quality product obscures the lack of useful information.

It is our contention that a major reason for the gap between potential and utilization is the lack of emphasis on the needs of users of these information systems. The applications development of GIS and other information systems has been driven by the science of information systems technology rather than the social science of information consumption. The culture of the empirical logical scientific paradigm has dominated the applications of the technology. All too often, systems designs are exported to organizations by developers whose professional training does not prepare them to understand decision-maker needs. A blueprint approach to information systems focuses on predetermining needed data elements and reports based on rational scientific analysis of information needs. Throughout health ministries, statistical offices are replete with data reports that are outdated and never used. Epidemiologists and computer scientists having limited understanding of management decision processes are frequently the architects of such health and management information systems. They often realize that the information they produce is irrelevant but have not developed systematic approaches to addressing the issue.

What is needed is an information system paradigm that unites the technology with the consumer. Design and implementation strategies that are keyed to the true desired outcome of information system activities; that is, improved management decision-making measured by impact. As mentioned earlier, we contend that GIS holds particular promise for accomplishing this objective because of its special emphasis on graphically summarized data presented spatially.

17.4 A sociopsychological paradigm

The need to integrate the process-oriented elements of information consumption and the technical product-directed bias of the information producer is not new. To direct the framework, the authors have turned to the experiences and approaches used by the architects of the Media Lab at the Massachusetts Institute of Technology (MIT) in the United States.

Concerned with how to understand the interaction between humans and machines capable of manipulating and presenting millions of data points, the Media Lab debated how to separate the idea of data: a series of signals, changes in patterns etc. from information: the final product judged useful to the user. They converged on a definition that, although seemingly simplistic, provides us with the base for exploring further the role of information systems and GIS. Turning to the work of anthropologist Gregory Bateson, the definition chosen was: "Information is a difference that makes a difference" (Brand, 1987).

At first consideration, "a difference that makes a difference" seems too all-encompassing to be useful. Yet such a definition implicitly identifies the importance of the user interface with information system technology. While historically, information availability was limited by a lack of methods for acquiring and processing data, the central problems today are essentially data reduction and presentation. Thus, the advances needed to make the technology more available to users include the development of guidelines to reduce data appropriately and improve its presentation.

Alan Turing, one of the intellectual fathers of electronic computing, provides further insight as to the importance of perception in the diffusion of technology in his article on the definition of an intelligent machine. As one of the most sophisticated mathematicians of his time, he defined the intelligent computing device in purely human terms. To paraphrase, an intelligent machine was defined by the ability to interact effectively with a human without the humans knowing that he or she was interacting with a machine (Turing, 1950). The engineers at MIT and Mr Turing are leading towards an understanding that, when data is turned into information, the perception and the reaction of the data consumer must be included in the definition of terms.

The important conclusion of the evolution of thinking in the domains of diffusion research and computer engineering is that formal consideration of the characteristics and needs of the data consumer must be made before GIS can be effective information systems. What must be considered is the technological equivalent of George Herbert Mead's (1934) "looking glass self", a sociopsychological paradigm explaining human development. The developing human gathers information from interpreting the ways others react to him or her. Likewise, a GIS must formally develop rules to enable it to learn from and adjust to the needs of its users. This feedback function must be incorporated into GIS projects and focus on two broad categories of issues: data reduction and presentation. Increasingly, standards against which to judge the quality of the visual display of quantitative information must be developed.

A point of departure for developing these standards can be borrowed from Edward Tufte's guidelines for graphic displays. In his epilogue for *The visual display of quantitative information* he makes the following statement: "The theory of the visual display of quantitative information consists of principles that generate design options that guide choices among options" (Tufte, 1983, p.190). He further elaborates what graphic displays should do (Tufte, 1983, p.1):

(i)   show the data;
(ii)   induce the viewer to think about substance rather than about method, graphic, design, the technology of graphic production or something else;
(iii)   avoid distorting what the data have to say;
(iv)   present many quantities in a small space;
(v)   make large data sets coherent;
(vi)   reveal the data at several levels of detail, from a broad overview to the fine structure;
(vii)   serve a reasonably clear purpose: description, exploration, tabulation or decoration;
(viii) be closely integrated with the statistical and verbal descriptions of a data set.

In addition to these prototype guidelines for the evaluation of GIS products, GIS project design needs to include explicit elements that link the perceivers and clients to the information system. Because the consumers positive or negative reaction to data presented through the technology of a GIS provides the "difference that makes a difference" or action response, this implicit element, seldom included in project design, is key to successful implementation. The authors experiences from the application of GIS to famine early warning in the Sahel serve to illustrate this point. Formal guidelines linking the consumer with GIS technology were developed from those experiences.

17.5 Lessons learned from designing and implementing a famine early warning system

The authors experiences in developing a famine early warning system for the United States Agency for International Development (AID) provide valuable lessons on both capitalizing on the appeal of GIS to introduce information technology and on developing GIS-based systems that make a difference. A major famine ravaged the Sahel during the period 1984-1985. AID, along with all other major development organizations, was caught in the embarrassing situation of having virtually no data as to the magnitude and distribution of the nutritional emergency. Congress placed great pressure on the AID to deliver basic information almost immediately so that emergency relief could be planned and managed on a rational and accountable basis.

Tulane University accepted a contract to gather, evaluate and transmit available data from six Sahelian countries to Washington, DC, where it was synthesized and reported to AID administrators. Given the emergency nature of the need for information, the project design process was severely restricted. Within two weeks after signing a famine early warning system (FEWS) contract with AID, Tulane representatives were to be in the field and transmitting data. But getting in the field and gaining access to data required that AID field missions and host country governments (the providers of data) would cooperate with this Washington-initiated project. Field missions and host country governments were initially very reluctant. Field missions, by and large, did not want Washington to interfere with efforts to manage the crisis. Data about the magnitude of loss of life and malnutrition were highly sensitive politically to host country governments. Thus, initiating the project activity itself was conflict-ridden and generally blocked by the field.

GIS-related products provided the major incentive to both parties for participation. In most cases Tulane was able to convince missions and governments to give the programme a trial period of a few weeks, during which existing data were gathered and mapped. Tulane also demonstrated how summarized satellite imagery data could be made available for analysis and presentation locally. The notion that host country governments could have some control over satellite data was extremely appealing to both missions and host country governments. These two GIS-related activities were a prerequisite to doing business in the Sahel during the famine.

The fact that FEWS had so many clients to satisfy made operation of the project very difficult. Missions, host governments and Washington all had different information needs. For example, Washingtons priority was on determining regional food assistance needs while host countries were concerned with internal food distribution. The information-gathering activity itself was difficult. Substantial quantities of data were being collected by host country organizations and private voluntary organizations in all country settings. However, in few cases were these data summarized or automated. Most data were in raw form. With a core staff in each country of one field representative, providing information products that met such diverse decision-making needs was not feasible during the first phase of the project. Because the locus of activity was the field, more emphasis was placed on developing country-level bulletins that satisfied missions and host country governments. As a direct result of this emphasis, Washington decision-makers were largely underserved by the project.

The project was further complicated by the changing information needs as the food security situation evolved. At first, the information most critically needed related to the magnitude of malnutrition and mortality. As emergency relief began to arrive more emphasis was placed on targeting and evaluating food distribution programmes. When the worst of the emergency was past, the focus of analytic activities shifted to predicting crop production sufficiency. The dramatic chaos created by the famine further emphasized the point that all information systems need to be dynamic as decision-making information requirements are and will be constantly changing.

In 1988, AID redesigned the FEWS project, making several strategic decisions that have had great impact on the utility of FEWS for decision-making purposes. First, the objectives of the system were redirected and made more specific. The primary objective is now to internalize a FEWS for the Sahel within AID. Secondly, great emphasis was placed on defining information products that were appropriate for clients in field missions and in Washington. Field missions need relatively detailed information pertaining to potential problem areas within their countries. On the other hand, Washington decision-makers are most concerned with highly digested graphic regional analyses.

There were two major approaches taken to ensure the development of this client-oriented information system. First, in designing the staffing of the project itself, the project director was to be a former consumer of the product. Indeed, a former high-level career administrator currently holds this position. This has brought the culture of the users to the technicians who produce the information products.

Secondly, great emphasis has been placed on the gathering of evaluative information from the clients. Regular user surveys are programmemed. The project is constantly informally assessing the utility of the products. Based on this monitoring of users' perceptions, the entire strategy for reporting was dramatically changed during the first year of the project. Originally, the major Washington-oriented products were to be relatively lengthy and technical trimestral reports. It was found that administrators were not reading these reports. Based on a combination of both in-depth qualitative techniques and the survey, one-page graphic decadenal bulletins produced during the agricultural season were perceived to be the most useful products by Washington decision-makers.

These bulletins have done much to raise the profile of food security problems in the Sahel, and they have effectively engaged Washington administrators in the analytic process that has been driving more intelligent policy-making with respect to these problems. As a primary indicator of the success of these strategies, consumer demand for information has increased dramatically over the years that FEWS has existed.

17.6   Improving the design and implementation of GIS: preparing and monitoring the information system users

Based on Tulanes experiences with FEWS and other information system projects, the authors believe that the spatial dimension of GIS brings great promise to improved health and environmental programme management; however, the efficacy of GIS could be increased dramatically by taking some practical steps to ensure user-oriented design and implementation of systems.

Conceptually, the key to successful GIS design is to focus on the role of GIS in supporting management decision-making. Mechanisms must be elaborated that assure close interaction between decision-makers and GIS technicians. The importance of project staffing has already been discussed. Other design options include articulation of linkages between information systems and user decision-making forums. The client forums as well as the types, frequency and periodicity of decisions for which information is needed should be identified. Ongoing evaluation of the impact of information products also should be pre-specified in the design of GIS projects. In the first stage of project implementation, sufficient time should be allocated to formulating information system outputs.

One useful tool for furthering user-oriented GIS is the information audit. An information audit combines elements of many types of information system design methods to address the central issue of what information is currently being used to allocate resources and to what extent information is being provided by the system. As a general rule, no information system should be developed that does not include a careful review of levels of information generation and use, resources allocated by the decisions made and some consideration of impact of the decisions taken. Steps appropriate to guide the application and development of GIS include:

(i) Determine the levels of information use and where data are generated in the system. In the FEWS example, information was utilized at the national and international level. Reporting was only provided to the international level and therefore, by making more information available at the local level through GIS applications, the system functioned more smoothly.

(ii) Determine the basic information needed from the GIS to first carry out and then improve tasks at each level of system operation. This is a formal or informal interview process where managers are queried as to what data are actually used in day-to-day operations or for planning and resource allocation. The information needs are then compared with the availability and timeliness of the GIS, and gaps in the information are identified.

(iii) Determine the quality and timeliness of information needed. In GIS settings this involves the often highly politicized task of choosing appropriate units for map display as well as the usual concern with the specificity and sensitivity of information. Indicators that change quickly or very slowly need to be reviewed with respect to their appropriateness to specific decision.

(iv) At a technical level, determine the degree of fit between information needs and the data currently being gathered. Do more or less data need to be collected? Again from the FEWS example, much more data were available than could be appropriately integrated into a decision-oriented GIS. Had researchers been more rigorous in pre-assessing this issue, a better use of GIS methods would have resulted.

(v) A specific review of the quality and type of presentation needs to included. Presentation technology is becoming more important in stimulating data use and the quality of presentations. General reporting format should also be explored at this point. Frequency of reporting, GIS format, colours etc. are all issues that deserve review once the prototype GIS data elements have been decided upon.

Although the first five steps outlined above are considered the classic elements of an information audit there are second stage activities which should be considered part of the same process and reviewed on a regular basis as part of the implementation of the GIS. While the information audit can be organized on a periodic basis the following steps should be part of the ongoing information system maintenance in most organizational settings.

(vi)   A prototype system should be designed and quickly implemented as a testing ground for technical ideas developed in the information audit and to assist in the design of the appropriate training. Prototypes should be placed into a well defined and representative user group early in the development process in order to discover problems and refine developments. This pretesting phase is critical before a GIS can be designed and overlaid onto a new environment. A culture of change, i.e., that change is normal and positive, should be instilled in the working group in order to assure that routine feedback will be an expected and important element of the new system.

(vii) Based upon the experience gained in the prototype stage, evaluate the need for more specific training. The goal of GIS, to make data available to consumers in time to make resource allocation decisions, should be built into the format of training and management of the new system.

(viii) Based upon prototype experience and after training, the system should be implemented. Since information systems are, by definition, dynamic in nature, there should be no fear to install partial systems or systems that may not be considered complete. The provision of GIS information will create new demands and new sophistication on the part of the users such that even a well designed first implementation will require constant revision and change. In the ideal setting, the user should be able to execute these changes. A real time evaluation function should be developed that examines and corrects problems or improvements in the GIS. By focusing information systems on improving resource allocation, the resultant evaluation of impact changes on formal organization can be monitored and evaluated.

The formal steps of the information audit and follow-up implementation can go far in assisting the GIS designer in respond appropriately to the needs of a specific client. The GIS designer or implementer can get the perspective needed to complement the more technical aspects of a GIS installation by focusing on:

(i)   the relationship between perceived information needs and actual information needs;
(ii)   user sophistication in data and information analysis;
(iii) cultural translations of quantitative analysis;
(iv)   establishment of direct links to consumer forums and feedback from those forums.

As is clear from the subject areas the skills involved in the information audit are only partly technical in the traditional sense. What is lacking in many environments is more specific training designed to provide to the GIS coordinator the needed theoretical perspective, methodological tools and interpersonal and observational skills to carry out the information audit and to manage the ongoing GIS. Further research should focus on how important the different elements of these systems are with relation to overall GIS impact in improving the efficacy of organizations that use the technology.

17.7 Conclusions

Adding the spatial dimension to health and environmental information systems greatly increases the promise of these analytic activities for research and management purposes. Spatial analysis and mapping are critical links in most health and environmental problems. To accelerate the adoption of GIS tools and to increase the utility of these tools to programme managers the authors recommend that a much more explicit consideration of user needs be more explicitly incorporated into the design and implementation of GIS. Specifically GIS planners should consider the science of visual presentation and the specific needs of GIS information consumers as part of the design of the system. The regular inclusion of an information audit into the GIS production and use cycle is a strongly recommended step to assure that this important method will continue to provide added value to decision-making in the human sciences.

References

Bertrand W., B. E. Echols & Khatidja Jusein (1987). Microcomputers and alternative data management techniques. *In: Management information systems and microcomputers in primary health care.* Lisbon, Aga Khan Foundation, pp. 123-134.

Brand, S. (1987). *The Media Lab: inventing the future at MIT.* New York, Biking Press.

Mead, G.H. (1934). *Mind, self and society.* Chicago, University of Chicago Press.

Rogers, E.M. (1983). *Diffusion of innovations.* New York, Free Press.

Rubin, A. & Z. Pearltone. (1989). *Art as technology.* Hillcrest Press.

Scholten, H.J. & M.J.C. de Lepper (1990). *The benefits of the application of geographical information systems in public and environmental health.* WHO Offset, Consultation on Epidemiological and Statistical Methods of Rapid Health Assessment. Geneva, World Health Organization.

Tufte, E.R. (1983). *The visual display of quantitative information.* Chelshire, CT, Graphics Press.

Turing, A.M. (1950). *Computing Machinery and Intelligence.* Mind

William E. Bertrand and Nancy B. Mock
Tulane University School of Public Health and Tropical Medicine
Department of Biostatistics and Epidemiology and
International Health Academic Programme
1501 Canal Street
New Orleans, Louisiana 70112
United States of America

18 THE LONG-TERM POTENTIAL OF GEOGRAPHICAL INFORMATION SYSTEMS FOR EPIDEMIOLOGY

Simon Raybould, Jacqueline Nicol, Anna Cross and Mike Coombes

Abstract
This chapter makes a brief assessment of the possible use of geographical information systems (GIS) in answering questions of health and the environment from the point of view of epidemiology. Health service planning is not dealt with, except as a baseline against which the needs of epidemiologists using GIS are compared. The initial discussion is about the contribution that spatial epidemiology itself can make, as a subject, irrespective of what actual tools are used. GIS is proposed as the most appropriate way of providing a tool-kit with the necessary features. The chapter goes on to suggest four key requirements of a GIS for epidemiologists, and looks at how well current GIS meet the needs as well as how easily the gap between the ideal and the (currently) feasible can be closed by judicious use of the existing features of GIS. Finally, the chapter outlines the necessary changes and developments in GIS technology that are needed to enable GIS to live up to their long-term potential.

18.1 The challenge of spatial epidemiology

In the present context of identifying the potential of geographical information systems (GIS), the arena of epidemiology of most interest is the investigation of spatial associations. It is important to stress at the outset that spatial epidemiology does not seek proof of causation. For example, to find that disease X is more prevalent under one set of social conditions than under another does not necessarily imply that the disease is an outcome of the conditions in question. There are two sources of uncertainty here - the first is that both may be caused by a third (and as yet unknown) covariate, and the second is that the actual direction of the chain of causation may possibly be in the counterintuitive direction. The task for GIS in spatial epidemiology, then, is unlikely to include the requirement to support formal hypothesis testing.

This leaves as a more limited field of application for GIS in epidemiology the urgent role of providing evidence of spatial association. Most epidemiological investigations have a common-sense approach regarding causality and so seek only cautious interpretation of results (such as the statement "X is spatially associated with Y" - or more rashly, "X is (in part) caused by Y"). The exact physiological pathway Y might take towards causing X is not within the remit of spatial epidemiology. It is therefore not possible to use the result of this form of analysis directly as a means of providing a medical cure. However, the result may provide a starting-point for looking at possible causes. Clinical experimentation (the process by which metabolic and/or physiological pathways for both pathogens and medicines are established) is both expensive and has a very long elapsed

M. J. C. de Lepper et al. (eds.), The Added Value of Geographical Information Systems in Public and Environmental Health, 277–284.

time (and indeed is morally undesirable where avoidable). There is, therefore, a role for spatial epidemiology prior to undertaking clinical work, to investigate a range of possible relationships. The majority will prove to be groundless and can be discounted from further work, but some will appear to be significant and prove to be fruitful areas for targeted clinical research.

The line of reasoning presented above leads to a conclusion that the spatial epidemiology for which GIS has greatest potential value is subservient to laboratory experimentation. There may remain, however, one or two particular forms of analysis with which GIS-based methods can more readily deal than more traditional methods. The most exciting area of opportunity may be that of cumulative causation or interactive etiologies. Many diseases have a compound etiology: a case is more likely to be observed in an individual if condition one is present, but condition two is absent, or condition three is present. With even a relatively small number of potential causes, this combination of uncertainties gives rise very quickly to a large number of permutations. In order to obtain robust results and simultaneously cope with high numbers of permutations, clinical experimentation would require prohibitively high numbers of individual experiments. Spatial epidemiology clearly has advantages over traditional methods of investigating compound etiologies when key environmental factors are among the potential contributing causes to be investigated.

18.2   The operational requirements for GIS in spatial epidemiology

A number of criteria can be seen to be vital when implementing a GIS for epidemiological applications. From the outset, there needs to be a design brief which will include the identification of system needs (and different users will have different requirements), the recognition of data requirements (balancing what is available and what could be required) and the organizational constraints to GIS implementation.

Underlying the strategic choices to be made is the potential problem that there are two distinct groups of potential users within the results of spatial epidemiology. The first is the health planner, who is concerned with where cases of disease occur, in order to organize the provision of health care facilities. The second is the research epidemiologist, whose concern is with the identification of spatial patterns inherent to the disease. The system needs are slightly different for the planner and epidemiologist. Both require information systems that enable the locational analysis of disease and the management of the requisite database. The health service planner places emphasis more on linking the analysis to data for facilities management (the management and coordination of the facilities and services to be provided, based on health needs). Optimized utilization of services and cost-efficiency are the key aims in responding to the health needs, leading to an emphasis on resource allocation. A much less normative spatial analysis is undertaken by the spatial epidemiologist and it is argued elsewhere in this volume that GIS are rather limited in their spatial analysis functionality.

The fundamental data-related requirement for an epidemiologists GIS is an information system able to handle the increasing quantity of data available from the many sources that are now geo-referencing health-related information. With the versatility of the GIS, not only can this be achieved, but it is also possible to update individual records regularly, enabling incidence to be monitored throughout a region or district. Unfortunately, major constraints for implementing a GIS for epidemiology lie in the common problem of poor-quality geo-referencing (and the use of incompatible systems across different datasets). This is particularly pertinent for epidemiologists, in that environmental etiology is often extremely spatially sensitive. For the (ill) health incidence data, provided as a result of planning or management activities, there is at least the advantage that the datas resolution is the appropriate individual (patient, operating room, etc.), allowing potentially very fine spatial referencing. In the longer term, these sources of data should be increasingly available in computerized form, so that neither the planner nor the epidemiologist will need to have recourse to published data, as this has generally been preaggregated and is therefore implicitly far less well spatially (or temporally) referenced. For the epidemiologists in particular, this will be a major advance that will unlock very real constraints.

The final prime difference between the design brief of a GIS for an epidemiologist, rather than for a planner, is the only one that is currently in the epidemiologists favour. This difference lies in the required response times. Except in very exceptional circumstances, an epidemiologist will tend to be in the relatively luxurious situation of not requiring the real-time answers that can be essential for the service planner. Fast answers will always be preferable, but accuracy and sensitivity are likely to be a more serious requirement for the epidemiologist.

In summary, therefore, one might take as the basic starting-point for a GIS to be used for epidemiology all the basic requirements of a GIS (mapping facilities, data manipulation functions and spatial manipulation capabilities such as buffering) and add the following:

(i)   access to health event data at the very finest possible level of resolution - preferably at the level of the appropriate individual event;
(ii)   the ability to handle very large quantities of data as a consequence of the above;
(iii)   ease of use, to enable subject specialists to access it themselves; and
(iv)   more analytical capacity than is commonplace in GIS currently.

The next section of this chapter seeks to flesh out this final point by looking at some of the higher-order questions that spatial epidemiologists need to tackle.

18.2.1 Analytical requirements of GIS for epidemiology

The functionality within a proprietary GIS is rarely, if ever, of a sophistication that could not be replicated by another method, such as a specially written program in FORTRAN or C++. The major advantage of using a GIS is that of convenience - less work should be needed, the algorithms have already been tested and advanced programming skills are not necessary. This should leave the user free to concentrate upon the substantive issues - but the spatial epidemiologist is unlikely to find a GIS that supports a standard list of analytical functions, even of a quite straightforward kind.

For example, an epidemic is defined as occurring when the observed amount of a disease statistically significantly exceeds the expected rate within specified time or space limits. Providing maps of reporting areas shaded accordingly should be a basic function of an epidemiology GIS, but they are not. Instead it is often necessary to calculate rates in the database associated with the GIS, compare observed and expected and create maps accordingly. Admittedly this is not a difficult operation, but nevertheless the fact that it has to be done reduces the very *raison d'être* of the GIS by re-introducing technical complexity. The GIS solution to this is to write programs using Macro languages, which would allow the function to be performed at the touch of a button. Such a solution may not be feasible if the mathematical solution lies beyond the practical limits of the database. Algorithms to recognize clusters of relatively rare diseases (e.g. childhood cancers) are notoriously complex and can require huge quantities of microprocessor resources. The same is true of almost any statistical function, including tests of the significance of an apparent association between a diseases pattern and other diseases or environmental phenomena. Such apparent associations are made easier to spot visually by coverage overlay in GIS, but a more robust evaluation of the pattern is surely essential for an epidemiology GIS application.

The remainder of this section will highlight specifically the questions GIS can potentially help epidemiologists to tackle when seeking to investigate spatial associations. The first of these can be phrased as "What is at ....?". This is a locational query, which in GIS terms would simply involve an interrogation of the relevant databases to deduce which factors are present in which areas. This may be useful in providing an overview of the distribution of patients, locating the patients with a particular disease or establishing what environmental factors affect an area. The environmental aspect is in fact becoming more and more influential in the search for causes of rare diseases of unknown etiology. For example, a pilot study in northern England adopted a GIS route in 1987 to highlight possible causes of childhood cancers in the region (Cross, 1991).

Even this apparently simple type of question is not straightforward to the epidemiologist, unfortunately. Generic questions of how to represent patterns and ratios visually are still unresolved: simple point maps take no account of the distribution of the underlying population, for example, while choropleth maps of rates are prone to problems associated with arbitrary and inappropriate boundaries (or class intervals). Other problems are associated with the simple visualization and presentation of epidemic data. First, the view portrayed is unlike an epidemic because it is essentially static, even though a GIS can store information for several time periods and produce a series of maps for desired time intervals. Each map represents a snapshot of events and the notion of developing and or or evolving disease patterns may be masked by their separation. With the development of computer movies and a possible move towards dynamic GIS, these two aspects may be overcome to some extent.

The second of the epidemiologists extra questions is easily stated as "Is this an important spatial pattern?". Such analysis requires far more functionality than is currently offered by any proprietary GIS; this may be the most important failure of contemporary packages. This question can cover a range of ideas, from a general spatial query on features to a more inquisitive level that asks whether the distributions observed are unusual and if so, whether they are deserving of or amenable to remedial action, on a practical level. Thus, although the production of relevant maps and the visualization of points or areas within the selected data may be sufficient to satisfy the epidemiologists' initial curiosity and support their opinions on a particular health problem, the more probing question as to whether the observed patterns are real or spurious is little more elusive, requiring some degree of intelligent investigation of the databases concerned that GIS cannot conduct without some human input.

The third of the epidemiologists questions is "What has changed since....?". This question exploits the temporal element of databases by studying the changes in incidence patterns over time. The potential of this approach lies in the fact that, should an obvious change in the rate of incidence occur in a particular time interval, then this may be an indication of some influential event or change in the surrounding environment (which has had a strong impact on the health factor being reviewed). Thus, the ability to interrogate databases according to the time element is likely to prove an important enhancement for a future generation of GIS packages, which are very limited in this respect at present.

The final extra question is the deceptively simple "What if .....?". This is probably one of the more useful methods of analysis that any software can offer. The sophistication of the relevant techniques available varies widely between software packages. However, the increasing number of health and environment issues will surely result in growing pressure for the monitoring and assessment of critical pollution and or or siting policies, both of which would benefit from this type of scenario testing. Such analyses require a variety of health, environmental and socioeconomic factors, which will be best stored and linked within the GIS framework. This final question, in fact, finds an echo in the planner's concerns, which will often include the need to look at the effect on response times of trauma units if the question is asked "What if this unit closed down - or was upgraded?".

18.3 Future prospects

The widespread enthusiasm for GIS is undoubtedly justified by the potential attractiveness of the basic toolbox and the improved state-of-the-art technology for the visualization and interpretation of data. However, it is now also well recognized that there are problems and limitations, such that some of the early adopters are finding their usage restricted to basic projects and database management. A second generation of analytical tools appropriate for an epidemiological framework may be essential to prevent the use of GIS within epidemiology from becoming severely restricted.

One of the main requirements is a means of providing statistical evaluation of any spatial patterns observed - or better still, a tool that can automatically flag areas of concern and possible areas to be followed up. Some programs have recently been developed and continue to be improved, including a means of identifying clusters of unusual events: for example, the geographical analysis machine (Openshaw et al., 1987) and other work on similar lines (Besag & Newell, 1991).

Once a statistically significant pattern is spotted, the next logical step is to investigate the evidence that may help to identify the possible causes (whether genetic, viral, environmental or a combination of these). This involves searching for all possible relationships between the factors of interest and the issue under investigation. Although GIS supports the visual inspection of such patterns by mapping, there is no means of verifying any suspected causal relationships (while the combination of different types of coverage can become very cumbersome and difficult to decipher). The development of a geographical correlates exploration machine (Openshaw et al., 1990) attempts to extend the capabilities of GIS by overlaying relevant datasets and then exporting the information to the host systems in order to carry out a virtual overlay process. This not only reduces the demands upon computer resources and time but also allows some degree of significance to be put on the polygons created by the overlay process, so that only those of interest are reported and mapped within the GIS.

Other analytical contingencies that are ripe for GIS interaction in epidemiology are the provision of a synoptic view of an area, highlighting high-risk locations with reference to particular common diseases. This can be achieved by the use of kernel estimators (Silverman, 1986) and the automatic monitoring of data and reporting - in real time - of the start of an outbreak of an epidemic of, say, food poisoning. These functions can be at least partially achieved with customized programs but not within proprietary GIS at the moment.

These new developments underscore the obvious point that many of the limits to existing GIS packages are explicable simply by the fact that GIS is a relatively young technology. Much progress has already been made in recent years and many valuable opportunities are now emerging.

One model may be that research in spatial epidemiology will develop new tools, such as the geographical correlates exploration machine, which will then become advanced functions of later generations of the GIS packages that are already available. Even so, the current utility of GIS in the epidemiology field is great.

References

Besag, J. & J. Newell (1991). The detection of clusters in rare diseases. *J R Stat Soc A*, **154**: 143-155.

Cross, A.E. (1991). *Building a health and environment geographical information system: an evaluation. Looking at childhood cancer in northern England.* Thesis. Newcastle upon Tyne, Department of Geography, University of Newcastle upon Tyne.

Mollering, H. (1980). Strategies of real time cartography. *Cartogr J*, **17**: 12-15.

Openshaw, S., M. Charlton & C. Wymer (1987). A Mark 1 geographical analysis machine for the automated analysis of point pattern data. *Int J Geogr Inf Systems*, **1**: 335-358.

Openshaw, S., A. Cross & M. Charlton (1990). Building a prototype geographical correlates exploration machine. *Int J Geogr Inf Systems*, **4**.

Silverman, B.W. (1986). *Density estimation for statistics and data analysis.*London, Chapman & Hall.

Simon Raybould, Jacqueline Nicol, Anna Cross and Mike Coombes
Centre for Urban and Regional Development Studies
University of Newcastle upon Tyne
Newcastle upon Tyne NE1 7RU
United Kingdom

PART VI

IMPLEMENTING GEOGRAPHICAL INFORMATION SYSTEMS

PART IV

DIAGRAMS AND GEOGRAPHICAL
INFORMATION SYSTEMS

19 GEOGRAPHICAL INFORMATION SYSTEMS IN ORGANIZATIONS: SOME CONDITIONS FOR THEIR EFFECTIVE UTILIZATION

Ian Masser and Heather Campbell

Abstract
Any commitment to the development of environmental monitoring and risk management information systems must be long term in nature given the complexity of the issues that are involved. This has massive organizational and interorganizational implications that need to be taken fully into account before a commitment is made to implement such a system. In the process, it is necessary to recognize that many things that may be technically feasible may not automatically be worth the organizational and interorganizational effort that will be required to implement and maintain them. With these considerations in mind the chapter sets out three conditions for the effective utilization of geographic information systems in organizations of the kind encountered in the health and environmental fields.

19.1 Introduction

The starting point for this chapter is a discussion published in 1979 by Paul Sabatier & Daniel Mazmanian entitled "The conditions of effective implementation: a guide to accomplishing policy objectives". In that article the authors sought "to identify and explain a set of five (sufficient and generally necessary) conditions under which a policy decision that seeks a substantial (non-trivial) departure from the *status quo* can achieve its policy objectives" (p. 483).

Sabatier & Mazmanian reflected a general concern among both scholars and decision-makers about why so many policies apparently failed to achieve their objectives. Such concerns are vividly captured by the subtitle of Pressman & Wildavskys (1973) classic case study of *How great expectations in Washington are dashed in Oakland*: why it's amazing that federal programs work at all.

There are obvious parallels between these themes and the growing recognition among scholars and planners alike that a fundamental distinction must be made between the acquisition of computer-based technology such as geographical information systems (GIS) and its effective utilization. Unless such a distinction is made, it is impossible to explain how a system that has proved successful in one situation can turn out to be a disaster in another. It is interesting to note that even the terminology used by the evaluators of the impact of computer systems is the same as that used by some implementation researchers. For example, the Chorley report for the Government of the United Kingdom, *Handling geographical information* (Department of the Environment,

M. J. C. de Lepper et al. (eds.), The Added Value of Geographical Information Systems in Public and Environmental Health, 287–297.
© 1995 Kluwer Academic Publishers.

1987), also argues that, although the continued development of the technology is guaranteed, "We believe this to be a necessary though not a sufficient, condition for the take up of geographical information systems to increase rapidly." (paragraph 1.22). To ensure their rapid take-up and effective utilization it is also necessary to reduce a number of human and institutional barriers to change, some of which are administrative and personal while others are financial and technical in nature. The extent to which effective utilization is influenced by the personal and organizational setting in which computers are introduced is one of the most important findings of a series of studies on the utilization of computers in local government carried out by the Public Policy Research Organization at the University of California at Irvine (Danziger et al., 1982; King & Kraemer, 1985; Danziger & Kraemer, 1986). The Irvine group emphasizes the dynamic, political and context-specific nature of the environment in which local government computer usage takes place.

With these considerations in mind, the present chapter examines some of the broader issues involved in the utilization of GIS in organizations of the kind encountered in the health and environmental fields in the European Region.of WHO. Like Sabatier & Mazmanian's paper, it attempts to introduce a more prescriptive dimension into a discussion which has been largely dominated by the findings of descriptive analysis (Campbell, 1990). By trying to identify necessary and generally sufficient conditions for effective utilization it seeks to sharpen the focus of the current debate and also to provide some guidelines for users. Because of this it follows Sabatier & Mazmanian in addressing two different audiences: scholars interested in developing a conceptual framework for the analysis of computer utilization in organizations and practitioners seeking to maximize the chances of effective implementation.

The organization of the chapter is as follows. The first major section reviews the findings of the studies carried out by the Irvine group. It draws attention to the importance this group attaches to three main factors: the organizational context, the role of individuals, and the impact of change and instability. With these considerations in mind the second major section sets out three conditions for the effective utilization of GIS in the organizational environment. Like the conditions set out in Sabatier and Mazmanians paper, these are presented as "necessary and generally sufficient conditions". The final section of the chapter outlines various strategies for increasing the chances of effective utilization in situations where these conditions are not met.

19.2 Conceptual framework for analysis

19.2.1 Introduction

The work of the group based at the Public Policy Research Organization at the University of California at Irvine provides a systematic account of the impact of computerization on local government activities in the United States (Danziger et al., 1982; Danziger & Kraemer, 1986; King & Kraemer, 1985). The findings of these studies suggest that the adoption and effective utilization of information technology is largely dependent on organizational rather than technical considerations. As a result, the Irvine group has argued that computer technology should be regarded as a package involving not only hardware and software but also personal skills, operational practices and the need for substantial supporting resources for the operation of automated systems. The computer package is not, however, an independent entity. The studies undertaken by the Irvine group highlight the contribution of three sets of organizational factors to an understanding of the use of the computer package within a particular organization. These are the organizational context, people and the impact of change and instability. The manner in which these organizational factors interact with the computer package determines the processes that influence the development and use of automated systems. These factors will be examined in turn.

19.2.2 Organizational context

Studies focusing on computer equipment have tended to view the outcomes of automation to be universal. The first element of the conceptual framework questions this assumption, emphasizing the contribution of the organizational context to the initial adoption and subsequent use of computer-based systems. Contextual factors are subdivided into two levels, the internal organization and the external environment, which together provide the background against which computer usage takes place.

The internal organizational context relates to the attributes of the organization in which the information system is located. These include features such as the organizational structure, administrative arrangements and procedures for decision-making in general and specifically with regard to computing resources. In many instances two sets of internal organizational factors must be examined: firstly, those that relate to the planning agency and secondly, the characteristics of the wider organization in which it is set, such as a local or regional authority or national government. However, consideration of contextual factors must not be limited to the organization in which planning activities are located; features of the external environment also influence the development and utilization of automated systems. There has been a tendency for evaluations of external factors to be limited to the actions of computer suppliers and manufacturers. The Irvine group questions this perspective, stressing in addition the importance of examining the political and administrative influence of other local authorities and central government. Socioeconomic characteristics, professional opinion and the attitudes and level of technical awareness within the society served are also included under this heading.

It will be clear from these comments that, by emphasizing the need to consider the organizational context, it is not assumed that the conditions into which a computer package is located are universal. The important implication is that studies must start by examining how organizations are, not as many prescriptive studies, how they should be. It is inevitable that the pertinent details will vary between and also within an individual organization over time. The stress placed on the constraining as well as facilitating role of the internal and external organizational contexts should not be assumed to suggest rigid determinism. These contextual factors are regarded as the background against which the activities of the second element of the conceptual framework should be set.

19.2.3 People

The Irvine group, in contrast to traditional technically oriented studies, does not presume that the interests of individual members of staff coincide with the goals of the organization in which they work or that the benefits of computerization are shared evenly. It is emphasized that individuals have differing values and motivations and that computerization tends to challenge traditional interests, threatening some and offering opportunities to others. As a result, it is argued that the adoption and development of computer-based systems must be viewed in relation to the underlying political processes of an organization. The political allocation mechanism is perceived to act as the means through which negotiation takes place between interested parties. In most planning contexts, bargaining over the allocation of benefits emanating from computer-based systems takes place both within the agency itself and with other units in the same organization. Individuals can perform an important role in this process with the acquisition of resources frequently associated with a single member of staff who possesses the necessary ability, willingness and probably experience to fight the inevitable political battles.

In the face of these activities, the rest of the staff are not passive. There appears to be no necessary link between the existence of an information system and its use. A great deal would appear to depend upon user characteristics and the relationship between personalities. Individuals within organizations possess very different skills and views of the type of activities for which information systems are useful vary in their willingness and inclination to use technology. Danziger & Kraemers (1986) work indicates that staff confidence in their own computing capabilities and the interpersonal relationship between users and the technical specialists have a marked impact on the utilization of information systems.

The activities of individuals can substantially affect the development and utilization of computer-based systems. The outcome of automation is therefore influenced by the interaction of a complex set of human and contextual factors, but conditions are not static, as the third element of the conceptual framework emphasizes.

19.2.4 Change and instability

The Irvine group points to instability as a key factor in understanding the difficulties organizations encounter and the tendency for unexpected outcomes. Computer-based systems tend not to be designed once and for all, as circumstances change which require alterations to be made. Development therefore appears to be ad hoc and incremental, with amendments to existing systems favoured over the potentially greater disruption of an entirely new system. No organization is static. Staff, in particular, change but as far as is practical, it appears that a stable internal setting enhances computer usage. A highly volatile technical, social or political environment is liable to impede the effective development and utilization of automated systems.

19.2.5 Summary

The conceptual framework developed from the work of the Irvine group attempts to take account of the social and political processes that underlie the utilization of computer-based systems in organizations. Three sets of factors are highlighted that will either facilitate or constrain the activities of the computer package. The detailed factors that need to be considered in relation to each element of the conceptual framework will vary between organizations and in time. However, once the pertinent factors and the manner in which they interact once are distinguished, understanding of the varied outcomes of the introduction of GIS is possible (Dangermond, 1990).

19.3 Conditions for the effective utilization of GIS in organizations

19.3.1 Introduction

The contention of this chapter is that three necessary and generally sufficient conditions have to be met for the effective utilization of GIS in organizations. These are:

(i)   the existence of an overall information management strategy based on the needs of users in the agency and the resources at its disposal;

(ii)   the personal commitment of individuals at all levels in the organization with respect to overall leadership, general awareness and technical capabilities;

(iii) organizational and environmental stability with respect to personnel, administrative structures and environmental considerations.

19.3.2   The existence of an overall information management strategy based on the needs of users in the agency and the resources at its disposal

In effect this condition involves three sub-conditions relating to the development of information management strategies in general, the expression of user needs and the availability of resources respectively. The importance of developing an information management strategy is central to the distinction made by Cartwright (1987) between collecting data and creating an information system. Without an overall information management strategy that takes account of data availability, computing capacities and management and user requirements, it is likely that major problems will arize in relation to computer utilization. This is particularly the case with respect to strategic monitoring activities, where a high degree of selectivity is required to ensure that the right information gets to the right people at the right time (Masser, 1986). Without a well developed information management strategy, it is likely that there will also be mismatches between information needs and data availability as well as between data collection and information processing.

The chances of failure are likely to be increased where there is no clear expression of management needs. All too often, information system designers are faced with circumstances that reflect the findings of a study of monitoring practices in United States Government agencies. This concluded that many system developers had "to design and supply information to a management structure which may not know what information it wants or how it would act if it received particular types of information" (Waller et al., 1976, p.19).

Operational difficulties will be further aggravated where unrealistic assumptions have been made about the resources at the disposal of the planning agency to promote effective utilization. There is often a tendency to assume that the bulk of the costs are incurred in the process of setting up the computer system. In the process, the recurrent costs of data collection associated with maintaining and updating systems are seriously underestimated and little if any provision is made for adapting the system to meet changing circumstances. In order to develop an incremental learning approach to computer utilization, it is essential that adequate resources be set aside for these purposes from the outset.

The chances of effective utilization are likely to be seriously inhibited when a number of agencies are involved, even when there is substantial agreement about the desirability of the overall objectives, and resources have been made available for this purpose. The extent to which apparently minor differences in interpretation and relatively small differences in the emphasis given to operational priorities can affect the chances of successful implementation have already been demonstrated by Pressman & Wildavskys (1973) case study of the operations of the Economic Development Administration in Oakland.

The experience of the urban planning field indicates that problems of this kind are particularly likely to arize when computing tasks are carried out by other agencies as a service function. It has been argued, for example, that this was a major factor in inhibiting the effective utilization of computers in British planning agencies until the advent of microcomputers during the late 1970s (Masser, 1988). As a result of this separation of tasks, many planning agencies found themselves faced with not only problems of access to computer technology but also with the possibility of having to implement planning applications on hardware and software systems designed for very different types of operation.

19.3.3 *The personal commitment of individuals at all levels of the organization with respect to overall leadership, general awareness and technical capabilities*

This condition also involves three sub-conditions relating to leadership, awareness and technical capabilities. The most successful cases of computer utilization are those where there is clear leadership and a commitment from senior staff who are aware of the potential opened up by GIS. However, success also depends upon the commitment and enthusiasm of agency staff at all levels. Effective utilization is unlikely unless, as Calhoun et al., (1987) have pointed out, computer applications are seen to have a tangible effect on personal career development paths and remuneration. The prospects for effective utilization are likely to be further inhibited where the need for technical capabilities has been underestimated, particularly in relation to operations management experience. Specific skill shortages can be remedied relatively quickly by special programmes if the resources are made available, but extensive on-the-job experience is necessary for operations line management.

It should also be recognized that many users in organizations are essentially passive users whose primary responsibilities lie outside the field of geographical information management (Danziger & Kraemer, 1986). Effective utilization is therefore likely to depend on the extent to which these users see the execution of their primary responsibilities as being facilitated by GIS.

19.3.4 Organizational and environmental stability with respect to personnel, administrative structures and environmental considerations

Given the lead time that is required to implement GIS a relatively stable organizational environment must be regarded as a critical factor in promoting effective utilization. Failure to recognize the vital importance of key individuals, particularly those working in rapidly changing technological fields, and the need for continuity in order to build up operational experience can seriously undermine the chance of effective utilization, even in organizations with clearly defined information management strategies and a high level of staff commitment. These difficulties are likely to be further aggravated by the uncertainties created by the periodic internal reshuffling of tasks and personnel between sections and threats of transfers of responsibilities and staff to different agencies within the public sector as a whole.

The overall stability of organizations is also likely to be affected by a variety of broader environmental considerations. Political and economic instability must be regarded as an important factor in the external environment of many agencies, and it should also be noted that responsibilities are constantly evolving as the issues agenda itself changes. Under these circumstances, the chances are high that systems designed to meet one set of requirements may find themselves having to adapt to meet completely new configurations of policies and programmes.

19.3.5 Summary

Matters relating to stability, overall leadership, general awareness and the technical capabilities of organizations are likely to reflect prevailing attitudes, particularly those of the central government agencies responsible for gathering data and promoting information technology. As a result, there are likely to be important differences between the experiences of agencies in the countries that have given a high priority to the collection and dissemination of information and the development of information technology skills and the countries that have not (Masser & Blakemore, 1991).

19.4 Some advice for practitioners

Three conditions were set out in the last section as being necessary and generally sufficient conditions that must be met for the effective utilization of GIS in the organizational environment. It must be accepted that only a small majority of agencies satisfy these conditions in full and that the vast majority fall short in one or more respects. This raizes the question about what these agencies might do to reduce the risks of ineffective GIS utilization. With this in mind, the concluding section of this chapter outlines some strategies that may be of value to planning agencies that wish to promote more effective utilization despite failing to meet all three conditions.

Two general strategies underlie the whole discussion about conditions for effective utilization. These can best be summarized by the slogans "small is beautiful" and "one step at a time". The former strategy is a general response to the need to avoid over-ambitious plans, which carry with them high risks of failure. Where there is no clear information management strategy, a small-is-beautiful strategy should concentrate on limited applications that directly meet specific needs. Such an approach should also impose less demands on leadership, staff commitment and technical resources. It is less vulnerable to organizational and environmental changes because the lead time required for implementing projects of this kind is likely to be relatively limited.

The small-is-beautiful approach to effective GIS utilization has been given fresh impetus over the last few years by the advent of powerful microcomputers with a wide range of user-friendly software packages. The cost of the hardware has fallen dramatically while computer capacities have increased in an equally spectacular fashion to the extent that a typical desktop configuration today is likely to have similar capabilities to those of the typical mainframe computer a decade ago (Yeh, 1988).

The obvious limitations of a small-is-beautiful strategy in terms of its overall impact on the work of the agency as a whole can be overcome to some extent by following the one-step-at-a-time strategy. Such a strategy emphasizes the vital importance of learning by practical experience in a field that tends to be dominated by technological innovations. It is impossible to underestimate the impact that a modest demonstration project can have on raising the overall levels of awareness in an agency and its value in highlighting the potential costs and benefits of further developments. By building up experience in this way, many of the problems associated with a lack of an overall information strategy can be minimized, while awareness within the organization of both the potential and the limitations of computer technology is increased. Such an approach is also likely to help in exploiting the opportunities and countering the threat imposed by organizational and environmental change.

User participation in the design and ongoing development of GIS would appear to underlie these strategies and should be seen as the key to ensuring effective utilization. Users are most aware of the organizational limitations that exist and the strengths and weaknesses of their environment in terms of resources and personnel. As organizational factors rather than technological issues are increasingly recognized as the critical constraint on computer usage, it is vital that those with the fullest appreciation of their context and needs be involved. However, participation should be seen as very much more than the consultation of users by technical staff through formal working parties. In calling for greater user involvement reference is not being made to this type of symbolic participation but rather to a situation that places potential users at the centre of the process. This approach requires a significant commitment as well as willingness to take responsibility on the part of management and users. Management must encourage staff and be prepared to allocate time for their involvement in the development of automated systems. Flexibility must be built into the work programmes of staff to enable the valuable process of informal learning in the workplace to occur. This type of training is often more useful than attendance at official courses. The minimal strategy would be the appointment of an intermediary. This individual must have experience of user needs as well as basic technical knowledge. If handled well, user participation will help to diffuse the anxieties that lead to a lack of staff cooperation and, in some circumstances, to an active effort to impede system development. By involving staff at all levels, a supportive environment should be created in which technical staff act as facilitators serving the needs of users. Time spent at the early stages of development should save resources later as well as ensuring the utilization of systems by the individuals actually required to operate and use the technology.

There can be no guarantee that there will be effective utilization of GIS in organizations unless the three conditions set out above are fully satisfied. However, given the suggestions that have been made in this section, it can be seen that a great deal can be done to increase the chances of effective utilization even in agencies that cannot meet all these conditions.

References

Calhoun, C., W. Drummond & D. Whittington (1987). Computerized informationmanagement in a system-poor environment: lessons from the design and implementation of a computer system for the Sudanese Planning Ministry. *Third World Planning Rev,* **9**(4): 361-379.

Campbell, H.J. (1990). *The use of geographicalal information in local authorityplanning departments.* Thesis submitted for Ph.D. in Town and Regional Planning. Sheffield, University of Sheffield.

Cartwright, T. (1987). Information systems for planning in developing countries: some lessons from the experience of the United Nations Centre for Human Settlements (HABITAT). *Habitat Int,* **11**: 191-205.

Dangermond, J. (1990). How to cope with geographicalal information systems in your organization. *In*: H.J. Scholten & J. Stilwell, ed. *Geographical information systems for urban and regional planning.* Dordrecht, Kluwer Academic Publishers.

Danziger, J.N. & Kraemer, K.L. (1986). *People and computers: the impacts of computing on end users in organizations.* New York, Columbia University Press.

Danziger, J.N., W.H. Dutton, R. Kling & K.L. Kraemer (1982). *Computers and politics: high technology in American local governments.* New York, Columbia University Press.

Department of the Environment (1987). *Handling geographical information.* Report of the Committee of Enquiry chaired by Lord Chorley. London, HMSO.

King, J.L. & K.L. Kraemer (1985). *The dynamics of computing.* New York, Columbia University Press.

Masser, I. (1986). Monitoring and evaluation of local and regional planning in developing countries. *In*: *Information systems for urban and regional planning: Asian and Pacific perspectives.* Nagoya, Japan, UNCRD, pp.181-237.

Masser, I. (1988). Information management in British planning agencies. *In*: P. Nijkamp & M. Giaoutzi, ed. *Informatics and regional development.* Avebury, Aldershot.

Masser, I. & M. Blakemore (1991). The institutional setting. *In*: I. Masser & M. Blakemore, ed. *Handling geographical information: methodology and potential applications.* London, Longman.

Pressman, J.L. & A. Wildavsky (1973). *Implementation: how great expectations in Washington are dashed in Oakland.* Berkeley, University of California, Press.

Sabatier, P. & D. Mazmanian (1979). The conditions of effective implementation: a guide to accomplishing policy objectives. *Policy Anal,* **5**: 481-504.

Waller, J.D., D.M. Kemp, J.W. Scanlon, F. Tolson & F.S. Wholey (1976). *Monitoring for government agencies.* Washington, D.C., Urban Land Institute, (Report 783-41).

Yeh, A.H. (1988). Microcomputers in urban planning: applications, constraints and impacts. *Environ Planning B,* **15**: 241-254.

Ian Masser and Heather Campbell
Department of Town and Regional Planning
University of Sheffield
Sheffield S10 2TN
United Kingdom

20 BUILDING A GEOGRAPHICAL INFORMATION SYSTEM IN THE EUROPEAN COMMUNITY: THE CORINE EXPERIENCE

David J. Briggs

Abstract
The CORINE Programme of the European Community represents one of the most ambitious and extensive attempts to establish a geographical information system for policy applications that has yet been undertaken in Europe. As such, it has much useful experience to offer the various new international systems that are now being developed or are planned. This chapter outlines the experience of establishing the CORINE information system and shows some of the conceptual, technical and administrative problems involved.

20.1 Introduction

Since their initial development in the 1970s, geographical information systems (GIS) have become established in many areas of education, science, industry and research, and their range of applications continues to grow (Green, 1990; Worrall, 1990). Some of the most important opportunities and challenges for GIS, however, probably lie in the area of policy development and analysis. GIS has already become an almost essential tool for local government (Gault & Peutherer, 1990). At the same time, many national systems are being built, often dedicated to specific policy areas - for example, defence (Ball & Babbage, 1989), natural resources (Cara, 1991), urban planning (Geertman & Toppen, 1990), environmental management (Cocks & Walker, 1987), or social services (Whitehead & Hershey, 1991). Now attention is beginning to to be focused on the wider arena of international policy.

Experience of building international GIS is limited. Possibly the best established example to date has been the GRID system, developed by UNEP as an information source for global monitoring and modelling (Mooneyhan, 1988). As Hastings et al. (1991) conclude, however, efforts to develop global databases have tended to suffer from a number of problems, including poor source data, intellectual isolation, inadequate attention to data integration, an over-simplistic top-down approach to system design and insufficient collaboration.

In this context the experience of the CORINE Programme of the Commission of the European Communities has much to offer. Adopted as an experimental programme to "gather, coordinate and ensure the consistency of information on the state of the

M. J. C. de Lepper et al. (eds.), The Added Value of Geographical Information Systems in Public and Environmental Health, 299–314.

environment and natural resources" as a basis for Community policy applications (Commission of the European Communities, 1985), this involved the construction and the evaluation of a broad-scale, transnational and multithematic GIS. The purpose of this chapter is to review briefly the background to and history of the programme and to outline some of the lessons to be gained.

20.2 The CORINE Programme

20.2.1 Background

The origins of the CORINE Programme lie in the development of environmental policy in the European Communityunity. Since the adoption of the first environmental action programme in 1973, the European Community has played an increasingly influential role in environmental affairs at both the national and international level. During the last 18 years, over 200 items of environmental legislation have been adopted (Commission of the European Communities 1987) and the Community's Fifth Action Programme on the Environment started in 1993.

These developments reflect not only the strengthening of the Community as a policy agent but also a change in environmental perceptions. Over recent years, it has become progressively more clear that many of the major environmental problems that face western Europe are international in their origins and impacts and often require an international response. The European Community provides an obvious body to coordinate and lead this response and, through its environmental policy, set targets and standards that its Member States should achieve.

Action to resolve the problems of the environment, however, requires more than political will or administrative structures and machinery. It is, equally, a matter of information. This is required:

(i)   to ensure that problems are recognized and tackled before it is too late and while remedies are still cost-effective;
(ii)   to allow policy actions to be adapted to and targeted at specific problems more effectively;
(iii)   to provide objective evidence on the nature and magnitude of environmental problems and thereby enable a more rational prioritization of environmental actions;
(iv)   to monitor and evaluate the effect of environmental policy on the ground; and
(v)   to inform the public.

To meet these objectives, it is clear that information must itself satisfy a number of criteria:

(i)   it must be consistent both spatially and temporally to allow real comparisons to be made;
(ii)   it must be available in advance of policy formulation so that it allows policy to be proactive rather than reactive;
(iii)   similarly, it must be available in advance of policy implementation so that baseline conditions can be defined against which the effects of policy can be measured;
(iv)   it must be cheap to obtain so that the benefits of using the data outweigh the cost of their acquisition; and
(v)   it must be presented in a way that policy-makers can readily understand and apply.

Until recently, it must be admitted, information that met these specifications was scarce. To a great extent, the limitations were administrative. Organizations to direct, collate and standardize environmental data did not exist. Information-gathering thus tended to be sporadic and *ad hoc*. But the problems were also in part technological: systems were not available for the wholesale monitoring of environmental conditions nor the handling of the data that could be obtained. Policy formulation, as a consequence, had to rely mainly on expert opinion and political judgement. Today, the constraints on information supply are being removed. As has been noted, the European Community is taking a central role in policy development and - through this - in information gathering and exchange. Indeed, it has now agreed to establish a European Environment Agency, with the explicit role of providing information to underpin Community policy (Briggs, 1991a). Other international bodies (e.g., UNEP, IUCN Environmental Law Centre) are also following suit. At the same time, technological developments, most especially in the areas of remote-sensing an information technology, are providing the means for mass data collection, analysis and use.

20.2.2 *Implementation and approach*

The CORINE Programme was established against this backcloth of both growing need for information and increasing capability to supply it (Commission of the European Communities, 1985). Its objectives were broad: not merely to collect policy-related information on the environment but also to establish procedures and instruments (including GIS) for future data management and control. At the same time, in order to narrow its focus, the decision specified a number of priority areas to which attention should be given:

(i)   biotopes (i.e., sites of importance for nature conservation);

(ii)   acid deposition (including atmospheric emissions and impacts on biotopes and soils); and

(iii) problems of the Mediterranean region (including land resources, soil erosion, land use, coastal problems and water resources).

A major component of the Programme was thus the establishment of an information system able to provide and analyse information on these various themes. To this end, a series of project teams was established, under the auspices of the Directorate- General for Environment, Nuclear Safety and Civil Protection, each focusing on one of the priority areas (Figure 20.1). The role of these teams was to devise appropriate methods for data analysis and modelling, to collect the data required and to evaluate and validate the results. These groups were serviced and supported by two further study teams: one responsible for constructing the database and the other for data quality, consistency and overall scientific coordination of the work.

20.2.3 The CORINE GIS

The information system was devized to provide a series of datasets, each relating to one of the themes specified in the original Council decision and registered to a consistent geographical base. Several different computer systems were used to create the information system - depending on the hardware and software available in the collaborating institutions. The final, central information system, however, was compiled in ARC/INFO - initially running on a VAX 11/750 at Birkbeck College, London and subsequently, on a series of Tektronix work stations at CORINE headquarters in Brussels. The general system specifications are shown in Table 20.1. Further details of the CORINE information system have been given elsewhere (Briggs, 1991b; Briggs & Martin, 1988; Briggs & Mounsey, 1989; Wyatt et al., 1988). An outline of the CORINE database at a formative stage in its development has also been given by Whimbrel Consultants Ltd/Huddersfield Polytechnic (1989), and detailed reports describing the various projects undertaken within the Programme are currently being prepared (Briggs & Giordano, 1992). Once it has been set up, the European Environment Agency will take over management of the system as part of its routine responsibility (Briggs, 1991a).

Figure 20.1: Organization of the CORINE Programme

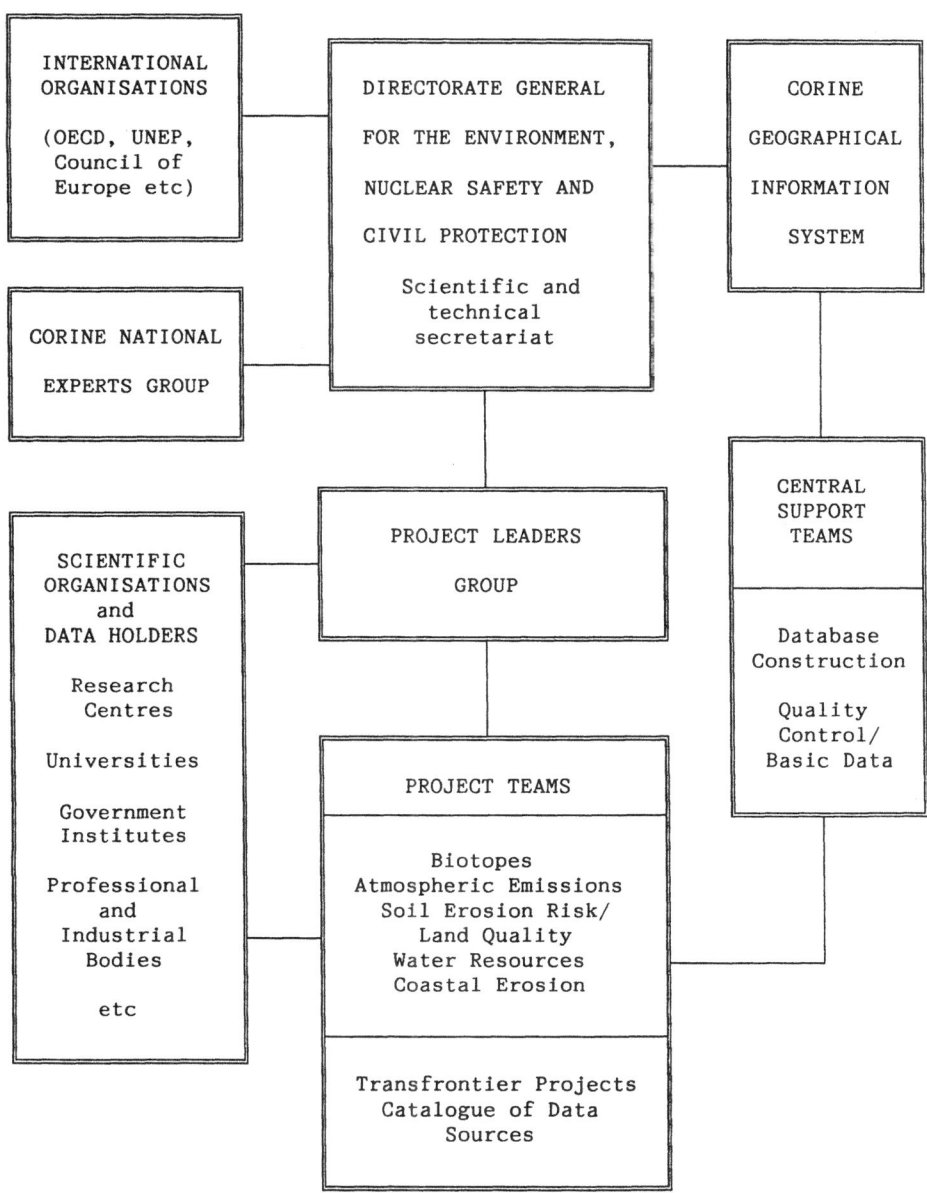

Table 20.1: Specifications for the CORINE Information System

Hardware	:	Tektronix work station
Software	:	ARC/INFO
Operating system	:	UNIX
Base scales	:	1:100,000, 1:1 million,
		1:3 million, 1:5 million
Base projections	:	Lambert Azimuthal
		Spheroid Int. 1909
		Radius of spheroid of reference : 0
		Longitude of centre of projection : 09 00 00 E
		Latitude of centre of projection : 48 00 00 N
		False easting : 00
		False northing : 00
Data volumes (estimated)	:	1500 MB

Main datasets:

Geography	- coastline
	- water pattern
	- settlements
	- national/regional boundaries
	- gazetteer
Climate	- temperature
	- precipitation
	- relative humidity
	- sunshine/cloud cover
	- windspeed
Nature	- biotopes
	- designated areas
	- vegetation
Land resources	- soils
	- soil erosion risk (Mediterranean area)
	- land quality (Mediterranean area)
Water resources	- water quality
	- stream discharge (Mediterranean area)
Air pollution	- emissions
	- air quality
Human	- population
	- socioeconomic activities
Land cover	
Coastal erosion risk	

20.3 The lessons learned

The design and construction of the CORINE information system inevitably faced many difficulties. Possibly the most fundamental of these related to the availability, quality and volume of data involved. Other problems, however, were largely administrative in origin. Together, these difficulties not only caused delays in the Programme but also imposed significant constraints on database design and on the analytical techniques and models that could be used - and thus affected the quality of the results obtained.

20.3.1 Data requirements

Fundamental to the CORINE information system - as any GIS - was the question of data requirements. As indicated earlier, these were to a large extent predefined by the Council decision: simply, the system had to provide policy-related information on the three thematic areas specified (biotopes, acid deposition and problems of the Mediterranean area). The implicit expectation of those who finalized the Programme was that such information readily existed in a relatively consistent form and would be freely provided by agencies in the Community Member States.

As will be seen, this was far from the case. Instead, to meet the objectives of the Programme, it was necessary to compile an extensive and often detailed, database. Topographic and administrative data had to be collected, for example, to form a geographical base on which the information system could be built and as a backcloth for output and display. Data on soils, climate, topography, socioeconomic activities and demography were needed as inputs to the various assessment methods and models (e.g., to assess soil erosion risk and land quality). Extensive ecological information was needed to allow the selection and description of biotopes. In addition, as the Programme proceeded, a range of extra information was requested to tackle short-term problems, to promote and publicize the information system itself and to support the day-to-day work of the Environment Directorate-General. The definition of data needs and the location of appropriate data sources, was thus a major - and to great extent underestimated - aspect of the work.

In this context, the issue of user requirements is also crucial. It is frequently stated that any GIS should be built upon a clear understanding of the user community and its specific needs. In the case of the CORINE Information System, initial attempts to survey potential users in other Directorates-General met with limited success and at times encountered considerable resistance to the very notion of GIS. Many potential users were unaware of what GIS could do until they had seen the system in operation. Even then, it was often perceived only as a substitute for existing operations rather than as a facility that might add new functions and generate new opportunities and areas of application.

Such wider and more liberal views tended to develop only when the users became relatively experienced and began to play with the system.

These problems of user awareness can perhaps be resolved by a clear strategy of promotion and education. In the case of the CORINE Programme, demonstrations and pilot exercises proved especially valuable in creating an understanding of the potential of the system. Even so, this itself tended to generate new problems. At times, it helped to encourage unrealistic expectations both about system capability and the expertise necessary to use it; more than one potential user said that he planned to arrange for his secretary to run the system on his behalf!

Since the early days of the Programme, fortunately, these problems have certainly declined, due to the wider awareness and acceptance of GIS. Nevertheless, the fact that users and uses of GIS typically germinate and grow as the system itself evolves means that it remains difficult to define the ultimate data requirements during the early stages of system conceptualization and construction. Yet, if policy development is not to be unacceptably delayed - and, indeed, if it is to be guided by prior understanding of the problems that actually exist - it is essential that data be readily available, more or less on demand. This, in turn, requires that data needs be generously provided for and that the range of data provided - the number of attributes and the level of detail - not be unnecessarily restricted. Some allowance must therefore be made in the system for data redundancy and data volumes may be expected to be inflated as a result: the attempt to pare the database to a minimum will almost certainly prove counterproductive. The original database also needs to be constantly updated and revized. To date, techniques for handling time-sequence data are limited and storage of serial data of this type itself results in much redundancy. Inevitably, therefore, any system must have substantial room to expand.

20.3.2 Data availability and quality

The wide range of data required by the CORINE Information System inevitably posed problems of data availability and quality. Systems such as this cannot realistically undertake primary measurement and monitoring; they must rely upon data that already exist or are being collected. Yet existing data are notoriously incomplete and variable. Gaps in data availability were encountered in almost all datasets required by CORINE and occurred for a wide range of reasons, including:

(i)   lack of primary survey or monitoring networks (e.g., water quality in Italy and Greece);

(ii)   equipment failure (e.g., many air quality monitoring stations);

(iii) administrative failure (e.g., gaps in stream discharge data in Italy due to failure to process the data at the regional or national level; gaps in climate data generally due to the interruption of the Second World War);

(iv) confidentiality (e.g., military restrictions on topographic data in Greece; commercial restrictions on emission data in the United Kingdom);

(v) cost (e.g., climate data in Germany; commercially published gazetteers; digital terrain models for France);

(vi) ignorance about the data that do exist.

Among the data that are available, considerable inconsistencies and errors may also occur. These, too, arise for a wide array of reasons, including differences or errors in:

(i) the parameters measured or surveyed, or in their detailed definitions;

(ii) the measurement methods or units used;

(iii) sample design (e.g., spatial distribution, timing, frequency, duration, number);

(iv) data processing (e.g., classification, aggregation, averaging, modelling);

(v) spatial referencing (e.g., projection, coordinate systems, size of spatial entities);

(vi) data reporting and representation (e.g., symbolization).

Examples of all these types of inconsistency were found in the course of developing the CORINE Information System. Definitions of designated (or protected) areas, for example, vary substantially between different Community Member States. The parameters used to monitor both water and air quality differ both in kind and measurement methods from one country to another (despite the adoption by the Community of legislation on data exchange). Data on stream and bathing water quality similarly show discrepancies in sampling frequencies, location and methods across the Community. Meteorological data vary in terms of the definition of the parameters, measurement method, timing and location. National models for estimating emissions from vehicles and industry vary. Methods of classifying soil, land use and even population are not consistent between Community Member States. Topographic maps differ in their symbolization, scale and projection. The administrative units used to compile socioeconomic statistics in EC Community Member States vary substantially in size. Even the 1:1 million Soil Map of the European Communities - one of the few ready-made sources available covering the whole Community - was found to contain serious cartographic inconsistencies that had to be resolved (Briggs et al., 1989).

Dealing with these various discrepancies and gaps in the available data was, without doubt, one of the major technical tasks involved in setting up the CORINE system. Particular attention had to be given to the process of data cleaning and integration. Data provided in digital form, for example, had to be reformatted and checked. Cartographic features typically had to be reprojected to the reference projection, either by using standard conversion functions (where projection details were known) or by the use of rubber-sheeting (where they were not). Each separate coverage had to be registered to the geographical base, either by using automatic fitting procedures or by manual editing.

The quality and consistency of attribute data similarly had to be carefully checked. Where data had been produced from different measurement systems, algorithms had to be sought to convert them to a common form. Where gaps in the input data existed, it was necessary either to fill them or to omit the variables concerned from the analysis. The former was clearly only feasible where suitable surrogates existed; in practice, however, surrogate data were often of uncertain quality, contained inconsistencies and tended to reduce the performance of the model or analytical procedure. Omission of the variable from the analysis was feasible when the variable concerned was relatively unimportant but was clearly unsuitable when it had a major influence on the outcome of the analysis. It must also be recognized that errors and inconsistencies in the source data may not be evident. Poor documentation, for example, often makes it difficult to compare the characteristics of data from different sources. In these cases, the inconsistencies and errors are liable to be incorporated into the information system and replicated in any subsequent analysis.

All the results generated had also to be validated for both positional and metrical accuracy. While plotting of results allowed broad-scale positional errors to be identified (e.g., sites falling in the wrong country, the wrong administrative region, or the sea), smaller locational displacements were difficult to detect. Similarly, while various statistical and logical tests could be applied to check for impossible outcomes, the quantitative accuracy of the results could not always be ascertained - not least because many of the phenomena being modelled are essentially conceptualizations (e.g., soil erosion risk, land quality) for which independent, measured data cannot be obtained. Validation was consequently often an essentially subjective process, in which members of the project teams visually examined the outputs and searched for inconsistent or unacceptable results.

20.3.3 Data volumes and system design

The spatial scale of the CORINE Information System was a major concern. The area to be covered was both large - some 2.3 million km² - and geographically complex and diverse. Many of the policy issues to be addressed also inevitably focused on small areas or entities (e.g., local biotopes or point emission sources). An inevitable conflict consequently arose: to be of value for policy applications, the scale of analysis (and thus the base scale for data collection) had to be relatively large, yet to be manageable it also had to be as small as possible. No single geographical scale could therefore meet all requirements. Instead, a scale of 1:1 million was chosen as a common base, with more generalized versions of the data (e.g., at 1:5 million or 1:10 million) being used for display purposes. In addition, a 1:100,000 scale database is being constructed from land cover mapping using remotely sensed imagery.

The varied data needs, the relatively large spatial scale required for basic data and the need to hold multiple copies of datasets at different scales all meant that the data volumes involved in the CORINE system were relatively large. The basic data held in the system probably amounted to about 500 MB; together with the derived data (results of overlays and modelling etc.), the total was about 1,500 MB. During processing, individual datasets often exceeded 300-400 MB in size. Despite the continuing improvements in computer power and performance, the implications of these data volumes for system performance and design are significant.

In the case of the CORINE Programme, for example, disk and core space were at times limiting, with the consequent need to download datasets onto tape whenever extra disk space was required for major operations. Conflicts of work also arose due to the need to tie up computers or plotters for long periods of time for routine tasks of compilation and checking. For the same reasons, some GIS tasks - such as rubber-sheeting the soils coverage to fit the base projection or overlay of the full datasets in the computation of soil erosion risk - could not be performed as a single operation. Instead, the relevant coverage had to be sectioned to provide smaller datasets, processed, then recompiled and edge-matched. All these procedures were inevitably time-consuming.

One important control upon data volumes and system performance is the choice of data structure. Much has been written about the relative merits and demerits of raster, vector and quadtree structures (Peuquet, 1984) and the general conclusion must be that each has advantages in specific circumstances. In the case of the CORINE system, however, it was clear that the vector-based system used was not always optimal for complex overlay procedures with large datasets. The data in Tables 20.2 and 20.3, for example, show disk-space requirements and processing times for overlaying various datasets in the analysis of soil erosion risk and land quality. As can be seen, both space and time requirements were relatively high under vector structures, especially when the data were spatially complex. In these cases, it proved more efficient to convert data into raster form for analysis outside the system.

Table 20.2:   Comparison of raster and vector overlays: disk space requirements for the
               analysis of soil erosion risk and land quality in Portugal

	Raster blocks	Vector blocks	Ratio raster:vector
Original data storage			
a. Erosivity	447	135	3.31
b. Erodibility	447	371	1.20
c. Vegetation cover	447	1,073	0.42
d. Slope angle	447	4,705	0.10
e. Climate	447	132	3.39
f. Soil quality	447	443	1.01
g. Irrigation	447	262	1.71
Total	3,129	7,121	0.44
Intermediate disk space	10,728	90,280	0.12
Storage of results			
Potential soil erosion (a+b+d)	447	3,831	0.12
Actual soil erosion (a+b+c+d)	447	3,884	0.12
Potential land quality (d+e+f)	447	5,128	0.09
Actual land quality (d+e+f+g)	447	4,214	0.11
Total storage of results	1,788	17,057	0.10

20.3.4 Project management

As the preceding discussion has indicated, technical problems of data quality, data
availability and system performance were major concerns throughout the CORINE
Programme. These, however, were not the only issues determining the success and
cost-effectiveness of the work. Equal importance was attached to the personnel
concerned. This was no trivial issue. Over 120 technical and scientific staff were
involved in developing the information system, working in some 30 institutes, in all 12
Community Member States. The opportunities for inconsistency and misunderstanding
were therefore considerable; effective project management and coordination were
essential.

Table 20.3 :   Comparison of raster and vector overlays: processing requirements for
                   the analysis of soil erosion risk and land quality in Portugal

	Raster minutes	Vector minutes	Ratio raster:vector
CPU processing time	31.1	589.5	0.05
Data transfer			
Rasterization	60.0	0.0	-
Import into raster	90.0	0.0	-
Total processing time	181.1	589.5	0.31

Source: Briggs & Giordano (1992))

The framework for management and coordination was provided by a series of
administrative and technical groups (Figure 20.1). To translate this structure into
effective operation, however, various methods were employed. In collaboration with the
scientific coordination group, for example, each project group compiled a manual
defining the minimum data specifications and the working procedures to be followed.
The project groups held regular meetings to exchange experiences and compare results.
New methods or procedures were tested in pilot studies before general application. To
standardize data transfer and exchange, a simple transfer format was designed (Hayes-
Hall, 1989). All data provided to the CORINE Information System were also checked,
either by the project leader or by the central management team.

Despite this effort, however, many discrepancies in method and data arose.
Computational procedures tended to deviate from those that had been agreed; protocols
for data description were not always adhered to; methods of digitizing were not always
consistent; and the formats used for data transfer frequently varied from those specified.

The reasons were largely institutional and personal and thus difficult to resolve. Many of
the organizations involved have their own protocols and procedures that they are often
reluctant (or unable) to abandon. Individuals often did not have the time to carry out the
necessary conversions or may simply have mislaid, misunderstood or ignored the
procedures required. No matter what was agreed at meetings, therefore, individual
contributors had a tendency to go their own way once they were back on home ground.

Nor should it be assumed that these problems are unique to the CORINE Programme or can be resolved simply by legislation. Similar discrepancies exist, for example, in almost all the data provided to the Commission of the European Communities by the Member States under the terms of existing environmental directives (e.g., bathing water quality, air quality, surface water quality), despite the existence of relatively rigid data standards (Commission of the European Communities, 1987). As was seen in the Programme, these discrepancies often feed through into the published datasets. Error and inconsistency in data are thus recyclable and transferable; in many ways GIS help to make them even more contagious. These errors and inconsistencies are not merely of academic concern. At the very least they raise the cost of constructing a reliable information system; at worst they may lead to inappropriate decisions and policy action. Quality control in GIS is therefore paramount. The ultimate lesson from the CORINE Programme is that, to achieve quality control, as much attention must be given to the human aspects of the systems as is normally lavished on the data, software and hardware involved.

References

Ball, D. & R. Babbage, ed. (1989). *Geographical information systems: defence applications*. Sydney, Pergamon Press.

Briggs, D.J. (1991a). Towards a European Environment Agency. *In*: Gilg, A., ed. *Progress in rural policy and planning*, London, Belhaven, pp. 157-174.

Briggs, D.J. (1991b). Establishing an environmental information system for the European Community: the experience of the CORINE Programme. *Inf Services Use*, **10**: 63-75.

Briggs, D.J. & D. Martin (1988). CORINE: an environmental information system for the European Community. *European Environ Rev*, **2**: 29-34.

Briggs, D.J., P. Brignall & A. Wilkes (1989). Assessing soil erosion risk in the Mediterranean region: the CORINE programme of the European Communities. *In*: H.A.J. van Lanen & A.K. Bregt, ed. *Application of computerized community soil map and climate data*. Luxembourg, Office for Official Publications of the European Communities, pp. 195-210.

Cara, P. (1991). The Italian geological database: in the beginning. *In*:: J. Harts, H.F.L. Ottens & H.J. Scholten, ed. *EGIS 91, Proceedings Second European Conference on Geographical Information Systems*, Brussels, April 2-5 1991. Utrecht, EGIS, pp. 876-883.

Cocks, K.D. & P.A. Walker (1987). Using the Australian Resources Information System to describe extensive regions. *Appl Geogr*, **7**: 17-27.

Commission of the European Communities (1985). Council Decision of 27 June 1985 on the adoption of the Commission work programme concerning an experimental project for gathering, coordinating and ensuring the consistency of information on the state of the environment and natural resources in the Community (85/338/EEC). Official *J Eur Communities*, **L176**: 14-17.

Commission of the European Communities (1987). *The state of the environment in the European Community, 1986.* Luxembourg, Office for Official Publications of the European Communities.

Department of Environment (1987). *Handling geographical information.* Report of the Committee of Enquiry chaired by Lord Chorley. London, HMSO.

Gault, I. & D. Peutherer (1990). Developing GIS in local government in the UK: case studies from Birmingham City Council and Strathclyde Regional Council. *In:* L. Worrall, ed. *Geographical information systems: developments and applications.* London, Belhaven, pp. 109-132.

Geertman, S.C.M. & F.J. Toppen (1990). The application of GIS for the allocation of housing in the Randstad Holland, 1990-2015. *In:* L. Worrall, ed. *Geographical information systems: developments and applications.* London, Belhaven, pp. 189-214.

Green, R. (1990). Geographical information systems in Europe. *Cartographic J*, **27**: 40-42.

Hastings, D.A., J.J. Kinemena & D.M. Clark (1991). Development and application of global databases: considerable progress, but more collaboration needed. *Int J Geogr Inf Systems*, **5**: 137-146.

Hayes-Hall, E. (1989). *The CORINE data transfer specifications. Version 4.0.* London, Birkbeck College, (CORINE Working Paper).

Mooneyhan, D.W. (1988). Applications of geographical information systems within the United States Environment Programme. In: H. Mounsey, ed. *Building databases for global science.* London, Taylor & Francis, pp. 315-329.

Peuquet, D.J. (1984). A conceptual framework and comparison of spatial data models. *Cartographica*, **21**: 66-113.

Whimbrel Consultants Ltd/Huddersfield Polytechnic (1989). *The CORINE database manual. Version 2.1.* Huddersfield, Whimbrel Consultants Ltd/Huddersfield Polytechnic.

Whitehead, C.D. & R.R. Hershey (1991). GIS in social services: child protection in the London Borough of Southwark. *In:* J. Harts, H.F.L. Ottens & H.J. Scholten, ed. *EGIS 91, proceedings Second European Conference on Geographical Information Systems.* Brussels, Belgium, April 2-5 1991. Utrecht, EGIS, pp. 1238-1240.

Worrall, L., ed. (1990). *Geographical information systems: developments and applications.* London, Belhaven.

Wyatt, B.K., D.J. Briggs & H.M. Mounsey (1988). CORINE: an information system on the state of the environment in the European Community. *In:* H. Mounsey, ed. *Building databases for global science.* London, Taylor & Francis, pp. 378-396.

David J. Briggs
Institute of Environmental and Policy Analysis
University of Huddersfield
Queensgate
Huddersfield HD1 3DH
United Kingdom

21 IMPLEMENTING A GLOBAL GEOGRAPHICAL INFORMATION SYSTEM FOR MODELLING SUSTAINABLE ENVIRONMENTAL QUALITY: THE CRITICAL LOAD EXPERIENCE

Evert N. Meijer, Jean-Paul Hettelingh and Paul Padding

Abstract

Air pollution has become an increasing threat to environmental quality over the past decades in all parts of the world. An example of this is known as acid rain: deposition of acidifying compounds, i.e., sulfur and nitrogen, which are emitted in the atmosphere by fuel combustion. The transboundary atmospheric transport of these acidifying compounds covers large areas, resulting in environmental effects, e.g., forest dieback in regions that may be distant from the sources. Recently, other environmental effects, e.g., climatic change, damage to the ozone layer and eutrophication of soils and waters are seen having global consequences. The assessment of the regional distribution of environmental effects has become an important issue for scientists and relevant input for policy-makers. One of the tools in evaluating abatement strategies is to compare regional acidifying deposition with regional patterns of acceptable acidification, so-called critical loads. In 1989 a Coordination Centre for Effects (CCE) has been established in the framework of the United Nations Economic Commission for Europe at the National Institute of Public Health and Environment Protection (RIVM) in the Netherlands in order to produce maps of critical loads and maps displaying the excess of critical loads by current acidification. The method used by the CCE consists of the application of a mathematical model to compute critical loads and the use of geographical information systems (GIS). In 1992 a project group financed by the World Bank, in which RIVM participated, has started the mapping of critical loads in Asia. In this chapter, the potential of using an environmental model in combination with GIS is presented based on the experience of air pollution research in Europe and China. It is shown that this combined method enables the computation of environmental quality indicators and the identification of regions where environmental quality is most at risk.

21.1 Introduction

Acid rain has become an increasing threat to environmental quality over the past decade. Acid rain is the term used to describe the deposition of acidifying compounds, i.e., sulfur and nitrogen, which are emitted in the atmosphere by fuel combustion. The transboundary atmospheric transport of these acidifying compounds cover large areas, resulting in environmental effects, e.g., forest dieback, in regions that may be distant from the sources. The decline of forest vitality in central and eastern Europe and the many dead lakes in Scandinavia and Canada are examples of damage that is attributed to acidification. Unfortunately, the protection of environmental quality in industrialized countries is often a reaction to degrading natural conditions that are difficult to reverse, especially in these countries. What is the situation in developing countries?

M. J. C. de Lepper et al. (eds.), The Added Value of Geographical Information Systems in Public and Environmental Health, 315–330.
© *1995 Kluwer Academic Publishers.*

Asian countries have rapidly growing economies leading to increased industrialization, energy consumption and the use of resources in general. There is a growing concern that the rapid growth in developing countries such as those in Asia does not sufficiently take into account the sustainability of the environment. Resources are depleted at a pace that cannot be kept up by nature, and the residuals of economic growth (e.g., waste and water pollution) may not be sufficiently taken into account.

Besides damage to ecosystems, acidification may in the long term reduce the quality of groundwater, damage materials such as buildings, fabrics and books and damage crops (RIVM, 1992). Recently other environmental effects, e.g., climatic change and damage to the ozone layer have been identified on a global scale. However indirect threats to environmental sustainability may take years to become apparent, which is the case for air pollution (Meijer et. al., 1993).

In Europe, it took until the OECD conference on Man and Environment in 1972 to recognize the air pollution effects that have evolved since the industrial revolution. Measures to reduce the emission of air pollutants did not start to become a common policy issue in European countries before the beginning of the 1990s. The reduction of acidifying emissions of sulfur and nitrogen are the subject of international negotiations in the United Nations Economic Commission for Europe (UN/ECE) under the Convention on Long-Range Transboundary Air Pollution (Economic Commission for Europe, 1985, 1988).

These negotiations have led to international agreements on the reduction of emissions of sulfur and nitrogen oxides. A sulfur protocol, which took effect in 1987, commits participating countries to reduce emissions by at least 30% as soon as possible and at the latest by 1993. A nitrogen protocol in effect since early 1991 requires that, by 1994, annual national emissions of nitrogen oxides should not exceed 1987 levels. This protocol also requires that average annual emissions of NO_x between 1987 and 1994 should not exceed the 1987 emission level. Both protocols are similar in that they do not explicitly include quantitative considerations about environmental effects. The abatement intentions are based predominantly on technical and economic considerations related to emission reductions. The revision of these protocols, currently underway, now also includes environmental protection as an explicit means of analysing the effect of emission reduction.

Also, among policy-makers concerned with developing countries, there is a growing awareness that economic development and policies to protect environmental quality should be considered simultaneously. Since 1992 a project group financed by the World Bank and the Asian Development Bank has started the analysis and mapping of critical loads in Asia.

An important aspect in the analysis of environmental problems is their transboundary character, which is not only limited to air pollutants. For instance, the pollution of the Rhine in France and Germany affects on the groundwater through the estuaries in the Netherlands; the Chernobyl disaster in the former Soviet Union caused fear in large parts of Europe. International collaboration between policy-makers and scientists has shown to be useful for the analysis of long-range transport of air pollutants. Scientists develop assessment methods allowing policy-makers to evaluate the effects of policy measures on varying temporal and spatial scales.

This chapter describes how geographical information systems (GIS) are used in combination with a mathematical model to quantify the effects of acidification. In section 21.2 the problem of acid rain is described briefly. Methodological issues are discussed in section 21.3. Finally, the results and conclusions are described in sections 21.4 and 21.5 respectively.

21.2 The origin and consequences of acid rain

The most important acidifying substances are sulfur dioxide (SO_2), nitrogen oxides (NO and NO_2) and ammonia (NH_3) and their reaction products. The anthropogenic emissions of SO_2 are largely due to the combustion of sulfur-containing fuels (oil and coal) used in the process industries (refineries) and power stations. NO_x is emitted by combustion processes associated with transport, power generation and heating. Most of the NH_3 in the atmosphere is due to the production of animal manure. NH_3 is a base gas that may lead to acidification after it, as a gas or as NH_4, reaches the soil and is nitrified. Total acid load is expressed as the moles of H^+ per hectare per year, allowing for the aggregation of varying acidifying compounds. It is assumed that each SO_2 molecule will eventually produce two H^+ ions and each NO_x or NH_3 molecule one H^+ ion (RIVM, 1992).

The concepts of critical loads and critical levels is a recent tool for quantifying deposition and concentration limits for various ecosystems (de Vries et. al., 1991a,b; Hettelingh & De Vries, 1992; Sverdrup & Hettelingh, 1990). The critical load is defined as a quantitative estimate of an exposure to one or more pollutants below which significant harmful effects on specified sensitive elements of the environment do not occur according to present knowledge (Nilsson & Grennfelt, 1988).

Critical chemical variables such as concentrations of aluminium and nitrogen, the ratio between aluminium and calcium, some nitrogen-base cation ratios and the acidity or alkalinity of the soil (moisture) or water and the nitrogen content of needles (De Vries et.

al., 1991a,b) are used as indicators for risk of damage. These indicators vary between ecosystems. For example a pH below 6.0 in surface water may lead to fish mortality, but this level of acidity in forest soils hardly affects the vitality of the forest. Using steady state models (steady state mass balance (SSMB) approach) that describe the production and consumption of acidifying compounds at equilibrium, it is possible to compute critical loads from such critical values.

In other words, when a critical load is exceeded for an extended period of time a violation of acceptable values for some or all of these critical chemical values will result (Hettelingh et al, 1992a) leading to an increased risk of damage to ecosystems in large geographical areas. Therefore, an appropriate assessment of (future) effects has to be modelled in combination with a GIS.

In 1992 a project group financed by the World Bank has started the mapping of critical loads in Asia as part of the RAINS-Asia project (regional acidification, information and simulation model for Asia; Hettelingh, 1991). Institutes in most Asian countries will gather information for the assessment of the vulnerability of the soil. A GIS forms the basis for the integration of all the data and the implementation of the models used. At the Research Centre for Eco-Environmental Sciences of the Chinese Academy of Sciences in Beijing (RCEES) a GIS Centre has been set up as the start of this network.

21.3 Methods behind the critical load concept

21.3.1   *Modelling the effect of acidification: the RAINS impact module*

The purpose of the RAINS impact module is to assess the environmental effects of emission reductions of sulfur. Environmental damage may occur when the deposition of sulfur exceeds critical loads (Hettelingh et. al., 1992a). By avoiding an excess of critical loads by acidic (sulfur-based) deposition, it is assumed that the aluminium concentration will not reach a critical threshold beyond which vegetation growth is jeopardized. This is the essence of the policy evaluation mode of RAINS via its impact module. The policy context in which critical loads are used in RAINS is iterative in that the sequence of (i) emission reductions, (ii) emission transport and resulting deposition patterns and (iii) comparison of deposition to critical loads can be repeated to find appropriate emission levels. These emission reductions can be optimized through RAINS by, e.g., minimizing emission reductions to a level where critical loads are not, or only marginally, exceeded. Another example is to minimize the area where critical loads are exceeded. Reduction of the area where particularly low critical loads have been computed can be given priority as a target for emission reduction as well. In short, the method of critical loads is a tool by which a great variety of emission reduction strategies can be evaluated in RAINS

(Hettelingh et. al., 1992a). The prerequisite for the application of this method is a geographical map of critical loads with a resolution similar to that of the deposition computed in RAINS.

Two methods have been used. First of all, the SSMB method was predominantly used for the computation of critical loads of acidity in forest soils. The other method, the method of relative sensitivity (MRV) assesses the sensitivity of forest soils and other vegetation soils to verify the results with SSMB in Europe (Kuylenstierna & Chadwick, 1989; Chadwick & Kuylenstierna, 1990; Chadwick 1990). MRV uses weighting factors to assess the influence of bedrock lithology, soil type, land use and rainfall on the sensitivity of a site to acidic deposition. The method by which critical loads are computed in Asia combines SSMB with MRV.

SSMB assumes a time-independent steady state of chemical interactions involving an equilibrium between the soil solid phase and soil solution (Sverdrup et. al., 1990, 1993, de Vries et. al., 1991a, b). Similar assumptions apply to chemical interactions in surface and ground waters. The SSMB computes the maximum acid input to the system that will not exceed the critical alkalinity value (Sverdrup & Hettelingh, 1992).

The equation used to compute critical loads is summarized as follows:

$CL = f$ (base cation weathering, base cation uptake, base cation deposition, vegetation soil leaching, base cation to aluminium ratio, Gibbsite coefficient, precipitation runoff)

The exact mathematical description can be found elsewhere (Hettelingh & de Vries, 1992; Sverdrup & Hettelingh, 1992).

21.3.2   *Modelling spatial variability of model inputs and outputs using GIS*

GIS is used in combination with the RAINS impact module for the following purposes:

(i)   for the computation of the regional distribution of critical loads using a large quantity of spatial data;

(ii)   as a spatial interface between the model output (sulfur deposition) of RAINS-Asia and the computed critical loads;

(iii)   for the assessment of areas at risk, including such characteristics as land use. The potential damage to rice fields, for example, may be of greater importance than the risk of damaged grasslands.

We will concentrate in this section on different GIS aspects of the computation of the critical loads. All variables in the equation described in the last section have their own spatial distribution. In order to compute the critical load for a geographical area we have to make a spatial overlay of all variables and solve the equation for each unique combination of values of each of the variables. Within GIS, the equation could be translated into a spatial model that is presented in Figure 21.1. The model input variables base cation weathering, base cation deposition and runoff are read from a quadtree map, while several other variables come from a table related to the map containing types of vegetation.

Figure 21.1: The SSMB critical load Spans model

```
E   Critical load model Beijing, May 1993
:   Spans Map model
:
: A. Read map values and table lookup
:
: BaseCation Weathering
bcw    = Table('wrval',  class('Weat1'),     'wr');
:
: Basecation deposition
bcd    = Table('bcdval',  class('BCdLow'),   'bcd');
:
: Basecation uptake, BaseCation / Aluminium ration, KGibb (from vegetation type)
bcu    = table('veg',     class('vegtype'),'bcu');
bcal   = table('veg',     class('vegtype'),'bcal');
kg     = table('veg',     class('vegtype'),'kgibb');
:
: Runoff
q      = table('q',       class('q'),       'q');
:
: B. parts of the expression
:
part1 = 1.5 * max((0.7*bcw) + bcd - bcu - (0.05*q), (0.002 * q));
:
part2 = ( pow(part1,(1/3)) / pow(300 * bcal, (1/3)) ) * pow(q,(2/3));
:
part3 = (part1 / bcal):
:
: Expression 1
cl1    = bcw + part2 + part3;
:
```

Both on the input and on the output side we have to fulfil certain conditions. First of all the selection of the projection used is important. Since part of the data are based on area-related values, it is necessary to choose an area-conformal projection and convert all data into this projection.

Second, on the output side the restriction is that the resolution of critical loads should be compatible with that of deposition, i.e., on the level of a grid cell. In general there will not be one value of critical load for a whole grid cell, because one grid cell may contain different ecosystems. Therefore there is a need for presentation through cumulative percentile distributions. The cumulative distribution of critical loads describes a relationship between ascending critical load values and the cumulative percentage (percentiles) of the area in which the critical loads occur. For example, a 5 percentile critical load of 550 equivalents of acid per hectare per year means that in 5% of the grid cell area critical loads occur with values smaller or equal to 550 equivalents of acid per hectare per year. For each cell a graph like Figure 21.2 can be constructed. The procedure of constructing a map of critical loads consists of (i) choosing a percentile critical load to be mapped in every grid cell and (ii) shading the colour assigned to the relevant range of critical loads. If the range of critical loads between 500 and 1000 acid equivalents corresponds to the colour light red, then a 5 percentile critical load of for example 550 will result in the grid cell being shaded light red. This means that, in 5% of the area of that grid cell, the critical load is between 500 and 1000 or lower (and thus worse). Decreasing percentiles correspond to lower critical loads and thus to more sensitive ecosystems.

Figure 21.2: Cumulative distribution of critical loads in a grid cell

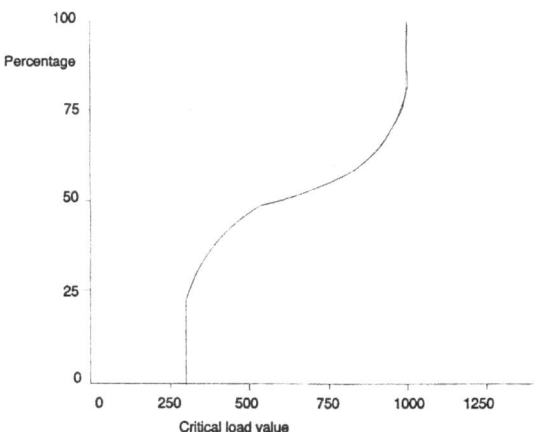

Furthermore, in the situation that data from different countries are used, the border between countries can give rise to conflicts. Two countries can have different critical load values for the grid cell that is partly in each of the countries. This can be solved by taking the overlay of the grid cells with the countries as the basis for shading: not whole grid cells but grid cells that fall within one country will be assigned critical load values.

Of course, not all areal data are available simultaneously and, in addition, not all countries provide data. It has been decided to use the following procedure: first a general dataset for the whole study area is composed, based on data from a variety of sources. This dataset, called the background data covers the complete study area.[a] The first results were made with these datasets and sent to the participating countries (the national focal centres) for comments, correction and updating with more accurate and detailed data. Many countries in Europe did send their own data for several reasons: (i) countries apply another, more detailed, geographical resolution, (ii) countries compute critical loads for other than forest ecosystems as in the background dataset and (iii) countries apply another approach for the assessment of the sensitivity of ecosystems. In Asia the network of national focal centres has just been initiated.

Finally, but far from the least important, input data come in different types of geographical objects: points, rasters of varying size and polygons. Rasters of different sizes are overlaid with each other to obtain unique combinations within the presentation grid cell. For the processing of point objects, we assumed an even distribution of ecosystems over a grid cell. The point data are weighted by the number of points within a grid cell.

In Europe, several levels of aggregation have been chosen for the presentation of critical load values. The highest level consists of the EMEP grid as defined by the Meteorological Synthesizing Centre - West (Norwegian Meteorological Institute) for the monitoring and evaluation of the long-range transmission of air pollutants in Europe. This grid consists of approximately 150km x 150 km cells in a projection with conformal areas (Polar Stereographic). The other level of aggregation consists of 1° longitude and 0.5° latitude (Lola) grid cells.

The format of the input data has been quite diverse, varying from irregular point sets, detailed grid raster, area percentages within a Lola cell to EMEP grid data. The following procedure has been followed in the European context. For the whole of Europe, critical loads on the Lola grid level have been calculated for forest areas. In a Lola grid cell there can be multiple critical load values, since each combination of soil type and type of forest (deciduous, coniferous or unmanaged) has a separate critical load. The area that is covered with each ecosystem is expressed as a percentage from the total area of the Lola cell.

Then all data for one EMEP grid cell are gathered. The Lola percentages are weighted by the relative share of the Lola grid cell in the EMEP grid cell. The resulting list of critical

---

[a] The data suppliers for the background data were the International Institute for Applied Systems Analysis (Austria) for Europe and the Stockholm Environment Institute (Sweden and United Kingdom) for Asia.

load values with area percentages are processed such that the cumulative distributions of critical loads within an EMEP grid cell are obtained. Finally, a percentile can be selected to produce a critical load map (Downing et. al., 1993).

In Asia also a conformal projection has been chosen (Lambert Azimuthal). Most of the data, the background data, were derived from small-scale maps. They came in polygonal format, which permitted the use of overlay techniques in the Spans modelling language to implement the full SSMB method within the GIS. Based on the critical load computation method summarized in 21.3.1, the model in Figure 21.1 was used to create a map of critical loads. The percentiles for each Lola cell had to be calculated out of the polygon critical load map, since 1°x1° Lola cells were selected for presentation.

For the southern part of China, the base cation weathering rates were produced locally and integrated in the model. The sulfur deposition data were derived from a Chinese air pollution model (Zhao et. al., 1991). The resulting point dataset was converted to a map with the surface interpolation method of potential mapping. Finally, the map with the areas at risk (excess of acid material) was calculated by subtracting the critical load map from the map with deposition. Area analysis assisted in the description of the land-use types in the areas at risk. In Asia the first round of sending the background data took place during summer 1993.

21.4 Results of RAINS impact modelled in a GIS environment

In this paragraph some preliminary results of the critical load model are presented. For Asia we concentrate on China, since the first calculations were made for this country.

Europe

Map 21.1 shows where acid deposition in Europe exceeded the critical loads in 1990. The map shows that areas most sensitive to acidification are located in Scandinavia and in central Europe. In central Europe, current acid deposition exceeds critical loads by a factor of 10 or higher (red shaded area) on the border of Germany, Poland and the Czech Republic (also known as the black triangle).

The next step in the GIS application of this research project will be the integration with other sources of information (e.g., atmospheric concentrations, depositions) and the use of point-interpolation methods such as potential mapping. The precision of maps displaying environmental effects and the flexibility in manipulating these maps will become increasingly important in scientific analysis and policy design for the protection of our environment.

China

The first (preliminary) results of the RAINS-Asia model applied to China are shown in Map 21.2 (Meijer et. al., 1993). The sensitive areas with critical loads of less than 500 equivalents of acid per hectare per year can be found in the eastern and southeastern China (south of the Changjiang river). Other sensitive areas are found south of Xian and in the upper northeastern part of China.

Map 21.3 illustrates excess sulfur deposition, which has been calculated by subtracting the map of critical loads for acidity from the sulfur deposition map. The following areas have excess sulfur in more than 50% of the area: around Beijing (Beijing, Tianjin, Hebei) and in the south-east (Jiangxi, Hunan, Guangxi, Guangdong). The highest preliminary category of excess (more than 500 equivalents of acid hectare per year) is found in Shandong (south of Beijing), covering 23% of the area. Other provinces with large areas in the highest category are Tianjin and Beijing with 59% and 46% respectively.

It is of interest to Chinese policy advisors to specifically be able to identify the kind of ecosystems that are at risk. An area cross-tabulation of the occurrence of excess and the land-use categories shows that irrigated farmland (rice fields) is the category where the excess occurs in the largest part of the area (40%). Also in the category other agricultural land and semi-arid hot scrub, 25-35% of the area is at risk.

The southern part of China has ecosystems that are vulnerable to acid components. When confronted with the deposition of sulfur, the areas around Beijing and in the south-east have large areas with a deposition exceeding the critical load of ecosystems. Rice fields and other agricultural land-use cover large parts of this area. Further research is needed to establish whether actual damage occurs to rice fields.

These preliminary results are currently linked to the deposition of sulfur. However, the effect of especially nitrogen oxide may also be important in China. Direct acidification effects (yellowing of needles) on coniferous forests have been monitored in the south of China (Ma & Shuwen, 1990).

Map 21.1:   Areas in Europe where acid deposition exceeded the critical load in 1990

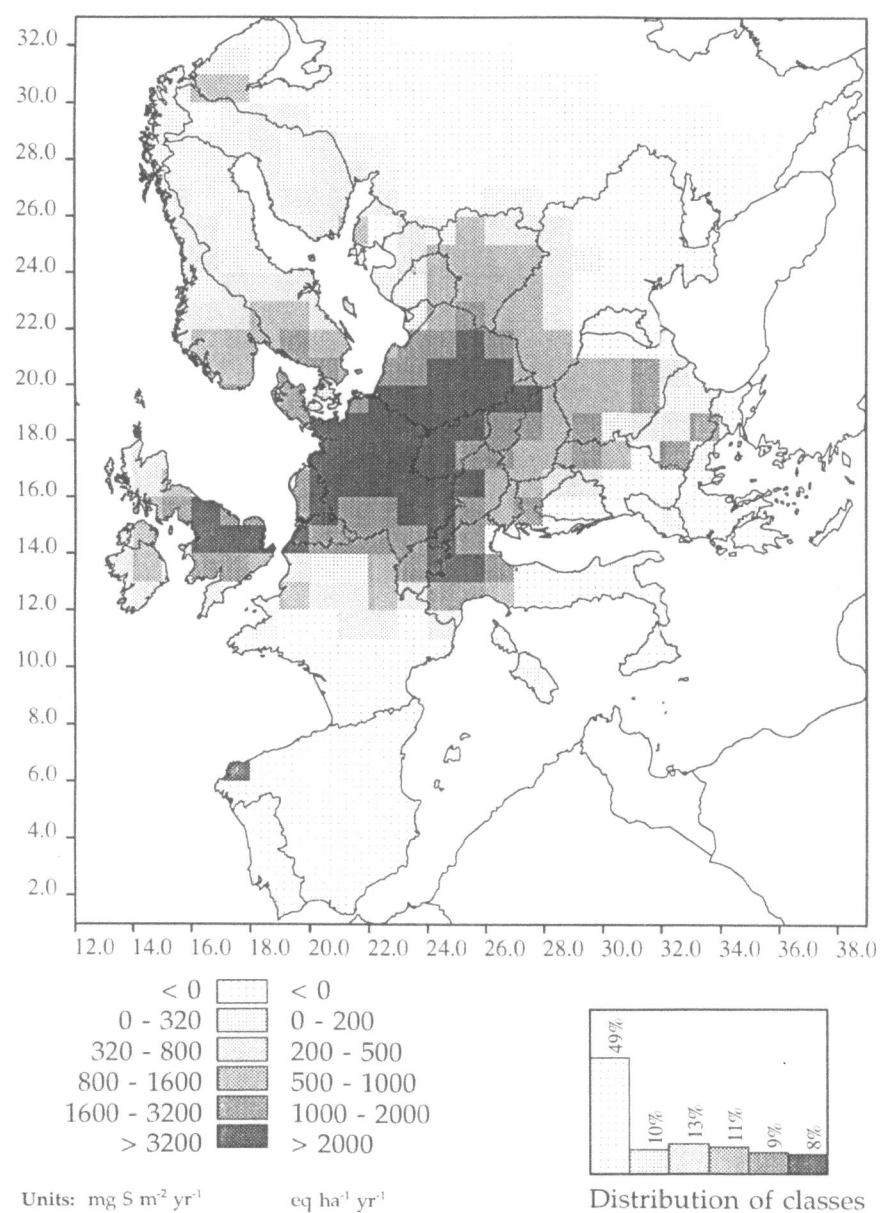

Source: RIVM / Coordination Center for Effects

Map 21.2: Critical loads for acid deposition in China

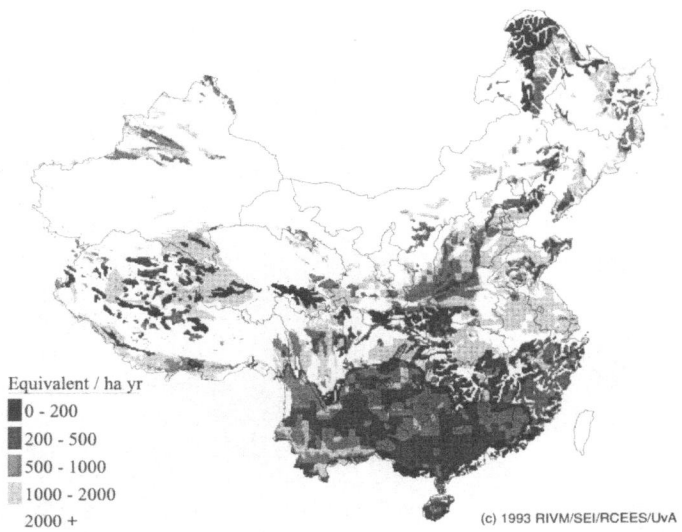

Equivalent / ha yr
- 0 - 200
- 200 - 500
- 500 - 1000
- 1000 - 2000
- 2000 +

(c) 1993 RIVM/SEI/RCEES/UvA

Map 21.3: Areas in China where sulfur depositon exceeds the critical load (in equivalents • ha^{-1} • year^{-1})

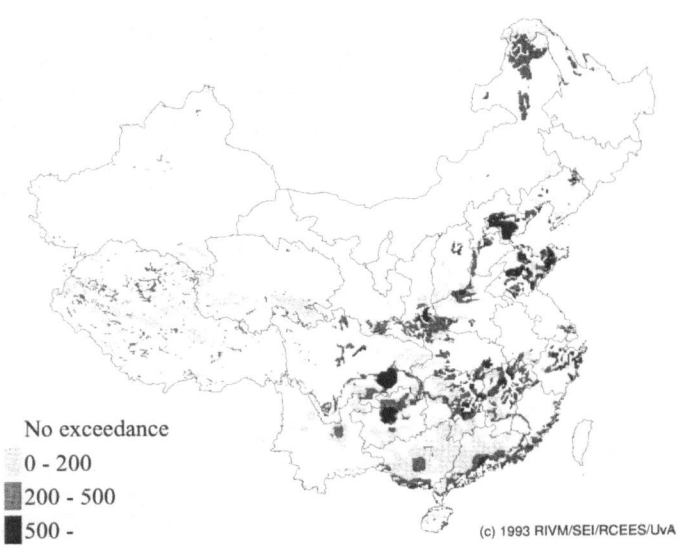

No exceedance
- 0 - 200
- 200 - 500
- 500 -

(c) 1993 RIVM/SEI/RCEES/UvA

21.5 Conclusions

The combination of environmental modelling and the application of GIS is an appropriate means for the analysis of changing environmental quality and the identification of areas in the world where environmental quality is most at risk. This chapter describes how this combined method has been applied to acid deposition in Europe and in Asia. GIS was not only integrated with the model for the computation of critical loads; it also served as a platform from which a great variety of data types and resolutions could be made compatible and consistent. The success of the method was not only due to the combination of tools. Other requirements include:

(i)   a network of international scientific participants in the field of environmental effects research;
(ii)   scientific consensus about the methods to be used for the computation of critical loads;
(iii)   the organization of frequent scoping meetings by the Coordination Centre for Effects involving all participants in Europe and Asia;
(iv)   policy interest: in Europe new protocols for the reduction of sulfur and nitrogen are upcoming, whereas, in Asia, consciousness about the link between economic growth and changing environmental quality is still emerging.

The combination of environmental modelling and GIS techniques will have an increasing potential for the analysis of environmental processes, the identification of environmental quality at risk and the simulation of effects of policy measures. This is especially true for the analysis of combined effects, e.g., acidification, tropospheric ozone and climate change. Synergistic effects of different pollutants under varying geographical and climate conditions possibly leading to effects for health, vegetation, crops and the cultural heritage can simultaneously be addressed with GIS. The reason is not only that GIS enables the overlay of maps displaying these effects but also that GIS allows for the application of comprehensive spatial statistics. GIS is a useful instrument for the support of policy development and scientific analysis of regionally distributed environmental effects that follow rapid economic growth in developing countries. The analysis of synergistic environmental effects using GIS in combination with environmental modelling is now becoming the subject of research.

Acknowledgements

The Environmental & Social Affairs division of the Asia Region of the World Bank and the Asian Development Bank are acknowledged for the support of the project "Acid rain and emissions in Asia", which enabled part of this chapter. J. Shah (World Bank), L.

Hordijk (Agricultural University, the Netherlands), W. Foell (Resource Management Associates, Madison, WI, USA), M. Chadwick, J. Kuylenstierna, S. Cinderby (Stockholm Environment Institute, York, UK), D. Streets (Argonne National Laboratory, IL, USA), G. Carmichael (University of Iowa, Iowa City, USA), M. Amann and W. Schoepp (International Institute for Applied Systems Analysis, Laxenburg, Austria) and Dianwu Zhao (Research Centre for Ecoenvironmental Sciences, Beijing, China) are gratefully acknowledged for the collaboration in the project Acid Rain and Emissions in Asia. The Dutch Ministry of Environment (DGM/LE, V. Keizer) and European national focal centres are acknowledged for the support of the computation and mapping of critical loads in Europe.

References

Chadwick, M.J. (1990). *Relative sensitivity, critical loads and the inclusion of damage components in acidic deposition abatement strategy models*. Presented at the Second Annual Workshop on Acid Rain in Asia, Bangkok, Thailand, 1990.

Chadwick, J.M. & Kuylenstierna, J.C.I. (1990). *The relative sensitivity of ecosystems in Europe to acidic deposition*. Stockholm, Stockholm Environment Institute.

de Vries, W., R.H. Hootsman, J. Kors, J.G. van Uffelen & J.C.H. Voogd (1991a). *Assessment and mapping of critical loads for potential acidity on Dutch forest soils*. Wageningen, Winand Staring Centre Report.

de Vries, W., M. Posch, J. Kämäri & W. Schöpp (1991b). *Long-term soil response to acidic deposition in Europe*.

Downing, R.J., P. de Smet, J.-P. Hettelingh (1993). *Mapping critical loads in Europe: status 1993*. Bilthoven, RIVM (in press).

Economic Commission for Europe (1985). *Protocol to the 1979 Convention on Long-range Transboundary Air Pollution on the Reduction of Sulfur Emissions or their Transboundary Fluxes by at Least 30%*. Geneva, ECE, 1985, (ECE/EB.AIR/12).

Economic Commission for Europe (1988). *Protocol to the 1979 Convention on Long-range Transboundary Air Pollution concerning the Control of Emissions of Nitrogen Oxides or their Transboundary Fluxes*. Geneva, ECE, pp. 13-20 (ECE/EB.AIR/18).

Hettelingh, J.-P. (1991). Guidelines for design of an acid rain policy model for Asia. *In*: W. Foell & D. Sharma, ed. *Acid rain and emissions in Asia*, Bangkok, Asian Institute of Technology.

Hettelingh, J.-P. & W. de Vries (1992). *Mapping vade mecum*. Bilthoven, National Institute for Public Health and Environmental Protection or Coordination Centre for Effects-West.

Hettelingh, J.-P., R.J. Downing, P.A.M. de Smet, ed. (1991). *Mapping critical loads for Europe*. Bilthoven, National Institute of Public Health and Environmental Protection (CCE technical report no. 1).

Hettelingh, J.-P., R.H. Gardner, L. Hordijk (1992a). *A statistical approach to the regional use of critical loads*. Presented at the Acidic Deposition Conference, Glasgow, 16-21 September, 1990.

Hettelingh, J.-P., R.J. Downing, P.A.M. de Smet (1992b). The critical load concept for the control of acidification. *In*: T. Schneider, ed. *Acidification research: evaluation and policy applications*. Amsterdam, Elsevier, pp. 161-174 (Studies in Environmental Science 50).

Kuylenstierna, J.C.I. & Chadwick, J.M. (1989). The relative sensitivity of ecosystems in Europe to the indirect effects of acidic depositions. *In*: Kämäri, J., Brakke, D.F. Jenkins, A., Norton, S.A. & Wright, R.F., ed. *Regional acidification models: geographical extent and time development*. Berlin, Springer-Verlag, p. 3-21.

Ma, G. & Y. Shuwen (1990). *Decline of Armand pine and acidic deposition in China*. Presented at the conference on "Acidic deposition, its nature and impacts", Glasgow, 16-21 september 1990. Beijing, Research Institute of Forestry and Shanghai, Institute of Plant Physiology.

Meijer, E.N. (1991). *Integration of European environmental data: mapping critical loads*. Presented at the Fifth URSA-NET Forum, Patras, 7-9 June 1991.

Meijer, Evert N, J.-P. Hettelingh, D. Zhao (1993). *A geographical analysis of air pollution effects in Asia*. Presented at the Second GISDECO Seminar, Soesterberg, June 1991.

Nilsson, J. & P. Grennfelt, ed. (1988). *Critical loads for sulfur and nitrogen*. Report from a workshop held at Skokloster, 19-24 March 1988. Copenhagen, Nordic Council of Ministers (Miljørapport 1988:15).

RIVM (1992). *The environment in Europe: a global perspective*. Bilthoven, RIVM (report no. 481505001).

Sverdrup, H. & J.-P. Hettelingh (1992). *Deriving the equations for calculation of critical loads with a simplified mass balance method*. Beijing, Research Centre for Eco-Environmental Sciences, Chinese Academy of Sciences.

Zhao, D., X. Jiling & M. Jietai (1991). *Mapping sensitivity, critical loads and exceedance for China*. Beijing, Research Centre for Eco-Environmental Sciences, Chinese Academy of Sciences.

Evert N. Meijer
University of Amsterdam
Department of Human Geography
&
Geodan bv
Jan Luijkenstraat 10
NL-1071 CM Amsterdam
The Netherlands

Jean-Paul Hettelingh
National Institute of Public Health and Environmental Protection
Coordination Centre for Effects - West
P.O. Box 1
NL-3720 BA Bilthoven
The Netherlands

Paul Padding
National Institute of Public Health and Environmental Protection
Department of Informatics Service Centre
P.O. Box 1
NL-3720 BA Bilthoven
The Netherlands

TOWARDS A HEALTH AND ENVIRONMENTAL GEOGRAPHICAL INFORMATION SYSTEM FOR EUROPE

PART VII

TOWARDS A HEALTH CARE
ENVIRONMENTAL GEOGRAPHICAL
INFORMATION SYSTEM FOR EUROPE

22   DEVELOPMENT OF A HEALTH AND ENVIRONMENT GEOGRAPHICAL INFORMATION SYSTEM FOR THE EUROPEAN REGION

WHO Regional Office for Europe

Abstract

A geographical information system combining data on both health and the environment would facilitate policy development, implementation and management, as well as research, in the fields of public health and environmental quality in Europe. A World Health Organization consultation therefore met to verify the usefulness and technical feasibility of creating such a permanent information management system. The issues discussed were its areas of application in the health and environment field, the administrative infrastructure required at both national and international levels, the usefulness of pilot studies, the data sources and requirements, the choice of indicators, the software and hardware specifications and the training of appropriate personnel. It was recommended that the European Centre for Environment and Health, Bilthoven, be considered as the coordinating point for launching the initial programme, to be based on pilot studies chosen according to well defined criteria, and that the Centre should seek the help of qualified technical collaborating centres and other national and international organizations engaged in the use of geographical information systems for environment and health management.

22.1   Introduction

The World Health Organization (WHO) Consultation on a European Environmental Health Database (held in Berlin, 22-25 November 1988) noted that:

> Geographical information systems are of value in the compilation and presentation of data at national and Region-wide levels, particularly environmental data and health outcome data related to the impact and use of health services.

In this context, and being aware of parallel developments in other multinational and nongovernmental agencies in Europe, the participants at the Consultation recommended that discussions be held with the Commission of the European Communities (CEC) to evaluate the possibility of using the CORINE Programme as a basis for a European health and environment geographical information system (GIS). They also recommended that the need for, and specifications of, such a GIS should be kept under review.

Later discussions with the CEC confirmed the potential of GIS in the area of environment and health, but also revealed the limited role (in terms of scope, timing and administration) that the CORINE system could play in such a development. At the same time, discussions were held with research institutions and national authorities in a

M. J. C. de Lepper et al. (eds.), The Added Value of Geographical Information Systems in Public and Environmental Health, 333–348.
© 1995 *Kluwer Academic Publishers.*

and health in the European Region and to evaluate the possibility of building on these systems. It appears that there are an increasing number of GIS centres in at least 16 Member States, a continuously expanding body of associated data, and a growing need for Region-wide coordination of the various systems.

As a result, the National Institute of Public Health and Environmental Protection (RIVM) in Bilthoven was asked to take the lead in assisting the Regional Office to develop a GIS programme for environment and health in the European Region. To this end, a planning meeting was convened at RIVM, in December 1989, to help verify the need for and define the scope of such a GIS system and to point to relevant information sources. This planning meeting recommended the convening of a consultation. This chapter presents the conclusions and recommendations of this consultation, which took place at RIVM, in December 1990, as reported to the Regional Office by D.J. Briggs and R.M. Stern in 1991.

The main aim of the Bilthoven consultation was to justify the need for and to define the potential uses and specifications of a spatial information infrastructure (i.e., GIS and its supporting facilities) to support the research and policy implementation of the Regional Office. Towards this goal, the Bilthoven consultation was requested: (a) to verify the usefulness of GIS, as a permanent information management system with the focus on the state of public health and environmental quality, for the WHO Regional Office for Europe and other European and national agencies dealing with public health and environmental policies and research; and (b) to evaluate the technical feasibility of creating a GIS that contains both health and environmental data in integrated databases.

The consultation brought together European professionals with wide experience in the fields of public health, environmental quality, spatial information systems and the development of suitable indicators of the state of environmental health and environmental quality. They consisted of 29 experts from 12 countries, 6 observers, 2 staff members from the Regional Office, 2 from WHO headquarters in Geneva, and one each from the United Nations Economic Commission for Europe, CEC, the International Institute for Applied Systems Analysis (IIASA) in Laxenburg, Austria, the Baltic Marine Environment Protection Commission, Helsinki Commission (HELCOM) and the United States Environmental Protection Agency.

22.2 Topics of discussion

Having established the mandate and role of the Regional Office in providing information on the local status of environment and health within the European Region and in developing and implementing policy guidelines for achieving the regional targets for health for all, the participants discussed the following issues:

(i)   information planning and analysis, including indicators on public health and environmental quality and the spatial resolution required;
(ii)   information infrastructure, including the functional and technical design of databases and user interfaces and the requirements for hardware, software and communications;
(iii)   meta-information systems as a catalogue of data sources for environment and health;
(iv)   the organizational aspects of a large-scale environment and health information system; and
(v)   the added value offered by a health and environment GIS.

Throughout the discussion, there was a consensus on the opportunities that separate programmes on environmental GIS (EGIS) and health GIS (HGIS) and a joint health and environment GIS (HEGIS) could offer to research and policy implementation in the European Region. In particular, it was noted that a combined HEGIS would add considerable value to such activities by allowing a wide range of relevant data to be integrated in a consistent spatial framework.

At the same time, the many difficulties involved in designing, establishing and using HEGIS were noted. Any successful GIS would require not only adequate computer hardware and software but also reliable data, an efficient organizational structure and well trained personnel. Each of these is a potentially limiting factor in the development of any GIS programme and especially a HEGIS programme. Experience with GIS programmes has established that special problems are likely to include: the availability of relevant and consistent data; the high operational and organizational costs of establishing and maintaining the system; the scarcity of adequately trained staff; the varied spatial scales of the data and applications involved; and the wide range of user needs.

To solve or minimize these problems, it was agreed that the system to be developed by the Regional Office should be kept as small and simple as possible, commensurate with its intended use and users, be developed step by step and involve potential users from the earliest possible stage. In this context, developing and using pilot demonstration systems to inform potential users was considered especially valuable. The development and use of meta-databases were also proposed as an effective way of improving access to and completeness of data. It was argued that the analytical capabilities of the system should be developed not only for mapping, spatial modelling and information retrieval but also for relevant statistical and relational analysis.

It was noted that the ultimate added value of HEGIS, compared with HGIS and EGIS, could only be established after HEGIS had been developed and tried; there was almost no national or international experience with HEGIS as such. It was clear that HEGIS would most likely be useful for answering "What if?" questions on public and environmental health, for seeking temporal and spatial patterns in health outcome and for the preliminary generation (and, under some circumstances, testing) of hypotheses about relationships between environment and demography and health. There was also, however, a significant risk of the misuse of such systems that must be guarded against. This potential risk is exacerbated by the persuasiveness of mapped output. A final understanding of health outcome and especially the origin of local variations thereof, using available spatially referenced public health data, can only be obtained from more detailed and appropriately designed ad hoc analytic epidemiological studies or exposure monitoring exercises.

Within these constraints, the clear consensus of the consultation was nevertheless that HEGIS would make a valuable contribution to health and environment research and policy implementation in the European Region. In particular, throughout the discussion, participants from the countries of central and eastern Europe expressed their wholehearted agreement on the appropriateness of the programme to their area, their enthusiasm for developing and taking part in pilot studies and their support for the Regional Office's HEGIS concept.

22.3 General conclusions and recommendations

The participants recognized that an urgent need exists for information on environment and health, at appropriate levels of disaggregation, to support policy development, implementation and management as well as research related to public health and the environment in the European Region.

They also stressed that the provision of this information at any level is likely to be most cost-effective where the system provides for multiple applications and meets the varied needs of different user groups. In this context, a clear prerequisite is to facilitate access to and improve the availability of relevant information (health, environmental, geographical, demographic, etc.). Likewise, it is important that the information be available in an integrated and comparable form, so that the links and relationships between different phenomena and different observations can be examined, analysed and illustrated. Processing and storing this information, facilitating its analysis and effectively communicating the results creates an overriding need to adopt modern information system approaches, using international databases and information management standards, built on a stable integrated database.

Much of the information required has an important spatial component, which is valuable in the analysis and explanation of spatial and temporal patterns of health outcome. It also provides the basic framework for the integration of the wide range of environmental and health information required. Incorporating this spatial dimension in environmental and health information adds considerable value to the information and greatly strengthens its use in policy implementation.

To provide the capability to collate and use environmental and health information in an effective manner, an information infrastructure must have the capacity for data capture, integration, storage, manipulation, spatial analysis, statistical analysis and presentation of results. GIS technology has now reached the stage of development where it can efficiently provide such an operational infrastructure.

The participants concluded that the Regional Office has a mandate to document and inform Member States about the local status of environment and health. They therefore recommended that the Regional Office encourage the development of national EGIS and HGIS as part of a wider environment and health information system, to support public policy implementation on issues of environment and health in the European Region. The Regional Office should also develop a HEGIS, with technical help from appropriate collaborating centres.

To develop an operational HEGIS, it will be necessary to:

- to be aware of and liaise with existing national and international agencies involved with health and the environment and to integrate their relevant activities, such as data collection sources and networks, GIS development and application and the need for and development of standards;
- to identify and recommend standards and guidelines specifically appropriate to HEGIS;
  to provide accessible training and educational facilities and staff;
- to create an organizational network of national focal points and specialized collaborating centres in the Region; and
- to rapidly develop a working demonstration or prototype that is capable of showing the potential of HEGIS, acting as a test case for system development and serving as a pilot model towards which other developing systems can converge.

Meeting these objectives will require the creation of a structured organization that can provide both central coordination and technical support as well as encourage the development of a collaborating network among the Member States in the European Region. On this basis, the participants recommended that consideration be given to incorporating the HEGIS project development into the programme of work of the WHO European Centre for Environment and Health. This Centre has been established with offices in Copenhagen, Bilthoven, Rome and Nancy with a view to strengthening collaboration in the health aspects of environmental protection, with special emphasis on information systems, mechanisms for exchanging experience and coordinated studies. To support the technical development of HEGIS in the Centre, especially at the Bilthoven office, the participants also recommended that RIVM in Bilthoven enter into a formal agreement with the Regional Office to become a collaborating centre for the development of GIS methods.

The criteria and procedures for implementing these general recommendations are described in detail below. They are divided into eight paragraphs: areas of application, administrative infrastructure, development of pilot studies, data requirements and indicators, hardware specifications, data access, storage and standards, software specifications and training and personnel. These recommendations were addressed to the Regional Office and the proposed GIS collaborating centre at RIVM. However, they also apply to the relevant activities that will be developed in different countries regarding the application of GIS in health and environment.

22.4 Areas of application

Conclusions

(i)   Environmental data usually have a spatial or geographical reference, are available with a wide range of spatial resolution and have been used in a variety of geographical information systems for various purposes.

(ii)   Public health data can be spatially referenced with varied spatial resolution. The WHO Consultation on Data Requirements and Methods for Analysing Spatial Patterns of Disease in Small Areas (held in Rome, 22-24 October 1990) recommended increasing the collection and availability of public health data with the highest possible degree of spatial resolution.

(iii)   Both environmental and health data show spatial variation. Numerous potential causal factors could relate local environmental conditions to health; a large number of non-environmental factors that are geographically related can also potentially affect public health. A spatial reference system provides a common reference framework for integrating these different types of data. In addition, analysis requires the use of non-spatial data, such as toxicological and chemical inventories and certain types of population-related data such as occupational exposures.

(iv)   The Regional Office has a mandate to document and demonstrate the extent of geographical variation in public health and environmental status in the European Region.

(v)   The purpose of integrating health and environmental data is to promote research, the monitoring and management of national environment and health policies and, in particular, the implementation of the WHO health for all policy.

(vi)   GIS can play a very useful role in making environmental and health information available to policy-makers, managers and the public in a more readily accessible form and thus in raising their awareness of the issues involved. A GIS programme can greatly aid policy implementation and management as well as research relating public health data to environmental data. Nevertheless, its scope and power is limited by the data available, the user's knowledge and the explanatory models that can be used.

(vii) GIS must be applied with care, however and the dissemination of results, especially for public information and political decision-making, must be conducted within a properly conceived context.

(viii) In this context, GIS is likely to be of particular value in the following types of application: risk management in both localized and large-scale disasters; the early warning of hazards; priority setting for health problems; the design of environmental sampling and monitoring systems; and the development of management support systems for the health for all campaign.

(ix)   Given these objectives, HEGIS can, with combined health and environmental information:
-    be a set of tools to manage the environment as a resource for health;
-    identify and highlight spatial patterns of health status;
-    help mitigate the consequences of catastrophic events on a local as well as on an international scale; and
-    assist in the optimal design of networks for the monitoring of environmental quality in the allocation of health care services.

Recommendations

A HEGIS should be developed under the auspices of the Regional Office in collaboration with the WHO European Centre for Environment and Health, specialized WHO collaborating centres, national focal points, subnational specialist institutions in the Member States and other supranational organizations such as the CEC, the International Agency for Research on Cancer and the United Nations Economic Commission for Europe.

22.5   Administrative infrastructure

Conclusions

(i)   National HGIS and EGIS and national and international HEGIS programmes could and should be used to aid Member States in implementing health for all policies.

(ii)   A clear statement of the tasks and responsibilities of the main participants in any international HEGIS programme will be required.

Recommendations

(i)   A clearly defined organizational structure should be developed that establishes the relationships between and responsibilities of all organizations involved in HEGIS.

(ii)   The WHO European Centre for Environment and Health at Bilthoven should be considered as the coordinating point for HEGIS.

(iii)   The Regional Office should develop one or more specialist international collaborating centres to provide technical support for HEGIS and its components.

(iv)   The RIVM in Bilthoven should be designated as the first WHO collaborating centre for GIS; additional, complementary collaborating centres should be identified and recruited with the help of RIVM.

(v)   Member States should be encouraged to develop national EGIS, HGIS and, if feasible and appropriate, HEGIS programmes with the help of national focal points (especially in the countries of central and eastern Europe).

(vi)   The WHO collaborating centres for GIS should have the prime responsibility for developing standards and keeping the HEGIS coordinating point up to date with developments in all aspects of HEGIS and its components.

(vii)   National focal points should have the task:
- of identifying specialist subnational institutes and coordinating collaboration between them;
- of providing support for national EGIS, HGIS and HEGIS programmes;
- of being prime movers in initiating national or subregional debate or research on policy issues;
- of acting as a point of address at the national level for the technical GIS service provided by the Regional Office; and
- of facilitating access to the training and educational programmes initiated by the Regional Office.

(viii)   The tasks of the HEGIS coordinating point should be:

- to develop and support the development of methods for policy implementation and policy-related research;
- to provide technical support for participating organizations;
- to foster the dissemination of relevant methods, standards and applications through training and educational programmes; and
- to provide support for the development of national infrastructures by the national focal points, especially within the countries of central and eastern Europe.

(ix)   The HEGIS coordinating point must maintain strong relationships with and assign specific tasks and responsibilities to, the supranational bodies, WHO collaborating centres and national focal points involved in HEGIS.

Advantage should be taken of subregional organizations (such as CORINE, EUROSTAT and HELCOM) to provide subregional GIS focal points for local GIS and HEGIS networks.

22.6   Development of pilot studies

Conclusions

(i)   Pilot studies or demonstration projects are likely to be extremely useful in ensuring the success of HEGIS. Such studies or projects would demonstrate the potential of HEGIS and would allow potential users critically to evaluate opportunities for their further application.

(ii)   In this context, the demonstration programme would provide an efficient and rapid illustration at the national level of the ability of HEGIS to carry out: data capture, supply and exchange; data structuring, cleaning and formatting; data integration; data storage; spatial modelling and data analysis; spatial querying and retrieval; and presentation of results.

(iii)   In addition, demonstration programmes would appear to be the most efficient way of addressing the critical and as yet unresolved issues of: data availability; data quality; maintenance and updating of the information system; data confidentiality and ownership; and data release and dissemination.

(iv)   Many environment and health programmes currently in existence at the international, national and subnational level throughout the European Region contain elements appropriate for incorporation in a HEGIS programme and have gathered experience that would be useful for the development of the Regional Office pilot HEGIS project.

(v)   Similarly, a great deal of experience in the field of environment and health has been built up in existing and developing GIS centres throughout Europe that can be used in the development of HEGIS.

Recommendations

(i)   The Regional Office should consider selecting possible candidates for pilot studies or demonstration projects on the basis of a subset of models that illustrate the following different features of HEGIS and their potential advantages:

-        GIS techniques can make better use of existing databases to deal with specific subregional problems, identified by networks of several local or national users with common interests (for example, environment and health in the Baltic states using the HELCOM network);
-        the development of the GIS methods for a particular local site or sites can then be extended to similar applications throughout the Region (for example, siting and monitoring of municipal waste incineration plants);
-        GIS technology can be applied to deal with different levels of spatial resolution (for example, the use of large-scale GIS to assist the countries of central and eastern Europe in developing priorities for environmental and health data collection); and

- HEGIS has the potential to observe clusters of disease in sensitive populations that might be environmentally related and for which an extensive data network already exists (for example, congenital malformations).

(ii) Critical criteria for the selection of pilot studies and demonstration projects should include:

- the relevance of the question to be answered to the spirit of the European Charter on Environment and Health;
- the extent to which interested parties are able to participate actively in the work programme;
- the existence of international, national or local programmes, preparatory studies and/or databases;
- the clear, demonstrable added value of a GIS application; and
- the possible additional benefits of such work in the form of sponsorship from other organizations.

(iii) Pilot studies and demonstration projects should be designed in such a way that their findings and results can be properly evaluated, especially with respect to questions of environmental health management and applicability at other sites.

(iv) A comprehensive work plan should be drawn up for each pilot study or demonstration project in consultation with the user communities. The work plan should contain a clear indication of what data input and evaluation criteria are required from the user communities.

(v) The Regional Office should ensure that proper use is made of existing health, environment and GIS experience within various agencies and institutes in the European Region.

(vi) The Regional Office should consider identifying a network of centres of excellence in the application of GIS to health and environment, to provide additional support for HEGIS activities at the European regional and subregional level.

22.7 Data requirements and indicators

Conclusions

(i) An overwhelming number of potential indicators and variables exist, from which a subset needs to be selected for incorporation into HEGIS. This subset should be based on a broad consensus about the usefulness of the indicators for the purposes of HEGIS.

(ii)   Not all the indicators in the initial subset will be universally available and additional ad hoc indicators will occasionally be needed for specific purposes.

(iii)   Local and national boundaries can create discontinuities of spatial data.

(iv)   For analytical purposes, data will be most useful at the lowest level of aggregation.

(v)   The dissemination of information to the public is an important function of HEGIS. Nevertheless, past experience has shown that maps showing disease incidence and other spatially referenced health and environment parameters may lead people to misunderstand the health risks.

Recommendations

(i)   The Regional Office should initiate the preparation of an inventory of existing data sources relevant to health and environment, including their content, format, quality and availability.

(ii)   The Regional Office should develop appropriate procedures to reach a consensus on the selection of a minimum core set of indicators from the available databases, based on the following guidelines: their relevance to the suspected impact on the health of populations; the potential for intervention or prevention; the availability of spatially referenced data at an appropriate aggregation level; and the degree of coverage of the population or area of the indicator.

(iii)   Some indicators should be included only for purposes of quality control.

(iv)   In the short term, only data from available registers for which there is a high degree of quality control should be used for indicators.

(v)   In the long term, additional indicators should be chosen or developed when needed for ad hoc purposes and as appropriate for any specific pilot programmes, taking into account the recommendations from previous WHO consultations. These could include indicators specific to sensitive population groups and indicators of ecological quality.

(vi) National HEGIS programmes, especially those participating in pilot studies, should be required to maintain the minimum core subset of indicators. Additional indicators of local interest may be developed at the discretion of the national programme managers. Procedures should be developed to harmonize and standardize indicators throughout the Region.

(vii) Care should be taken to minimize discontinuities at national boundaries and, where unavoidable, to consider their effects in interpretation.

(viii) Quality control measures and the definition of terms must be standardized among the Member States and introduced at the national level to facilitate the international use of data.

(ix) All efforts must be made to ensure that data are collected on confounding variables (such as smoking, nutritional and alcohol consumption habits and gender), and known risk factors must be taken into account before any attempt is made at interpreting patterns of disease.

(x) Member States should be encouraged to make data available at the lowest levels of aggregation, within the constraints of spatial referencing, data quality and confidentiality. Health and environmental data should, however, be at consistent levels of resolution, taking into account their respective scales of spatial variation.

(xi) Information from HEGIS that has sufficient detail to satisfy researchers should be digested and packaged before being presented either to policy-makers or to the general public.

(xii) The Regional Office should develop guidelines for the harmonization of national HGIS, EGIS and HEGIS databases and infrastructures.

(xiii) The Regional Office should develop stringent quality control requirements for any indicators incorporated into HEGIS. Member States should be encouraged to assist in implementing these controls.

(xiv) The indicators and quality control of the system should be re-evaluated periodically and modified at appropriate intervals.

22.8   Hardware specifications

Conclusions

(i)   Developments in GIS methods have reached a sufficient level of maturity for the concept of HEGIS, available at international, national and subnational levels, to be realizable.

(ii)   The use and development of HEGIS requires a level of computer hardware and software that is generally available in all European countries.

(iii)   The difference between international, national and subnational systems is essentially one of scale and aggregation, rather than one of principle.

(iv)   The operation of HEGIS in a multi-user environment relies on a network of work stations and servers to provide local processing and centralized data management. A local area network is therefore an essential requirement of HEGIS.

Recommendations

(i)   The HEGIS coordination point should have access to the hardware, software and personnel necessary to fulfil its required functions.

(ii)   Any HEGIS system should have an interface that is user-friendly to both the non-computer literate and the specialist and should be ergonomic, interactive and customizable, with on-line help and access control mechanisms.

22.9   Data access, storage and standards

Conclusions

(i)   The integration of data into HEGIS will involve bringing together data from many different sources and with different characteristics, requiring accurate and detailed documentation.

(ii)   A meta-database is also necessary and should be closely coordinated with HEGIS.

(iii)   Relational software of the database management system (DBMS) type, for the storage of attributed data in health and environmental applications, is widespread in European countries.

Recommendations

(i)   The database system should include the standard capabilities of storage, manipulation (including updates), retrieval, support for integrity constraints, access controls, distributed processing and remote access.

(ii)   Support for meta-databases should include data dictionaries for local information, data content, data quality, etc., with code lists, thesaurus components and a catalogue of relevant external databases. Meta-database coverage should go beyond the data contents of HEGIS and should extend beyond data in the strict sense.

(iii)   Any database that is part of the system should provide proper links between the digital-cartographic elements and their associated attributes and other related data.

(iv)   Datasets used in support of research, monitoring and analysis functions should be derived from the most disaggregated data available.

(v)   HEGIS should adopt standardized boundary files (both current and historical), where available. If not, they should be developed with the help of national focal points who have the prime responsibility for providing accurate updated source data at standardized projections, etc.

(vi)   The HEGIS coordinating point should provide access to meta-data and information-support materials. Access to meta-data should take into account developments in network communications and multimedia databases.

(vii)   GIS systems used as part of HEGIS should allow for the analysis of both raster and vector data and provide an integrated environment for the storage of vector data, which include the topology of digital cartographic elements.

(viii)   The development of HEGIS should take full account of the development of standards in a wide range of related areas, including: the ISO standard for user interfaces; cataloguing standards for digital databases and digital cartographic datasets; and the ISO/ANSI standard for SQL as a database query language. At the application programming level, the use of forthcoming standards on the import/export of data from databases, data dictionary interchange and interconnection of heterogeneous DBMS are expected to be most important.

(ix)   WHO should address the subject of standards in coordination with the various national, international and nongovernmental organizations working in the field.

(x)   Where appropriate, the Regional Office should produce new data standards and provide continuing support for them (in the form of documentation and convenience software) and, in particular, is encouraged to develop procedures to enable national organizations to reach a consensus on the standardization of the logical database design for time series and other kinds of HEGIS data, to facilitate data interchange.

22.10 Software applications

Conclusions

(i)   A general assumption is that a system for health and environment using geography would include the capability for statistical techniques and methods, exploration, simulation, spatial operation and visualization (including mapping).

(ii) The analytical part of a system is also expected to allow the preliminary testing of hypotheses (for example, in relation to pattern identification, hazard identification and risk prioritization) and the explorative analysis of the data.

Recommendations

(i)   Software for the compilation of spatial statistical measures, including the comparison of maps and for the modification of standard statistical tests in the presence of spatial autocorrelation, should be made more widely available. Researchers in the fields of health and environmental analysis should be made more aware of the effects of spatial autocorrelation and multiple testing on tests of significance.

(ii) Research should be undertaken to evaluate the role of expert systems in HEGIS.

22.11 Training and personnel

Conclusions

The development and implementation of HEGIS will require a large number of professionals with multidisciplinary expertise and training. Current personnel resources in this area are probably not sufficient in the Region to meet the immediate or long-term needs of an extended HEGIS programme.

Recommendations

The Regional Office, together with national focal points and collaborating centres, should promote multilevel training in the use of HEGIS and the interpretation of results from HEGIS. A task force should be created to examine the personnel resource requirements of the programme and to prepare support material and programme outlines so that the necessary training facilities can be developed and the required staff can be trained at international, national and subnational levels.

Annex 1   The European Charter on Environment and Health

Preamble

In the light of WHO's strategy for health for all in Europe, the report of the World
Commission on Environment and Development and the related Environmental
Perspective to the Year 2000 and Beyond (resolutions 42/187 and 42/186 of the United
Nations General Assembly) and World Health Assembly resolution WHA42.26,

- *Recognizing* the dependence of human health on a wide range of crucial
  environmental factors,
- *Stressing* the vital importance of preventing health hazards by protecting the
  environment,
- *Acknowledging* the benefits to health and wellbeing that accrue from a clean and
  harmonious environment,
- *Encouraged* by the many examples of positive achievement in the abatement of
  pollution and the restoration of a healthy environment,
- *Mindful* that the maintenance and improvement of health and wellbeing require a
  sustainable system of development,
- *Concerned* at the ill-considered use of natural resources and man-made products in
  ways liable to damage the environment and endanger health,
- *Considering* the international character of many environmental and health issues
  and the interdependence of nations and individuals in these matters,
- *Conscious* of the fact that, since developing countries are faced with major
  environmental problems, there is a need for global cooperation,
- *Responding* to the specific characteristics of the European Region, and notably its
  large population, intensive industrialization and dense traffic,
- *Taking* into account existing international instruments (such as agreements on
  protection of the ozone layer) and other initiatives relating to the environment and
  health,

The Ministers of the Environment and of Health of the Member States of the European
Region of WHO, meeting together for the first time at Frankfurt-am-Main on 7 and 8
December 1989, have adopted the attached European Charter on Environment and
Health and have accordingly agreed upon the principles and strategies laid down therein
as a firm commitment to action. In view of its environmental mandate, the Commission
of the European Communities was specially invited to participate and, acting on behalf
of the Community, also adopted the Charter as a guideline for future action by the
Community in areas which lie within Community competence.

M. J. C. de Lepper et al. (eds.), The Added Value of Geographical Information Systems in Public and Environmental Health, 349–355.
© *1995 Kluwer Academic Publishers.*

Entitlements and responsibilities

1.   *Every individual* is entitled to:

    •   an environment conducive to the highest attainable level of health and
          wellbeing;
    •   information and consultation on the state of the environment, and on plans,
          decisions and activities likely to affect both the environment and health; and
    •   participation in the decision-making process.

2.   *Every individual* has a responsibility to contribute to the protection of the
      environment, in the interests of his or her own health and the health of others.
3.   *All sections of society* are responsible for protecting the environment and health as
      an intersectoral matter involving many disciplines; their respective duties should
      be clarified.
4.   *Every public authority* and agency at different levels, in its daily work, should
      cooperate with other sectors in order to resolve problems of the environment and
      health.
5.   *Every government and public authority* has the responsibility to protect the
      environment and to promote human health within the area under its jurisdiction,
      and to ensure that activities under its jurisdiction or control do not cause damage to
      human health in other states. Furthermore, each shares the common responsibility
      for safeguarding the global environment.
6.   *Every public and private body* should assess its activities and carry them out in
      such a way as to protect peoples' health from harmful effects related to the
      physical, chemical, biological, microbiological and social environments. Each of
      these bodies should be accountable for its actions.
7.   *The media* play a key role in promoting awareness and a positive attitude towards
      protection of health and the environment. They are entitled to adequate and
      accurate information and should be encouraged to communicate this information
      effectively to the public.
8.   *Nongovernmental organizations* also play an important role in disseminating
      information to the public and promoting public awareness and response.

Principles for public policy

1.   Good health and wellbeing require a clean and harmonious environment in which
      physical, psychological, social and aesthetic factors are all given their due
      importance. The environment should be regarded as a resource for improving
      living conditions and increasing wellbeing.

2. The preferred approach should be to promote the principle of "prevention is better than cure".

3. The health of every individual, especially those in vulnerable and high-risk groups, must be protected. Special attention should be paid to disadvantaged groups.

4. Action on problems of the environment and health should be based on the best available scientific information.

5. New policies, technologies and developments should be introduced with prudence and not before appropriate prior assessment of the potential environmental and health impact. There should be a responsibility to show that they are not harmful to health or the environment.

6. The health of individuals and communities should take clear precedence over considerations of economy and trade.

7. All aspects of socioeconomic development that relate to the impact of the environment on health and wellbeing must be considered.

8. The entire flow of chemicals, materials, products and waste should be managed in such a way as to achieve optimal use of natural resources and to cause minimal contamination.

9. Governments, public authorities and private bodies should aim at both preventing and reducing adverse effects caused by potentially hazardous agents and degraded urban and rural environments.

10. Environmental standards need to be continually reviewed to take account of new knowledge about the environment and health and of the effects of future economic development. Where applicable such standards should be harmonized.

11. The principle should be applied whereby every public and private body that causes or may cause damage to the environment is made financially responsible (the polluter pays principle).

12. Criteria and procedures to quantify, monitor and evaluate environmental and health damage should be further developed and implemented.

13. Trade and economic policies and development assistance programmes affecting the environment and health in foreign countries should comply with all the above principles. Export of environmental and health hazards should be avoided.

14. Development assistance should promote sustainable development and the safeguarding and improvement of human health as one of its integral components.

Strategic elements

1. The environment should be managed as a positive resource for human health and wellbeing.

2. In order to protect health, comprehensive strategies are required, including, *inter alia*, the following elements:

- The responsibilities of public and private bodies for implementing appropriate measures should be clearly defined at all levels.
- Control measures and other tools should be applied, as appropriate, to reduce risks to health and wellbeing from environmental factors. Fiscal, administrative and economic instruments and land-use planning have a vital role to play in promoting environmental conditions conducive to health and wellbeing and should be used for that purpose.
- Better methods of prevention should be introduced as knowledge expands, including the use of the most appropriate and cost-effective technologies and, if necessary, the imposition of bans.
- Low-impact technology and products and the recycling and reuse of wastes should be encouraged. Changes should be made, as necessary, in raw materials, production processes and waste management techniques.
- High standards in management and operations should be followed to ensure that appropriate technologies and best practices are applied, that regulations and guidance are adhered to, and that accidents and human failures are avoided.
- Appropriate regulations should be promulgated; they should be both enforceable and enforced.
- Standards should be set on the basis of the best available scientific information. The cost and benefit of action or lack of action and feasibility may also have to be assessed but in all cases risks should be minimized.
- Comprehensive strategies should be developed that take account of the risks to human health and the environment arising from chemicals. These strategies should include, *inter alia*, registration procedures for new chemicals and systematic examination of existing chemicals.
- Contingency planning should be undertaken to deal with all types of serious accident, including those with transfrontier consequences.
- Information systems should be strengthened to support monitoring of the effectiveness of measures taken, trend analysis, priority-setting and decision-making.
- Environmental impact assessment should give greater emphasis to health aspects. Individuals and communities directly affected by the quality of a specific environment should be consulted and involved in managing that environment.

3. Medical and other relevant disciplines should be encouraged to pay greater attention to all aspects of environmental health. Environmental toxicology and environmental epidemiology are key tools of environmental health research and should be strengthened and further developed as special disciplines within the Region.

4. Interdisciplinary research programmes in environmental epidemiology with the aim of clarifying links between the environment and health should be encouraged and strengthened at regional, national and international levels.
5. The health sector should have responsibility for epidemiological surveillance through data collection, compilation, analysis and risk assessment of the health impact of environmental factors and for informing other sectors of society and the general public of trends and priorities.
6. National and international programmes of multidisciplinary training, as well as the provision of health education and information for public and private bodies, should be encouraged and strengthened.

Priorities

1. Governments and other public authorities, without prejudice to the importance of problem areas specific to their respective countries, the European Community and other intergovernmental organizations, as appropriate, should pay particular attention to the following urgent issues of the environment and health at local, regional, national and international levels and to take action on them:

    *   global disturbances to the environment, such as the destruction of the ozone layer and climatic change;
    *   urban development, planning and renewal to protect health and promote wellbeing;
    *   safe and adequate drinking-water supplies on the basis of the WHO *Guidelines for drinking-water quality* together with hygienic waste disposal for all urban and rural communities;
    *   water quality in relation to surface, ground, coastal and recreational waters;
    *   microbiological and chemical safety of food;
    *   the environment and health impact of
        - various energy options;
        - transport, especially road transport;
        - agricultural practices, including the use of fertilizers and pesticides, and waste disposal;
    *   air quality, on the basis of the WHO *Air quality guidelines for Europe*, especially in relation to oxides of sulfur and nitrogen, the photochemical oxidants ("summer smog") and volatile organic compounds;
    *   indoor air quality (residential, recreational and occupational), including the effects of radon, passive smoking and chemicals;
    *   persistent chemicals and those causing chronic effects;
    *   hazardous wastes, including management, transport and disposal;
    *   biotechnology, in particular genetically modified organisms;

- contingency planning for and in response to accidents and disasters; and
- cleaner technologies as preventive measures.

2. In addressing all of these priorities, the importance of intersectoral environmental planning and community management to generate optimal health and wellbeing should be borne in mind.

3. Health promotion should be added to health protection so as to induce the adoption of healthy lifestyles in a clean and harmonious environment.

4. It should be recognized that some urgent problems require direct and immediate international cooperation and joint efforts.

The way forward

1. Member States of the European Region should:

    - take all necessary steps to reverse negative trends as soon as possible and to maintain and increase the health-related improvements already taking place. In particular, they should make every effort to implement WHO's regional strategy for health for all as it concerns the environment and health;
    - strengthen collaboration among themselves and, where appropriate, with the European Community and other intergovernmental bodies on mutual and transfrontier environmental problems that pose a threat to health; and
    - ensure that the Charter adopted at this meeting is made widely available in the languages of the European Region.

2. The WHO Regional Office for Europe is invited to:

    - explore ways of strengthening international mechanisms for assessing potential hazards to health associated with the environment and for developing guidance on their control;
    - make a critical study of existing indicators of the effects of the environment on health and, where necessary, develop others that are both specific and effective;
    - establish a European Advisory Committee on the Environment and Health in consultation with the governments of the countries of the Region; and
    - in collaboration with the governments of the European countries, examine the desirability and feasibility of establishing a European Centre for the Environment and Health or other suitable institutional arrangements, with a view to strengthening collaboration on the health aspects of environmental protection with special emphasis on information systems, mechanisms for exchanging experience and coordinated studies. In such arrangements, cooperation with the United Nations Environment Programme, the United

Nations Economic Commission for Europe and other organizations is desirable. Account should be taken of the environmental agency to be established within the European Community.

3. Member States of the European Region and WHO should:

   • promote the widest possible endorsement of the principles and attainment of the objectives of the Charter.

4. European Ministers of the Environment and of Health should:

   • meet again within five years to evaluate national and international progress and to endorse specific action plans drawn up by WHO and other international organizations for eliminating the most significant environmental threats to health as rapidly as possible.

The GeoJournal Library

1. B. Currey and G. Hugo (eds.): *Famine as Geographical Phenomenon.* 1984
   ISBN 90-277-1762-1

2. S.H.U. Bowie, F.R.S. and I. Thornton (eds.): *Environmental Geochemistry and Health.* Report of the Royal Society's British National Committee for Problems of the Environment. 1985   ISBN 90-277-1879-2

3. L.A. Kosiński and K.M. Elahi (eds.): *Population Redistribution and Development in South Asia.* 1985   ISBN 90-277-1938-1

4. Y. Gradus (ed.): *Desert Development.* Man and Technology in Sparselands.
   1985   ISBN 90-277-2043-6

5. F.J. Calzonetti and B.D. Solomon (eds.): *Geographical Dimensions of Energy.*
   1985   ISBN 90-277-2061-4

6. J. Lundqvist, U. Lohm and M. Falkenmark (eds.): *Strategies for River Basin Management.* Environmental Integration of Land and Water in River Basin.
   1985   ISBN 90-277-2111-4

7. A. Rogers and F.J. Willekens (eds.): *Migration and Settlement.* A Multiregional Comparative Study. 1986   ISBN 90-277-2119-X

8. R. Laulajainen: *Spatial Strategies in Retailing.* 1987   ISBN 90-277-2595-0

9. T.H. Lee, H.R. Linden, D.A. Dreyfus and T. Vasko (eds.): *The Methane Age.*
   1988   ISBN 90-277-2745-7

10. H.J. Walker (ed.): *Artificial Structures and Shorelines.* 1988
                                                                        ISBN 90-277-2746-5

11. A. Kellerman: *Time, Space, and Society.* Geographical Societal Perspectives.
    1989   ISBN 0-7923-0123-4

12. P. Fabbri (ed.): *Recreational Uses of Coastal Areas.* A Research Project of the Commission on the Coastal Environment, International Geographical Union. 1990   ISBN 0-7923-0279-6

13. L.M. Brush, M.G. Wolman and Huang Bing-Wei (eds.): *Taming the Yellow River: Silt and Floods.* Proceedings of a Bilateral Seminar on Problems in the Lower Reaches of the Yellow River, China. 1989   ISBN 0-7923-0416-0

14. J. Stillwell and H.J. Scholten (eds.): *Contemporary Research in Population Geography.* A Comparison of the United Kingdom and the Netherlands. 1990
                                                                        ISBN 0-7923-0431-4

15. M.S. Kenzer (ed.): *Applied Geography.* Issues, Questions, and Concerns.
    1989   ISBN 0-7923-0438-1

16. D. Nir: *Region as a Socio-environmental System.* An Introduction to a Systemic Regional Geography. 1990   ISBN 0-7923-0516-7

17. H.J. Scholten and J.C.H. Stillwell (eds.): *Geographical Information Systems for Urban and Regional Planning.* 1990   ISBN 0-7923-0793-3

18. F.M. Brouwer, A.J. Thomas and M.J. Chadwick (eds.): *Land Use Changes in Europe.* Processes of Change, Environmental Transformations and Future Patterns. 1991   ISBN 0-7923-1099-3

The GeoJournal Library

KLUWER ACADEMIC PUBLISHERS – DORDRECHT / BOSTON / LONDON